AF543011

Systematics of Fruit Crops

scan for catalogue

Browse, Search, Read & Buy...

Subject Catalogues

eChapters

Publishers

Print Books

Forthcoming

eBooks

New Titles

NIPA
GENX

ONLINE RESOURCES

- Publishing Books and Journals
- Ebooks and e articles, Current Affairs
- Reasoning, Logic and Aptitude
- Competitive Examination Preparation
- Language Learning and Development Programme
- Document Quality Checker and Improvement Tool
- Effective Public Speaking, Presentation and Interpersonal Skills
- Online Programmes for Professional Development
- Personality Development and Human Values

PAY USING UPI PayPal

Systematics of Fruit Crops

Girish Sharma
Professor
Department of Fruit Breeding and Genetic Resources
Dr. Y.S. Parmar University of Horticulture and Forestry
Nauni, Solan-173 230 Himachal Pradesh, India

O.C. Sharma
Subject Matter Specialist (Pomology)
Regional Agricultural Research Station
(Sher-e-Kashmir University of Agricultural Sciences Technology of Kashmir)
Leh, Jammu & Kashmir, India

B.S. Thakur
Senior Scientist, Department of Fruit Science
Dr. Y.S. Parmar University of Horticulture and Forestry
Nauni, Solan-173 230 Himachal Pradesh, India

NEW INDIA PUBLISHING AGENCY
New Delhi – 110 034

NEW INDIA PUBLISHING AGENCY
101, Vikas Surya Plaza, CU Block, LSC Market
Pitam Pura, New Delhi 110 034, India
Phone: + 91 (11)27 34 17 17 Fax: + 91(11) 27 34 16 16
Email: info@nipabooks.com
Web: www.nipabooks.com

Feedback at feedbacks@nipabooks.com

© Author: 2009

ISBN 978-93-80235-06-6

All rights reserved, no part of this publication may be reproduced, stored in a retrieval system or transmitted in any form or by any means, electronic, mechanical, photocopying, recording or otherwise without the prior written permission of the publisher or the copyright holder.

This book contains information obtained from authentic and highly regarded sources. Reasonable efforts have been made to publish reliable data and information, but the author/s, editor/s and publisher cannot assume responsibility for the validity of all materials or the consequences of their use. The author/s, editor/s and publisher have attempted to trace and acknowledge the copyright holders of all material reproduced in this publication and apologize to copyright holders if permission and acknowledgements to publish in this form have not been taken. If any copyright material has not been acknowledged please write and let us know so we may rectify it, in subsequent reprints.

Trademark notice: Presentations, logos (the way they are written/presented) in this book are under the trademarks of the publisher and hence, if copied/resembled the copier will be prosecuted under the law.

Composed and Designed NIPA

PREFACE

India is regarded as a horticultural paradise due to prevailing rich diversity of agro-climatic and socio cultural conditions present in the country and appropriate steps are needed to increase the quality horticultural crop production through recent techniques. For crop improvement fundamental knowledge of botany and systematic pomology of crop is essential.

Various taxonomists dealing with fruit crops have rated systematic pomology as an advanced horticultural subject and takes into consideration the basic aspects of taxonomy i.,e identification, naming of fruit plant species and varieties, besides, placements or logical classification of each fruit type under specific units of classification. For sound horticultural systematic knowledge primarily those of taxonomy, morphology, genetics, cytology and plant breeding is essential. For good reading material it is essential for systematic pomologists to use information of the associated sciences with appropriate explanations, and applications. The present work in book form is prepared to provide elementary knowledge to the students who have started studying systematic pomology.

Introduction (three sub-heads) presents the more academic elements of taxonomy related to the theories, hypothesis, basic principles pre-requisite of systematics which are required for a minimum working knowledge of systematic pomology. The detail regarding general origin and distribution, flower and fruit structure is given so that students with this background knowledge are in a position to cope with problems related both to varietal description and testing. Significance of systematic pomology to varietal improvement, new variety sources and methods of synthesis are detailed. The section 1-4 comprises of systematic enumeration of 58 fruits, discussed under tropical and subtropical, small fruits, nut fruits and temperate fruits. The text is supported by drawings that are accompanied with captions. Under each crop, historical background, origin and distribution uses, pomological traits of fruit, important species and cultivars are briefly dealt with. List of cultivars is restricted to popular cultivars as the cultivation status is ever change one. The glossary and annexures are designed with thrust on clarity and brevity. Most of the words are described in the text. The annexures provide detailed information of fruit crops, fruit types, genera, species and tribes their number and status in fruit crop families. Botanical terms chart provides morphological description of leaf, floral structure and form, inflorescence root and rootstock for easily understanding by the students. The authors have drawn required material from number of texts, monographs, books, periodicals and floras for compilation of this work. We acknowledge the material information retrieved from these sources.

Authors

ACKNOWLEDGEMENT

We would like to express our appreciation and gratitude to our colleagues for all the helpful advice and encouragement given during writing and illustration of this book. Special thanks to S Sharma who prepared the rough and final diagram. Our thanks to one and all who have been associated with book preparation and some deserves special mention like Dr O. C. Sharma, Roshan Anand, Shailander Yadav and Rubiqua Bashir.

It is hoped that the book will be useful to the students and also to amateurs of natural history and horticulture. My sincere thanks are also due to the entire team of DPT computers for formating, composing and printing of the first print of the manuscript and to M/S New India Publishing Agency, New Delhi, the publishers for their effort and keen interest throughout the production of this work. Errors or omission notified by the users of the book would be appreciated to improve the book value.

Authors

CONTENTS

Introduction

Systematic Pomology, systematic botany/plant taxonomy is a branch of botanical sciences, which primarily is concerned with identification, nomenclature and classification of fruit crops and their resemblances and differences between and among various fruits and nut crops.

The taxonomy study of fruit plants i.e Systematic Pomology is very useful in increasing man's knowledge of evolution of plant kingdom and also reveals the manifold diversities present in plant kingdom and is useful

- From teaching point of view of different seeds of plant, presently available on earth their different characters, names geographical distribution, similarities and dissimilarities etc.
- Provides scientific knowledge of world's plant resources, which highlight the natural raw material of the plant kingdom that is needed for the everincreasing world population.
- Lastly it establishes the correct scientific name(s) of plants in nomenclature which is of international significance.

The prime aim of Systematic pomology, which is an important part of general horticulture is to form a practical basis for the recognition and classification of plants. Systematic means orderliness and through systematic pomology using units of classification, plants especially fruit trees, small fruits and nut crops of commerce are placed under well recognized categories of plants like division, order, family, genus, species and variety.

Systematic Pomology is a specialized science that deals with the study of various fruits and their botanical and horticultural relationships and it further identifies names and classifies fruits and nut crops. When a new taxon is found it has to be established whether it is identical or similar to one, which is already known whose identity is than determined by matching it with a known type. If it fails, then authorities of special plant groups are asked to make or establish its correct identity.

Nomenclature is primarily concerned with the determination of correct name of the identified taxon. In short nomenclature is system of names/ naming. Plants have number of direct and indirect uses and on the basis of utility, their presence in abundance, plants are often recognized by their vernacular names. These names though important and useful are not recognized world over by the people who speak different languages. For the sake of uniformity it is of utmost significance that plant(s) be given a correct name. The naming of plants (scientific names) is done by following the rules of nomenclature. While naming the fruit crop and varieties, systematists follow the rules of nomenclature formulated by the American Pomological Society, whereas for other plants, Code of the International Botanical Congress known as International Rules of Botanical Nomenclature are followed. In general Latin names have been adopted and have become officially accepted all over the world.

Classification follows nomenclature and is involved in placing of a plant or a group of plants or varieties in definite groups or categories after following some internationally accepted scheme or procedure. At present number of well accepted systems of classification are in vogue, important ones being those of Bentham and Hooker, Bessey, Engler, Hutchinson and Tippo. The significance of all these systems of classification are that they are relatively simple, has regional and wide reorganization, and all have tried to associate every species as a member of a particular genus, every genus to a family, family to an order and so on. In short aim is to associate the specimen with an appropriate units of classification.

Systematic Pomology and its significance can be gauged from the fact that it is intimately related to number of sciences. Hence it is essential to establish its relationship to other related subjects or branches of Horticulture or Pomology. The Systematic Pomology in its objectives encompasses the learning of various kinds of fruit crops which are considered significant in horticulture. The study involves their history, habitat, distribution, resemblances and differences, names and classification, structure and above all their relationship among each other, as well as to other members of the plant kingdom.

The term systematic means orderliness, meaning doing some work in a particular well-defined manner or through some established or recognized way/arrangement. The basic aim of systematic pomology is to formulate a practical platform for recognition and classification of plants, especially concerned with fruit trees, nut and berry crops which usually are grown commercially. As part of plant taxonomy, systematic pomology primarily deals with identification, nomenclature, and classification and is also involved in establishing resemblances and differences between and among various fruits and nut crops.

Prior to further knowing in detail the subject of systematic pomology and its significance it is important to know its relationship to other subjects and the usual terms employed need to be very well understood for clear understanding of the subject. In short systematic pomology which broadly falls under horticulture is part of agriculture. Agriculture further is defined as a science of raising plant and animal products from the soil which are regularly supplied to fulfill human demands. On the other hand horticulture forms an important segment of agriculture and usually it deals with fruits, nuts, vegetables, flowers and ornamentals from their general development, use and production. The exact boundaries of horticulture are largely confined to their general use and popular opinion and is not linked with any natural or botanical relationship. There are large numbers of examples which indicate variation like lawn grass is horticulture (landscape gardening) whereas field grass is part of agronomy, sweet potato is horticultural entity and Irish potato is regarded as an agronomical or field crop. Likewise the garden pea and sugar and starchy corn are differentiated as per their usage.

Horticulture in its very early days was considered more of an art, as many of the horticultural practices such as pruning, grafting were more artisan in their outlook. However today this view does not hold good any more as horticulture is strongly connected with number of branches of science. Sciences like physiology, chemistry, genetics, breeding are intimately and inseparably linked with horticulture. Horticulture comprises of four important branches, namely pomology, olericulture, horticulture and landscape gardening.

Of these branches, pomology is further subdivided into (i) practical pomology (ii) commercial pomology (iii) systematic pomology. As these names suggest they are largely self-explanatory. The practical pomology is of prime importance as it deals or is the 'science of fruits'. The name has been derived from Latin world pomace meaning fruit and a Greek term logos or logy referring to science or discourse.

The practical pomology mainly deals with the practices of growing fruits whereas commercial pomology is concerned with marketing and fruit disposal and takes into consideration the necessary steps required for the preparation of fruit for marketing, appropriate storage and transportation so that it reaches the consumer in a fine form. Lastly systematic pomology deals with learning various kinds of fruits that are of significance to horticulture, their resemblances and differences, names, classification, distribution, natural habitat, their history and various structure. The information so gathered provides a strong base for the formation of fruit inventory.

Systematic pomology is a branch of plant taxonomy and hence primarily is concerned with plant identification, nomenclature and classification of fruits. These three essential taxonomic components are very useful in determining their relationships both between and among various fruit plants considered important in horticulture.

Plant Identification

In botanical usage when a new taxon or a type / specimen has been identified, it now becomes essential to determine whether the new type is similar or identical with another which already is a known and established plant. Hence identification is primarily the comparison of the unknown with known specimen. There are a number of methods or combination of these methods that provide the procedure for determination of the identity of the new taxon. The most accepted method of plant identification is through reference to the (i) plant manual (ii) latest flora/check lists (iii) monograph. The first and most established and reliable method in identification of an unknown type is through referring a plant manual. Plant manual(s) is a complete reference book, regarding the placement of various orders, families alongwith their detailed botanical or pomological description. Plant manual usually is a wholesome reference book which highlights in specific order the plant or fruit crop families.

There are number of manuals that are recognized world over, most popular among students of horticulture are "Manual of Cultivated plants" (Bailey). Manual of Cultivated Trees and Shrubs (Rehder), Manual of Botany (Gray). Besides these, states within a country, also have manuals that deal with the local flora and could be effectively used for identification. For appropriate identification of the unknown type the procedure in using the manual is to first determine the family name. For this purpose the help of keys is used which is an essential segment of almost all manuals. For effective use of system of keys for identification one must be well versed with the terms used for describing various features of plant parts. Flower features are regarded to be most stable. Through the keys and checking the same with the flowers of the unknown type the family name is located. After this the section where family is described is found out from the key to the species, the plant identity is determined, then through key to genera, specimens generic name is determined. Later from species key the specific identity of the plant is determined. At this point of identification besides specific features, usually both scientific-binomial and common name(s) is known.

Second way to identify the new or unknown plant is determined through referring it to the floras and check lists of the region where the plant has been growing. The plant information in this case is not prepared in key form, but information covered is limited and mainly confined to areas such as valleys, peaks, mountains and countries. Important features of the system is that 'what plants are not known to the area' in short is a process of elimination, and by this approach number of plants may be left out. Hence, then one works with limited number of plants. In this way the unknown plant can be associated with one or more species. However the exact identity could only be established by making a reference to the plant manual or to some standard work on the plants carried out in some specific area.

Reference to some new monograph or related revised work apart from identifying through manuals is often said to be very reliable system for plant identification. The basic requirement for use of such monograph(s) is that one should have a good knowledge of plant family and its generic name, because these monograph(s) through detailed study and analysis present the family and genus information. Identification through monograph is very important as they are regarded to be the most reliable source of information and are invariably published by state / national institutes or by some recognized botanical societies.

Prior to use of manuals or monographs one must also know what the plant keys are. Keys are said to be some sort of simple arrangement usually in an outline work which is very effective and useful in identifying the unknown or the new type. In this arrangement special importance is given to each step between two contradictory choices or characters. The head for characters / categories of the key are used in number of ways, in lettered or numerical, but of these lettered head keys are more common all over the world.

As the species number and varietal wealth of horticultural crops is very vast, horticulturists all over have frequently found it useful to develop keys for identification. Keys have been devised using leaf flower and fruit characters. Flower petals in peach before they start bearing are helpful in identification of varieties. Cultivars like Early Elberta, Hiley, Sullivan have large showy petals in comparison to cvs. Elberta and Belle of Georgia, but have dished petals. 'Cucumberland' and 'Raritan Rose' the two white fleshed peaches can be identified as 'Cumberland' has large showy flower initially that later turn non-showy. Likewise 'JH Hale' and 'Elberta' can be distinguished as flowers are self sterile in 'JH Hale', but in 'Elberta', petals are longer than those of J H Hale. Raspberry presents wide variation in number of characters like presence or absence of prickles or bristles on the young ones. Glands on the petiole, peduncle, pedicel, glands sessile or stalked, form of leaflets, length and width relation of terminal leaflets. If in a genotype number of characters are variable and if all are taken to form a key for identification, the key then becomes very lengthy and complex and then the very purpose of identification is lost and it becomes very difficult and cumbersome to separate several hundreds of named cultivars on the basis of limited characters which otherwise than would be overlapping.

The leaf characters i.e. arrangement of leaves, venation, form, their outline, margin characters have been invariably used by the systematists or horticulturists in systematic works for identification. The leaf characters have been very useful for identification of fruit varieties especially in the nursery stage before they start flowering. J K Shaw (University of Massachusetts) has been a pioneer in systematic work (1902-1943) and according to him the leaf characters were very useful in identifying 92 apple varieties that have been detailed in a descriptive catalogue. Besides leaf, other characters like form and tree vigour, shoot, bark lenticels, serrations, surface and leaf blade are also useful in identification. In peach, leaf shape, size, colour are important for identification. The identification of fruit crop varieties through leaf characters is of paramount significance especially to the nurseryman as well as for the orchardists in a way that the buyers get true to type plants. Apart from these usefulness, identification has some limitation also in a way that only some few varieties can be identified. Vast experience gathered through regular observation is generally required for appropriate identification. The hybrid varieties are very difficult to be identified.

Nomenclature

Nomenclature is defined as the system of name(s) or naming, and usually is concerned with awarding or assigning a correct and befitting name to a known plant. The most significant function of nomenclature is to give a most appropriate and correct name to the plant. According to a well renowned systematist 'science can make no real progress without a regular system of nomenclature'. In respect of nomenclature principles, each entity, be it an object, species or variety has to be given a distinct and separate name. Many attempts were made in the early days to classify and name the plants and animals present on the earth. Aristotle (334-322 BC) the Greek scientist and rightly called as father of 'National History' attempted and classified animals and plants according to their morphological and anatomical similarities. Presently the scientific names of the plant are usually based on the Latin words or language and are recognized universally by these names all over the world. Further the Latin or latinised words are very specific and exact in their meaning and for their preciseness and exactness they fit in very well in otherwise keeping in mind the very descriptive work that one comes across in the natural sciences. As number of languages are spoken i.e. Greek, Chinese, Sanskrit or Hebrew, use of these words would result in far more confusion in naming.

Plants as such have no names of their own. The names by which we know them are the names given by us to them. The naming of plants serve two important functions, firstly it becomes convenient to refer to any particular plant and secondly they indicate relationships. A number of plants have been given a large number of popular names on the basis of their utility direct or indirect use to mankind in different languages. Under such conditions it becomes all the more important to assign a definite and particular botanical name which is used and recognized all over the world. The use of scientific names rather than common names is important. It is very often seen that many English names have been applied to the same plant e.g. Pansy has been given about more than 50 common English names and nearly the same number is given in German, French and Spanish languages. Likewise plantain has some 46 English names, 75 Dutch, 100 German and 11 French. The word 'Foxtail' has been employed to designate different grasses and 'cedar' was used to refer to a group of trees for quite long period of time. These popular names are most unsatisfactory and often confusing as some plants have too many names whereas ironically several have no common or popular names at all. The common names have a limited or restricted usage and is essentially restricted to the people who know the language, and showed no inherent genetic relationship. Hence the need for an acceptable system was felt which was international in its characteristics. Linnaeus (1707-1778) a Swedish scientist worked as Professor of Natural History at the university of Upsala. Linnaeus introduced the binomial system of nomenclature wherein every type of animal and plant shall have a particular name compounded of a substantive and an adjective. The former represented the genus and the latter the species. The generic name in general is common to all the species of a particular genus, for the simple reason that they all show common features of resemblance with one another. On the other hand the adjective or the specific name primarily refers to certain definite and specific characters which results in differences of the individuals of a particular genus from one another. The generic names to a large extent have been used arbitrarily and without any generally accepted principle. At present large segment of generic names are based upon combination of Latin and more so to the Greek words indicating some special or peculiar characters possessed or supposed to be possessed by the plants. This however lacked accurate knowledge and generic names had small characteristic.

As the number of plants known to man started increasing it was apparent that some uniform and generally accepted set of principles had to be adopted in the resulting confusion in names/ system of nomenclature was to be avoided. Professional botanists gradually adopted a definite system of naming plants and plant groups according to international agreements reached at meetings. These international meeting are called as International Congress and the rules adopted and published by them are known as International Code of Botanical Nomenclature. The prevailing International Code of Botanical nomenclature is the result of many years of trial and error. When nomenclature of fruit(s) is in question, systemists follow the rules of nomenclature formulated by the American Pomological Society. Some important features of the Code of Nomenclature of the American Pomological Society are,

- The main purpose of the Code is to establish a simple system of pomological nomenclature that is apt, stable and dignified. Name should be small usually one word, expressive of some trait like quality, place or person.
- The Code has wide scope and it applies to all cultivated fruits and nut crops.
- Within the Code limits, discoverer, introducer and originator has the right to name a new variety.
- Under priority head, the name published first is usually accepted and recognized. Variety name published shall not be applied to another variety in the same group.
- Names of new varieties shall preferably be of one word, but two words would also be accepted.

- Possessive noun to be avoided
- Variety name derived after a person place or a substance shall be spelled and pronounced in the same way.
- An imported variety to retain its name
- Titles and initials not to be part of variety name.
- Hyphen within variety not permitted
- Avoid use of general terms like seedling, hybrid, pippin etc
- Person name not allowed without his consent till he is alive
- The name of the published variety should be distributed in printed form along with description and illustration
- Variety description should cover all plant characters like plant, foliage, flower, fruit, growth habit etc.
- Varieties under test should be assigned specific letters, numbers code etc.

Systematic is a small part/segment of and is very closely associated with the taxonomic botany the systematic pomologist hence should be well versed with various systems and rules of botanical nomenclature. In the early days the plants were named and classified as per the convenience, but with the passage of time various efforts were initiated to establish some sort of universal code that would govern the botanical names. Perhaps the most popular and widely accepted or used was that given by Linnaeus. His system is referred to as binomial or Linnaeus system of classification and according to this system name of every species shall comprise of two words or epithets. The first word refers to genus and first letter is always written in capital and the second indicates the name of species. The two names together form the scientific name and results in a binomial. This system of naming came into operation with the publication of 'Species Plantarum' in 1753, the classical work of Linnaeus. According to the guidelines or concept of Linnaeus it was said that no two genera could have the same species name. Under the head priority, systematists avoided giving different names to same species and identical / similar names to different species. One significant problem which the systematists faced at early times was the large number of new species and genera which were being added to the existing ones at a rapid pace and these had to be described and classified, which resulted in duplication and added to the confusion that was already existing. Stendel (1821) published an index (Nomenclator botanicus) that had Latin names of all known flowering plants at that time. This work resulted in the formation of the present days most popular index called as Index Kewensis. After a lot of deliberation which occurred at several International meetings, minor codes were formed (Paris Code 1867, Rochester Code 1892). The Vienna Congress, resulted in the Vienna Code in 1905 and this turned out to be the most useful as regards the formulation of the existing international set of rules. By 1930, set of Rules of Nomenclature were formed which was International both in name and function. Some important provisions of the Rules of Nomenclature are listed.

- Author name must accompany the plant names which he finds or identifies i.e. Prunus persica L. (Linnaeus abbreviated to L.) *Malus soulardii*, Bailey etc.
- The name of the author is in no way part of plant name, and name(s) of the authors after the scientific name are usually abbreviated.
- A botanical name must comprise of 1) species name 2) variety name and 3) name of the author.
- Priority of publication. According to this rule the first published name is the official name and those published later are referred to as the synonyms. Descriptions are given in Latin and publication should be in printed form.
- Names in capital are made clear in botanical names e.g. Generic names should begin with a capital letter e.g. *Vitis, Passiflora, Macadamia, Mangifera* etc. Species names should began with small letter except those which held generic rank e.g. Prunus Persica, species name is written in capital because Persia was once the generic rank.
- Type specimen are the original herbarium specimen on which the identity of the plant is based.
- Plants that are brought into cultivation and those which do not differ much from the parent type are given the same name which are applied to the species and subdivision of species in nature.

Plants that arise as a result of hybridization, polyploidy or mutation or through any other means and those which show distinct perceptible differences from the parent specimen are often indicated with species epithets indicating some special feature.

Classification

The history and evolution of plant classification is very interesting subject as it provides the information not only about the person or persons concerned with it and their contribution but that from their contribution has developed the classification of plants based on biological facts and figures. The arrangement of plants into groups and sub divisions is usually spoken of as classification. The classification of Angiosperms presented to the taxonomists the greatest problem in systematic botany, as this group is very large and is inherited with many variations that at first it appeared impossible to arrange this enormous heterogeneous collection in some suitable and satisfactory order. Some people at very early time gave name and classified some plants that were directly suitable for human food. From this time plant taxonomy or systematic botany took its origin. Later certain practices, rules or principles of classification have gradually developed during the past many centuries. Now today we have a well developed and recognized principles of classification. The term taxonomy has been derived from two Greek words (Taxis+ nomos) meaning arrangement/ classification and laws governing the arrangement or classification of plants. The prime aim of classification is to arrange plants in such a way so as to give an idea about the sequence of their evolution starting from very primitive, simple and earlier to more recent, developed and complex types.

The classification of flowering plants is usually based upon a number of characteristics. Features like geographic distribution, anatomy, life-history, flower colour and many more are used as basis of classification. After constant observation over several years it has been found that reproductive or generic characters of a plant pertaining to the flowers are of more permanent nature and are not or little affected by the changing environment than the vegetative characters relating to shape, size, division, colour etc of vegetative parts. Except for phyllotaxis, venation and the presence or absence of stipules (vegetative characters) others are seldom used or are discarded as they are easily influenced by the changing environmental conditions. From time to time number of plant classifications have been proposed and are classified into one of the three systems namely 1) artificial 2) natural and 3) phylogenetic. The artificial classification is the most primitive and classifies plants for the sake of convience by taking one or few important characters that are used in classification. This is primarily used for identification and does not show any relationship and is based on one or a few characters such as fruit and flower colour, ripening season or some other superficial characters. The major drawback of the system is that closely resembling plants are often placed in widely separated groups while those quite different from one another are often placed in same group.

The natural classification makes use of all the important characters. The most peculiar and prominent morphological structures or characters are considered with the intention of grouping together those which are most similar in number of traits. This classification is based on the assumption that plants will fall into the same group (genus or family) that are closely related to each other in their evolution. It utilizes all informations that are available and reflects the situational knowledge as it exists under natural conditions.

The phylogenetic, system classifies plants according to their evolutionary and genetic relationships. Through this classification one is able to find out the ancestors or derivatives of any taxon. However the most limitation of this system is the limited knowledge available to the man to study the earlier forms. As the evolutionary picture is not very clear, hence the present knowledge is insufficient to construct a perfect phylogenetic classification. Despite serious problems and the records available are of doubtful nature yet the present day systematists with the help of sciences like genetics, cytology, anatomy, chemistry, paleobotany are sincerely involved to reconstruct the phylogeny of the present day plants. Artificial system of classification at present are not in use especially for the classification of higher categories such as genera and families.

The entire history of classification in short can be summarized and placed under distinct four groups. Each period is significant in its own ways and related botanists/scientists and systematists have contributed immensely towards plant classification.

Systems of classification are grouped into four type as below,

1. Classification based on plant habit
2. Classification based on an artificial system on numerical classification.
3. Classification based on plant forms
4. Classification based on phylogeny.

Classification is the natural end of systematic pomology. Most authorities dealing with fruits are in agreement that fruits deserve as careful study and as accurate classification as for other organism including fungi and mosses etc.

The pomological classification differs from the botanical classification in a way that botanical classifications are usually complete when the plants have been associated with species, genera and higher categories whereas pomological classification begins at this stage. The number of varieties in different fruits is quite very large, further the new varieties are regularly being added to the existing ones. The source of new varieties come from chance seedlings as well as from varietal improvement programmes. Some difficulties are often encountered in classification, major being that most characters or traits do not show pronounced taxonomic importance. Differences in varietal characters exist and at times it is quite distinct in a physiological rather than distinctly morphological in nature. These variations are often influenced by soil, irrigation or climate. Apart from these several new varieties are added every year to the existing ones mainly from hybridization between varieties of different groups. This results is the blending of varietal characteristics and lines of demarcation or their inherent variation is often broken down resulting in increased confusion among varietal groups.

From time to time pomologists have developed and used variation for different fruits, most of them being arbitrary in nature and some were build on natural basis, however none appears to be completely satisfactory. Hence on number of occasions, he makes a selection of, that is a combination of both natural and arbitrary. Today pomological classification pays less attention on characters that show high variability and over lapping continuity. Much emphasis is now laid on genetical characters that either show more discontinuity or contrast differences. The genetics/genetical tool are more frequently applied, as variability is that phase of genetics that consider the study of uniformity and variation or resemblances and differences. Genetics is therefore the exact study of variation as it takes not only morphological characters but also the chemical and physiological processes of plants. The geneticist invariably studies the frequency with which the characters constituting these resemblances and differences appear from generation to generation. Genetics along with cytology usually provides a reliable platform for the establishment of variation their resemblances and differences, relationships that are useful in classification of plants both cultivated (also varieties) and as well as the wild ones.

The vegetative characters like leaf margin (serrations), surface shiny or pubescent, number shape, size and colour of lenticels, straight or reflexed and mid rib are under the control of a single gene and discontinuous is nature hence are of greater value in classification compared to the fruit traits like size, shape which are polygenically controlled and quantitative. The use of plant characters in identification have been acknowledged and are by far more used in classification. A number of authors have used them extensively for the classification of fruit crop varieties, species as well as their wild relative types. In peach classification of small leaf characters, shoot and flower traits has resulted in reasonably suitable classification of varieties, Rolfs (1907), Connors (1919-21), Black and Edgerton (1946), Moore and Flory (1947). On the basis of leaf glands, form, shape size and type, calyx form and size, stone shape, pits and grooves (Sculpturing) and fruit characters, Bunyard (1938) an authority on peach classification has classified large number of peach cultivars. Characters like, small or large flower, gland type, pubescence present or absent, fruit flesh yellow or white has monogenic (single gene control) inheritance and hence showed discontinuous variation. Winters(1925) said that varietal characters of primary importance for identification of raspberry cultivars were color and structure of the spines (Prickles) on the cane and on the leaf petiole, presence or absence of glands, number of leaflets on each petiole, foliage colour and its intensity, cane height, glaucous or pubescent.

In apple the big or giant sized apples have been classified by the application or use of cytology i.e. determination of chromosome number and polyploidy. In cherries variation is more limited with distinct discontinuity in skin and flesh colour, and in other fruit characteristics. Plums show great genetic variability, hence classification could be based on number of characters, e.g. seeds may be smooth or rough, winged or not wide or narrow, apex pointed or blunt, oblong or round, necked or without neck, thin or plumb. Growth could be erect spreading, or spreading-drooping and drooping. Stone usually cling stone. Fruit shape usually ovate, bloom thick, red purple or black traits are dominant to oblate shape, thin bloom and yellow fruit characters.

In recent years besides classification primarily based on morphological characters, efforts are ahead for classification which is based on chemical and physiological processes and is of great practical significance to the orchardists. Cherries are being classified according to their incompatibility or pollination reaction, apples, strawberry, pears blueberries, Brassica according to their chromosome number (polyploidy), and in doing so it may show fertility relationship and their inherent potential for intercrossing. Classification based on chromosome number would be usefulness in determining the origin and the method of origin of species and varieties.

Classification of flowering plants may be based upon number of characteristics possessed by them, like life history, geographical distribution, environmental conditions, anatomy, uses in daily life, color of flowers etc. The Rules of Nomenclature define the categories into which the plants should be classified. The categories in taxonomical studies are called as units of classification which are stable and place fixed according to their status or hierachy by the well defined Rules. Any deviation from the set sequence amounts to the violation of rules and hence no credit is given to such classification which do not follow the rules. The units of classification are placed below in a descending orders from the units of greatest magnitude to the units of least magnitude.

Kingdom-Planta	
Division-Spermatophyta	
Sub-division-Angiospermae and Gymnospermae	
Class- Monocotyledoneae and Dicotyledoneae	
Subclass-Archichlamydeae	
Order- Rosales	(ales)
Suborder-Rosineae	(ineae)
Family-Rosaceae	(aceae)
Sub family-Rosoideae	(oideae)
Tribe-Roseae	(eae)
Subtribe-Rosinae	(inae)
Genus	(Rosa)
Subgenus	(Eurosa)
Section	(Gallicanae)
Sub-section	
Species	(gallica)
Sub species	(ssp)
Varieties	(var.)
Sub-varieties	(subvar.)
Forma	(f)
Clone	(cl.)

The categories of classification are further divided into major and minor units. The categories of taxonomy i.e. family genera and species etc. show their inter relationship. The categories in their development are highly variable and each unit is very subjective and is primarily based upon human concept. The major categories like divisions, orders and classes are established, whereas the minor categories like genera, species and varieties are small in magnitude but very variable in nature, hence provide a suitable platform to the taxonomists for extensive study work. The plant kingdom is divided into division and according to the rules of nomenclature it usually represents category of highest magnitude. Spermatophyta refers to seed producing plants and it includes most of our essential crop plants. The next category that follows or is next to division is sub division. Under this category the seed plants are divided in two groups, the Gymnospermae and Angiospermae. Special features of first group is that they are wind pollinated, ovules (seeds) are not enclosed in a chamber or ovary. Group comprises of conifers, yew trees etc. The group Angiospermae is very large group and comprises of all the flowering plants. The next unit that follows subdivision is class. The class name ends with 'eae'. The Angiospermae is further division into two distinct groups called monocotyledaneae and dicotyledoneae. Characteristic features of monocotyledous are, flower parts are in three's or six, parallel veination, vascular bundles many and scattered, stem section mostly herbaceous, seeds with only one cotylelon. Whereas dicotyledons have leaves with reticulate veination, stem woody with concentric rings, flowers pentamerous or in fours, cotyledon two in number.

The category class is divided into orders, and the order name ends with suffix 'ales'. Sometimes when orders are large in number it is divided into suborder the name ending with 'ineae'. The category order in turn is comprised to number of families, but sometimes there is only one family in an order. The family name usually ends with 'aceae' but some families end with 'ae'. Of the major units, family is the most smallest but most in usage. As in case of orders the large families like Compositaceae and Rosaceae, the category is sub-divided into sub families, the names end with suffix 'oideae' like Pomoideae, Rosoideae or Riberoideae. Sub families which are still large are in turn sub divided into tribes. The Latin name of tribe usually ends with 'eae' e.g. Pomeae, Pruneae, Rubeae Potentilleae etc. The category tribe may be further divided into sub tribes, that have Latin names ending with suffix 'inae'

The minor categories of classification include the units of classification like genera, species, varieties etc. Though termed as minor and smaller in magnitude but are of great concern to the systematists compared to the major categories because of their greater functional significance and usage. Consequently over years of extensive research and exploration huge data has been assembled and widely interpreted compared to the higher categories which are less variable. The Genus is the subunit of family. The generic name is the first name applied to the plant when a scientific name is given to it and the second name refers to the species name. The basic concept prevalent during Linnaeus days was that a genus is a distinct category whose species have more similar characters or resembles each other than they do with the members of other genera in the same family. However the drawback in this aspect is of assigning values to the selected characters when a genera is described. Some character(s) may be very effective or useful in organizing one genera, but may not be of any use to other genera with the same family. If the family component is very large, genus is sub divided into sub genera, which further are divided into sections, sub sections and series.

Species is the basic unit of classification and is the character on which binomial system is based upon. Individuals resembling one another in all important morphological characters both vegetative and reproductive constitute a species. The individuals of a species resemble so closely that they may be regarded as having derived from the same parent. However differences in shape size, colour of plant structure may not always mean a different species. Any variation below the unit of species is subdivided into subspecies, varieties, sub-varieties, form and individuals. Sub-species are individuals with very little variation in magnitude, the morphological features are not that distinct and clear. In contract to this some sub species have distinct morphological variation as well as distinct geographical distribution.

Systematic Pomology - Literature

The botanical variety is division of sub-species and is based on very small morphological variation in relation to growth, colour, ecological or geographical distribution. The last two units of classification 'form' and 'clone' represents the smallest categories that reveals variation in the individuals of a population. The term 'form' is frequently employed in describing many colour sports, petal or fruit colour. On the other hand the term 'clone' is applied universally to individual plants that are propagated only by asexual means. Hence all the individuals of apple, peach, plum, walnut, chestnut, mango, citrus, sapota, pomegranete, grapes, avocado etc. and most other fruits comprise a clone as they do not come true to type when propagated from seed.

Taxonomy encompasses identification nomenclature and classification and each aspect in itself is very elaborate and descriptive in nature. Systematic pomology deals with the same aspects relating to the fruit crops as the minor categories of classification are also taken seriously, the science therefore is more descriptive in nature. The science of systematic pomology is as old as mankind, and as man's interest in fruit crops increased since they were of direct use in number of way i.e. food, fiber, fuel, medicine, he took to write in detail about them. These initial writings, although improved from time to time resulted in some very important literary works in the form of documents, indexes or books. The pertinent literature is voluminous in nature and is one of the significant aspects of this science of fruits. World over many volumes relating to local flora have been published which in very simple and systematic way have given detailed description of plant species and their fruits and varieties. To familiarize with the subject one ought to know some of the classical standard works. The literary work of systematists is published in several languages worldwide. Some significant developments in systematic pomology that brought this science of fruits earlier considered as an art on a more recognized scientific lines is detailed below.

Confucius. 479 BC	Mentioned 'tao' the peach in his book
Aristotle. 384-322 BC	Father of Natural History. Transformed the biological data into a science
Theopharatus. 370-286 BC	Foundation of botany was laid. Described grafting, different types of apples and pears
Cato, MP. 234-149 BC	Described seven different apple cultivars
Varro, Marcus. T. 116-27BC	His publication 'Roman Farm Management' contained description of grapes, olive, apples, pears, dates figs, pomegranate and nuts
Camerarius, R.J. 1961	For the first time indicated sexuality in plants
Linnaeus, C. 1707-1778	Father of taxonomy
DeCandolle, A.P. 1778-1841	French, contributed enormously to systematic pomology
Van Mons, J.B. 1785	Practical plant breeding was started
Bentham, G. and Sir J. Hooker. 1817-1911.	Published 'Genera Plantarum' in 3 volume. The work comprised of names and description of all genera of seed plants known at that time. System of classification being simple was accepted world wide.
Hooker, William. 1818	Published an English book that had high quality painted pictures of the most esteemed fruits cultivated in British Gardens.
Diel, AFA. 1756-1839	'Systematic Nomeclature or Description of the Finest Fruits of Germany', considered to be classics of pomology in Germany
Thomas. A. Knight. 1823	Applied plant hybridization to a practical and systematic breeding of fruit crops
Engler, Adolph. 1844-1930	Published 20 volumes illustrating the 'Natural Plant Families' which provided the guidelines for the identification of all known genera from algae to highest seed plants. Published number of encyclopedias also.

Darwin, Charles. 1859	Published 'Origin of Species' is considered to be an outstanding work that gave new evolutionary viewpoint on the origin and variation of species. The work was based purely on natural selection and this indicated new search all over to find evolutionary evidences in all branches of mankind learning.
DeCandole, Alphonse. 1886	Published a book 'Origin of cultivated plants'. Formulated set of rules that later became the basis of present International Code of Botanical Nomeclature.
Hedrick, UP. 1900	Published 'A Laboratory Manual in Systematic Pomology'.
Beach, SA. 1862-1922	Placed the systematic pomology teaching on a scientific basis and also started apple improvement work through breeding.
Viala, P and Vermorel V. 1905-1910	Amelography. An outstanding work on the grape varieties of the world. A comprehensive grape work.
Schamel, A.D. 1909	Outstanding work on citrus improvement initiated through selection, introduction and breeding.
Bailey, L.H. 1914	Published 'Standard Cyclopedia of Horticulture'.
Bailey, L.H. 1910-1935	Published classified work on different fruit crops (temperate) like peaches, plums, cherries, pears, small fruits, blackberries of New York.
Bailey, L.H. 1925	Published a text book 'Systematic Pomology'.
Vavilov, N.I. 1936	Established eight centres of origin.
Bailey, L.H. 1941	Published 'Hortus Second'.
Webber, H.J. and Bachelor L D. 1943	Citrus Industry. The book contains wide description of all citrus species and various citrus fruits.

Plant Characters used in Systematic Description

The flowering plants exhibit enormous variability in different characters. The systematic botanists or systematic pomologists after studying vegetative and floral characters in detail have found them to be useful for classification purpose. A brief outline of different characters for identifying and describing flowering plant is detailed below. Without adequate and appropriate knowledge of these characteristic, description becomes a tedious job especially for graduating students.

Habitat	The natural place or locality where it has been growing of its own-wild type as food crop, ornamental or for any use
Habit	The plant may be annual, biennial or perennial Herb, shrub or tree Special features i.e. epiphyte, xerophyte, hydrophyte, parasite etc.
Root	Tap or adventitious Branched or unbranched Any other specific feature like aerial, fibrous, fleshy, tuberous nodulated etc.
Stem	Herbaceous or woody Circular, angular or compressed Shiny, hairy, spiny or waxy Branched or unbranched. If branched indicate branching mode Erect, or climbing. In case branching indicate their mode i.e. tendrils, hooks, spines etc. Solid or pithy Any modification i.e. bulb, tuber, rhizome, corm, phylloclade
Leaf	Evergreen or deciduous Alternate, opposite, whorled Sessile or petioled Stipule present or absent Simple or compound. For simple leaf, description of lamina, margin its nature, apex, surface venation. If

	compound then whether palmate or pinnate, latter may be pari or imparipinnate leaflets number, arrangement. For compound leaf, leaflets should be described in same way
Inflorescence	Are of different types like raceme, racemose, cymose, mixed compound etc. For each detailed information should be given
Flower	Without or with pedicel Complete or incomplete Regular or irregular Bracteate or abracteate. Bract to be described if bracteate Sex - uni, bisexual, hermaphrodite Ovary position, hypo, peri or epigynous Colour of flower
Calyx / Sepal	Number of sepals Green or coloured / petaloid Inferior or superior Folding of the calyx Poly or gamosepalous. If former, shape apex and its outline and if gamosepalous, forms like tubur, campanulate, spurred or any other form be described
Calyx/Corolla	Number of petals Aestivation / folding of corolla Petal colour. Inferior of superior. Poly or gamopetalous. If polypetalous shape apex and outline to be described. If gamopetalous, form like tubular, companulate, bilabiate etc.
Perianth	The terms gamophyllous and polyphyllous should be used and described in a way similar to calyx and corolla
Androecium	Number of stamens Filament nature short, long or any other form Colour of anther, dehiscence, fixation etc. Epipetalous or free from the petals

	Anthers introse or extrose Polyandrous or adelphous
Gynoecium	Number of carpels, mono, tri, penta or polycarpellary Ovary superior or inferior Syncarpous or apocarpous Number of loculi Number of ovules in each loculus or on each placenta Shape of ovule Placentation Style, stigma, their form and nature Nectary/nectar gland if present should be described
Fruit	External characters-shape, size, colour, ground/over colour, Lenticels- form, colour, cavity, stalk-thin thick, peel- smooth, thick Internal characters, flesh colour, texture consistency etc.
Seed	Cotyledon number, straight or folded Endospermic or non endospermic Position, shape and size of embryo etc.

1.1 Origin and Development of Fruit Crops

Study of fruit crops have always fascinated the scientist, pomologist, and breeders all over the world. Systematists have said that classification is the backbone of horticulture and provide a very stable and suitable platform on which varietal improvement is based upon. Variability is the primary need of the breeder and before the initiation of any breeding programme of substance, a breeder ponders and asks himself as what is the usually descent of plants with which improvement has to be carried out, in what form and where does this variability exists in nature and finally the extent, range and distribution among its wild relatives. This is essential because in nature there still are present a large number of wild types that hold potential to provide man with food clothing, drugs, fiber and other things of his use, the extent to which the related group breed together and finally the distribution, habitat and the variation in form do they express in nature. The systematists by giving definite generic, specific and varietal names to the varied type and have described their ecological and geographical distribution for better understanding of the plants.

As variation is the final tool employed by the breeder in crop improvement and the original habitats are the natural and valuable source of required variation. Hence the first step to develop a better or a superior variety or type, one should have collection of material from the wild native population or stands that represents wide range of variability within the species. In recent times, breeding programmes all over have been successful in collecting range of wild types from their original habitat that exhibited range of genetic variation. To broaden our knowledge about genetic variability one must know the natural habitat and centers of origin of our economic plant-fruit crops.

N.I. Vavilov through one of the most complete and extensive surveys of crop plant carried out on a very large scale is regarded as one of the monumental work in exploration and collection of tremendous mass of plant material. His sole efforts resulted in the establishment of eight world centers in which most of the plant wealth of present day major crops was present. He further emphasized that these primary centers were enriched with wide genetic variation as these were separated from one another by large mountain ranges, oceans, rivers, and even deserts which in a way restricted human activity and acted as natural isolation barriers. In these isolated places far away from human activity the process of evolution took place at a very rapid pace. Growing of their own, selection and natural hybrid occurred at a rapid pace, new types were evolved and selected / preserved or rejected, or may be left alone in the mixed population to survive of its own. These new types evolved provide horticulturist - new opportunities to work on hence they would like to be evolved in gene centre(s) which are more dynamic. Out of the eight basic centers of origin, the following centers are recognized as dynamics gene centers where fruit crops have evolved in plenty. These are

1. *The Chinese Centre:* This centre of origin is considered to be the oldest and largest of all centres and encompasses the mountainous regions of central and western China and adjoining lowlands. The important fruit crops that have been evolved here are pear, peaches, apricots, plums, oranges, citrus, litchi etc.

2. *The Asia Minor Centre of Origin:* The centre is known as Near East or Persian Centre of origin. This centre comprises interior of Asia Minor whole of Transcaucasia, Iran, highlands of Turkemistan is considered to be important centre as number of fruits have originated here i.e. fig, pomegranate, pyrus and malus and number of species of almond, grape, chestnut and pistachio nut.

3. *The Hindustan Centre of Origin:* The centre includes Burma, Java, Assam, Malaya Archipelago, Borneo, Sumatra and Philippines. Fruits that have

originated are mango, orange, sour lime, banana, coconut, citrus species rambutan, breadfruit, citrus, lime, grape fruit etc.

4. *The Central Asia Centre of Origin:* This centre includes the states of north west India, Kashmir, Afghanistan, Uzbekistan, Tadjikistan and Tian Shan province of China. Pistachio, apricot, plum, pear, almond, grape, apple, walnut have originated here.
5. *Central American Centre of Origin:* This is called as Mexican Centre of origin and includes south Mexico and central America. Papaya, guava and avocado are the fruits of this centre.
6. *South American Centre:* The centre includes high mountain regions of Peru, Bolivia, Ecuador, Columbia, parts of Chile and Brazil and whole of Paraguay. Pineapple, guava, passion fruit have originated in this centre.

Hawkes (1983) on the basis of early farming sites discovered by archaeologists in Thailand (11,000 BC), Near East (9000 BC) and Mexico (6000 BC) agriculture began several times and not once in different parts of the world and he proposed the concept of 'Nuclear centres and regions of diversity'. He gave the following nuclear centres and regions of diversity.

Nuclear Centre	Regions of diversity	Minor centres
A. Northern China	I China	1. Japan
	II. India	2. New Guinea
	III. South East Asia	3. Solomon Islands, Fiji and South Pacific
B. The Near East	IV. Central Asia	4. North Western Europe
	V. The Near East	
	VI. The Mediterranean	
	VII. Ethiopia	
	VIII. West Africa	
C. South Mexico	IX. Meso America	5. United States, Canada
		6. The Caribbean
D. Central and Southern Peru	X. Northern Andes (Venezuela to Bolivia)	7. Southern Chile
		8. Brazil

All over the world there are number of places/regions where the crops in fact did not originate, which is evident from the fact of the absence of wild types. In these regions, crop species were brought mainly from the nuclear centres at very early times and primarily due to spatial isolation in time and enormous human selection pressure played a very decisive role in increasing the present genetic diversity.

Important centres of fruit crops have been briefly described along with these there are some prime regions of diversity and some important are detailed below:

1. *Chinese – Japanese Region :* This region is very diverse in general climatic conditions ranging from below zero in the northern parts to warm temperate in the south. The region is spread over a very large areas covering vast areas of eastern Asia including parts of Iran, Afghanistan, Siberia, besides other adjoining places. A large number of fruit crops are part of this centre namely peaches, sweet orange, cavandish banana, cherries (Chinese Bush type), plums (Japanese), ber/jujuba also called as Chinese date, mulberry, loquat, litchi, kumquat, pears (sand pears), persimmon (Japanese persimmon). Among nuts, chestnut is not wide spread, apricot kernel is usually referred as Chinese sweet almonds. This region is remarkable in the sense that many fruit trees and ornamentals have their origin in this centre.
2. *Indo-Malayan Region:* This region is regarded to be very old and agriculture was in practice some 6000 years back, but very less is said to be known about it. Indo-Malayan region includes India, Indo-China, East Indies, Malay Peninsula and large number of small and large islands of the south seas that form or are part of Polynesia. This region in most parts usually experiences tropical or subtropical climate and heavy to very heavy rainfall is characteristic feature. The most significant contribution of this region to the world is large number of tropical and subtropical fruits along with some very useful condiments which are of prime significance from economic standpoint. Fruits and nuts that are part of this region are mango, banana, grapefruit, jackfruit, breadfruit, limes, lemons, citron, oranges, grapefruit, durian, rambutan, tamarind and carambola. Besides these fruits, nut crops that have originated are coconut, macadamia, Indian almond, walnut (*J. regia*) only in parts.
3. *South American Centre:* The topography of this centre is in geographically contrast to most other regions. The large geographical area is very diverse and comprises of highest mountains of the western world and some of the peaks may rise over 20,000 feet and decreases to sea level along the Pacific coast. The gradient of the slopes is such that from the base one can see tropical fruit and other crops cultivation and gradually the

scenario changes with the mountain heights and the top with absolutely different crops i.e. temperate fruits could be seen being cultivated and all this happen within a distance of few kilometers may be three or four. Many important economic plants have been originated and given to the world, namely cashew, pineapple, cherimoya (*Annona cherimoya*), strawberries, tree tomatoes and tomato. Some nut crops of importance evolved are cashew, coconut, black walnut, peanut, Brazil nut and palm nuts.

4. *Central American Centre:* According to Hawkes (1983) the nuclear centre southern Mexico, has Meso-America as the region of diversity and minor centres comprises the territories of United States, Canada and the Caribbean. This region is characterized by mountains, deserts and heavily watered areas extended to extreme southern portion and parts of central America. This area mostly lies with the warm temperate and tropical regions. The mountain elevation though high but ecologically ends/ results into only cool temperate zone. Important fruits that have originated are papaya, custard, apple (Annona), strawberry, persimmon (*D. virgianana*), Mexican cherimoya (*Annona longiflora*), Mexican plum, Crategus, gooseberry, raspberry, blackberry and among nut crops walnut (*Juglans* spp.), hickory, pecan.

If Indo-Malayan and South American Centre has resulted in evolution of tropical and subtropical it is Asia Minor/Near East, Central Asia and Chinese centre which are regarded to be the store house of temperate and warm temperate fruit crops from origin standpoint. In these centres certain prominent areas have been demarcated which over centuries have accumulated immense genetical variability. A brief outline of geographical location and fruit crops that have originated is described. Russian scientists have made detailed studies of Caucasus, Turkestan, Tian-Shan, Siberia and Far East in this regard.

Caucasian Centre: According to Russian scientist after very fine investigation of this region have said that 'wild pear and apple trees, plums *Prunus divaricata, P. spinosa, Crataegus* spp. and *Corylus avellana* plants grew in large number covering huge areas, and grew in the form of forests. The maximum concentration of these species were found on the northern slopes of Caucasus, in the mountain zone and as well as in the foot hill regions. In an another provenance Petak, these species of various fruit crops grew in the form of woodlands. In the Transcaucasia mountains of Armenia and in Georgia the same plant species are available but these species and forms are very divergent. In this region of diversity most of the wild fruit trees occur mainly at an height ranging between 900-1300 m, and usually the apples and pears are found considerably at higher elevations. In Savanetia, these wild forms of fruits are located at higher altitudes i.e. apple are found upto 2,130 m, wild pears 2,050 m, Sorbus, Ribes and Rubus at even higher altitudes of 2,500 m, and wild sweet cherry at lower elevation of 1,980 m. This region also represents principal centres of varietal diversity of such important fruit trees as *Cydonia oblonga, Punica granatum, Prunus divericata* and *Prunus avium* may be found. The form of such species as *Diospyros lotus, Pyrus eleagrifolia, Cerasus incana, C. microcarpa, Amygdalus georgia, Ficus carica, Corylus colurna* and *C. colchica* originated in Transcausasia. *Vitis vinifera* cultivated and its wild forms have originated in Transcausasia, as large number of indigenous types are wide spread in Georgia, Armenia and Azerbaijan, showing wide variation in size, colour, seeds, and very often the grape vines are seen twined around wild quinces and pears. The colour of wild grape is usually black with sour taste. This region represents huge diversity in the number of crops that have evolved and also in the number of species available here. Fruits that have originated are pears, apples, plum, quince, cherry, almond, medlar, hawthorn, walnut, fig, persimmon, chestnut, hazelnut, jujube, grape, pomegranate, currants, gooseberries, blackberries and blueberries.

Tian-Shan : In the Tian Shan (eastern area) of China very thick plantation of wild almond in the form of huge grooves are found growing in the thinly populated mountain gorges. The plants are found growing in almost xerophytic conditions at an elevation of 11,000 to 11,500 ft. Both sweet and bitter type of almonds are present as depicted by morphological variation between the wild and the cultivated ones. In some wild almonds the kernel quality was as good as the cultivated ones. Apricot is another fruit of this region that has shown wide genetic variability. Some wild apricot types found in the forests have several forms, varying from the cultivated being smaller in size and stones, have bitter kernels but are perfect edible. In wild apricot, the stone comprised nearly 20 per cent by weight, and also indicated variation in number of associated endocarp traits i.e. shape, size, flatness, apex being sharply pointed or slightly pointed, shape orbicular or broadly orbicular or rounded, surface rough or smooth, thin or thick endocarp ventral suture and its associated characteristics.

Siberian areas: The cold deserts of Russia, the Siberian area is often said to be densly populated with the wild apple especially the crab apple i.e. *Malus baccata* Borkh. The most striking feature of wild form of apples found particular in this region is their winter hardiness. Three species *M. baccata* var. siberica, *M. baccata* var. mandschurica and *M. baccata* var. himalaica are found in abundance and on account of its free hybridization with the cultivated apple, many hybrids have been obtained, and the hybrids have adapted well to the northern most region where the common apple has failed to establish due to lack of adequate winter hardiness.

The Far East Area: This area is considered to be rich and diverse in wild pear and most important hardy species is *Pyrus ussuriensis*, the apricots namely Siberian apricot (*Armeniaca siberica* Pers) and Manchurian apricot. The original wild cherry specie *Prunus mandschurica* Koehne, the Manchurian walnut and the wild hazelnut. Apart from this very hardy grapes *Vitis amurensis* are of particular interest because of its variable forms, whose fruits over long period of time have been consumed by the local inhabitants.

Turkestan: The area exhibits tremendous variability in different types of fruit crops. As in the case of Caucasus the major wild fruit types were invariably found in the lower or foothills and some even in the higher reaches of the mountains. The species and varieties exhibit variation from those available in Transcaucasia and even in Caucasus. The area is highly enriched with the wild forms and species of *Pyrus, Prunus, Vitis, Rubus* and *Mespilus*. In this region also wild forms of apples and walnuts are present in dense woodland types. In the eastern mountain areas of Turkestan wild apples, pears, almond, apricots, Persian walnuts, pistachios and raspberries are found in plenty, varying enormously in morphological forms. The Pistachio are typical endemics of Turkestan. In fact the fruit belongs to semi-desert, or to the dry slopes of low mountains and foot hills with loose soils. The apple types available here show huge variation in form and type. *Malus pumila* var. Niedzweizkayana is remarkable in the sense that intense red colour of skin enters the fruit pulp, the bark and veins are intense red in colour so are the flowers and seeds of the fruit. Further in this area whole lot of variable forms small, sour, low in production to large sized sweet in taste intense red coloured and high in productivity type are found in abundance. Some wild trees bear fruits which in no way are inferior in quality to the cultivated types. 'Alma-Ata' – 'City of apples' is said to be surrounded by thick forests of wild apples. The most striking features of wild apples of Turkestan area is that they are usually bigger in size as compared to wild apples of Caucasus region.

The origin and manner in which the fruit crop varieties have evolved is still not very clear, because they have been in cultivation from prehistoric times, almost about 3000 to 4000 years prior to the beginning of Christ era. In these early times the only method for improvement used was selection, which was based on external or phenotypic appearance or characters or entirely based on its direct use. At this time no knowledge about male (pollen) and female parent existed. The early improvements in varieties were mostly of chance seedling. The findings of sex in plants by Camerarius in 1691 was a major breakthrough in plant history. With this the plant improvement was placed on more scientific basis. Later Mendel's work provided the requisite boost and selection, hybridization and selection procedures were employed in a much refined manner. Through Mendel's law of inheritance it was possible to know or predict the nature or inheritance mode of a character if the genetic make up of the plant was assessed earlier. With this discovery breeding / improvement work in number of fruit crops was initiated especially in Europe by very large number of workers like Thomas Fairchild, Van Mons, Thomas Knight and a whole lot of new aspiring fruit breeders were growing thousands of hybrid seedling. Appropriate data of both the parents were also recorded for comparison of the resultant hybrid or their deviation from the parents. To improve plants, modern methods of research have been applied, especially a combination of number of sciences like genetics, cytology, pathology, chemistry, biochemistry have been used for early and effective attainment of breeding or improvement objectives. The magnitude of information obtained from supporting or allied sciences like cytology, genetics, geography, plant morphology and floral anatomy provided breeders with useful information that was essential in plant improvement work. Most widely used methods of plant improvement are

1. Selection of chance seedling.
2. Selection from hybrid progeny.
3. Selection of gene mutation or bud sports.
4. Polyploidy-chromosome doubling of varieties and species.
5. Species hybridization-wide hybridization.

1. *Chance seedling selection:* Most of the fruit crop varieties that are cultivated commercially have all originated as chance seedling. It was this raw material on which man implied his inquisitive nature and keen power of observation has made some selections which later became the backbone of commercial orcharding system. These seedlings, numbering in thousands were found growing along high ways, deserted fields. This method is perhaps the most simple form of breeding, which only requires keen observation alongwith the assumption of the potential worth of the new selected plant. Chance selection of seedling has resulted in selection of thousands of varieties in fruit crops and quite a large are even cultivated commercially. According to an estimate, considering all fruits, nearly 95 per cent and sometimes even more in some fruit crops have resulted from this simple seedling selection method. Some of the selections which were made two to three hundred years back are still in cultivation mainly due to the accumulation of desirable genes over years

of growing under particular environmental conditions. A few selections have made an everlasting impression on the fruit industry.

2. *Hybrid progeny selection* Improvement is through controlled hybridization. Through this process desirable genes for many characters from a plant can be transferred from one variety to another. Donor and recipient are selected after their proper evaluation. Desirable pollen is allowed to grow on the stigma. After fertilization, when fruit formation and development processes are completed, hybrid seeds are collected, later grown to assess the potential of the seedlings. Crop improvement through hybridization especially with fruit crops is long and continuous, time consuming and costly affair and is often said to be never ending process, unless the breeding objectives are very well and clearly defined. Successful breeding work usually requires adequate assistance to grow and evaluate the hybrid seedlings till the fruiting stage. Baring few fruit crops the juvenile period is quite long where breeder can do nothing but watch and wait till the seedling starts fruiting. Much attention currently is directed to shorten this period. The other prime objective include many characters like improved plant and fruit characters, large and uniform colour and size of fruits, better storage and transportation qualities, hardiness to climatic conditions, resistance to major attacks of insects pests and diseases and above all higher in productivity. Despite tremendous progress made in varietal improvement through hybridization, still a variety high in all dessert quality is yet to be developed. The task appears to be a difficult one as quality traits are usually controlled by polygenes. In the present varietal or crop improvement scenario there is no fruit crop which does not require improvement in one or the other respect such as better resistance to insect pest and diseases (to reduce input cost incurred for their control), resistance to low and high temperatures, improved fruitfulness and quality, adaptation to varied soil and climatic conditions etc. In spite of number of constraints and inherent problems associated with fruit crops the prime aim of the breeding is to combine as many good traits as possible in a single genotype as just one single trait like better storage, very early or late ripening, intense red coloured or uniform colour character(s) can revolunize the fruit industry.

3. *Gene Mutation:* Means sudden heritable change in the genes and one which is transmitted to the offspring. A large number of varieties of horticultural crops have arisen through mutation which are also called as bud sports. During growth the cells, undergo rapid division. One single cell divides into two daughter nuclei and so does the chromosomes

in a well defined and systematic manner. However sometimes due to errors in division of chromosomes and genes on chromosome also undergo re-arrangement and these changed patterns are transmitted to the new cells. These changes may then accumulate at specific places. It is interesting to note that if an axillary bud is formed from the changed locations, new growth in the form of shoot or branch may develop varying in growth habit, leaf morphology and if the changes are large it may effect characters of flower and fruit significantly. These changes are known as mutations and may occur in the cells of leaves, flowers, fruits or in any other part of the plant. The mutants are easily spotted from the main variety as it may be differing in size, colour, texture or leaf morphology. These mutations or mutants are of immense value especially in vegetatively propagated plants. Isolate the variant or mutation or the mutant segment and propagate either through budding or grafting. If the changed characters are reproduced the mutant type is immediately recognized as a new variety. The process of gene mutation is very specific i.e. some fruit crops are more prone to such changes and some do not respond. Apples, peaches and roses are known for their high frequency and number of bud sports that they have produced. Gene mutations have played significant role in differentiating crop plants. Nectarines have arisen as recessive mutation from the peach in that it is devoid of pubescence on the exocarp and for all other traits it is a peach. This and a large number of new varieties of various fruit crops have resulted spontaneously from gene mutation. Ever since man took selection as a way of improvement, gene mutations have attracted the breeders attraction as it has contributed largely as a viable source of new variability.

Mutations that appear in the sex cells i.e. pollen or egg are not detected until the next year and the variability is enclosed in the seed. The seed after germination shall show the changed characters. Further if the parent variety is cross pollinated and the change occurred is recessive, the variability may be expressed after several generations. Gene mutation also result in chimeras and they are the individuals that contain both mutated and non-mutated tissues or an individual composed of two or more genetically distinct tissues. Mutations may occur in number of plant characters but those that influence the fruit colour, its intensity, early or late in ripening are regarded of greater significance to the fruit industry. Apart from desirable, undesirable changes also occur in high frequency e.g. abnormal fruit shape and ribbing, skin russeting and roughness, pollen sterility etc.

4. *Chromosome doubling or polyploidy :* Is a phenomena in which the plant has more than double the basic number of chromosomes. Improvement of fruit varieties through natural or artificial doubling of chromosome has in recent times been recognized as a method to enhance the existing variability for better utilization. Most of our economic plants are natural polypoids and to this list number of fruit crops like apples, plum, cherries, strawberries and other are also added. The hereditary information is located on the genes which are present on chromosomes, hence any charge in chromosome bring about marked changes in the plant. Sometimes the sex cells are not reduced and are formed with all the chromosome or the chromosome number may be doubled in the simple cells of the stem root or axillary buds. The resultant type with double chromosome number when further divides, all the daughter cell will have doubled chromosome number. When the phenomena of polyploidy was known, chromosome doubling was attempted by number of ways i.e. heat and shock treatments and some drugs were also tried. Later a poisonous product called colchicine was extracted from *Colchicum*. This potent chemical inhibited spindle fiber formation, the daughter nuclei remained within one cell and resulted in chromosome doubling.

   With this technique polyploidy was attempted in number of species and varieties of different fruit crops with the hope to develop bigger and better plants compared to the normal diploids. In some cases induced polyploids had bigger flowers, fruits and plant size whereas in some tetraploids the plants were of smaller size, with high degree of sterility.

   In some species and varieties the fertility was rated as good and was restored through inter-crossing with different tetraploid lines. Despite some short coming in polyploids they are very valuable for horticulturists in hybridization programmes and also of prime importance to study genetics. The breeder uses four different diploid plants when he crosses two tetraploids, large combinations may occur among the four different sets of chromosomes which were part of the original species or variety. The pollen sterility observed is mainly due to the fact that chromosomes fails to pair effectively during meioses. However when two tetraploid are crossed, pairing occurs with parental set of chromosome and the resultant plant shall be fertile. It has been observed that effects of polyploidy are very distinct in the form of large sized leaves, flowers and fruits. In England and United States of America bud sports of pear have been designated as 'giant' because of their large sized fruits. Cytology studies showed chromosome number to be sixty eight instead of thirty four of diploid types. The tetraploid apples were also larger than diploid ones. The stone fruits respond very less to chromosome doubling.

5. *Species hybridization:* With the finding of new crops and later their detailed studies have shown that they have evolved through interspecific hybridization which was different from the methods described so far. This resulted in the fifth method of improvement. The development of cultivated strawberry clearly reflects how new, different and important plants and cultivars have been selected from off springs resulting after hybridizing two species. The present day strawberry never existed in the form available today. The large fruited strawberry was obtained when three species with small fruits i.e. *Fragaria vesca, F. virginiana* and *F. elatior* were crossed with *F. chiloensis* a large fruit bearing species. It is interesting to know that plants of *F. chiloensis* are dioecious i.e. male and female flowers are borne on different plants. Further by chance only female plant were introduced to Europe. At the planting site *F. chiloenisis* did not bear any fruit for long time as there was no pollen source from any other species. As time passed, some of its plants through some error were planted near other European species that had small fruits. The pollen of these small fruited species pollinated female flowers of *F. chilensis,* which resulted in good yields consistently with large fruits. This provided new vistas for the development of better strawberry. The seeds from large fruits were collected, seedlings thoroughly evaluated were later crossed with *F. virgianana* to induce better flavour and quality traits. Slowly and gradually other useful traits like better shelf life, day neutral etc traits were transferred and strawberry industry was established on a firm footing. In a short time, from unknown, poor in quality, strawberry has now established itself as a prime fruit.

Another fruit crop currant, especially the Red currants belonging to genus Ribes is said to have been evolved from three species *Ribes vulgare, R. rubrum,* and *R. petraeum*. The species show wide variability in number of characters like, pubescence, bristles or prickles, deciduous or rarely evergreen, leaves simple or palmately lobed, flowers very small to missing, solitary, few flowered or many flowered raceme etc. Besides this, Duke cherry has been obtained from after systematically crossing sweet (*Prunus avium*) with sour (*P. cerasus*) cherries. Apple also has resulted from inter specific hybridization. Plum cultivars from *Prunus cerasifera, P. spinosa* and *P. domestica* are inter-specific hybrids.

India is a vast country and has been bestowed with varied agroclimatic conditions which are highly suitable for cultivation of different fruit crops ranging from tropical, subtropical to temperate fruits. Presently, India is ranked second in world fruit production, next only to China. India is a natural genetic reservoir for number of fruit crop species. About 20 genera such as Artocarpus, Carissa, Dio, Emblica, Ficus, Juglans, Grewia, Mangifera, Musa, Morus, Malus, Prunus, Punica, Pyrus, Ribes, Rubus, Syzygium, Vitis and Zyzyphus provides enormous fruit variability. These and other large number of fruit plants grow in various regions of India give fruits a special status that are of great economic value both as fresh fruit or after processing.

Indian agriculture so far has been hinged around production of annual crops particularly the cereals to provide food to the huge population. This approach has resulted in severe ecosystem imbalance and also lack in balanced diet. Today fruit crop trees have become an integrated part in plantation programmes especially on the marginal or degraded lands both in hills and plains. The fruit crops if selected and planted judiciously would definitely enrich the ecosystem and on the other hand would provide nutritive fruits to the population. At present there is limited land available for expansion of agriculture purposes, but plantation of fruit trees on the wasteland comprising of rocky, sandy areas, undulating terrain, saline and sodic soils could be effectively exploited for cultivation of hardy fruit trees like karonda, ber, datepalm, Indian gooseberry, datepalm, guava, jackfruit, bael, custard apple, fig, pomegranate etc.

Fruit crops exhibit unique wealth of our country. Presently crops grown include mango which is native of India, and apple which is an introduced species. For adequate growth flowering and fruiting, fruits require specific, agroclimatic conditions i.e. apple performs best under cold temperate region while mango does well under subtropical and tropical conditions, while date prefers hot and dry conditions–desert. The fruit crops show divergent anatomical and morphological variation in growth, flower and fruit characters. The plant growth varies from low spreading in strawberry, to perennial herbs like rasbhari, vines like grapes, kiwifruit, passion fruit, low bushy shrubs i.e. fig, pomegranate and tall trees like mango, jackfruit, date, coconut, walnut etc. Floral structure is variable and size varies from very small in Indian gooseberry, mango ber, avocado, to very large in passion fruit. As regards the sex of the flower(s), most fruit crops have bisexual flowers, but crops like papaya produce male and female flowers on separate plants, persimmon, kiwi also produce male and female flower on same or separate tree (vines). Fruit, the economically important plant product also reveals structural variation. Thalamus forms the edible part in pome fruits, are the outgrowth of seed is edible part in litchi, pomegranate, rambutan. In mango, ber, karonda, phalsa fruit is a drupe. Fruits like cashew, almond, walnut, pecan filbert and chestnuts are designated as nut crops, seed/kernel is the edible fruit and passion fruit, sapota, grape, guava, citrus, passion fruit, cape gooseberry are classified as berry fruit, varying in their edible parts.

Another interesting fact after studying various fruit crops is revealed that staple grain crops like wheat, maize are obtained from single family Graminaceae, but fruit crops are derived from number of varied families belonging both to monocots and dicots. Further the history of fruit cultivation too is very old i.e. mango, banana and coconut are prehistoric, to crops like strawberry and karonda have very short history of cultivation. Fruits have always fascinated mankind by ways of its delicate flavour, taste, flesh consistency, colour, shape etc.

1.2 Principles of Plant Taxonomy and Botanical Nomenclature

Plant taxonomy (syn. Systematic botany) as well as systematic pomology is regarded to be a functional science and is mainly involved with the identification, nomenelature and classification of plant of various kinds and forms which are available to be human being. Identification basically is concerned with determining that a particular plant is similar to some other known type or individual, whereas nomenclature is involved in application of correct name or names to the individual or group of individuals. As such usually plants have no names of their own, but are known by the names which are given to them by the people concerned with plant taxonomy. Giving a correct and an appropriate names is very essential and it serves two very essential purposes, one they are convinent in referring and secondly it shows/ indicate relationship. Classification the last aspect of taxonomy deals with the placing of the plants or groups in some specific categories based upon some specific procedures or plan.

In the 19th century with the initation of descriptive taxonomy the principles of taxonomy were formed on scientific lines, and the prime aim was to establish the similarities and differences in the total morphological characters of the plants which were present all over the earth at that time. This classical and outstanding descriptive taxonomy work was started with the works of taxonomy greats like Tournfort, deJussieu and Linneaus. This work was then followed by other workers like Robert Brown, John Lindley, Hookers, deCandoles, George Bentham and many more in this line. At this time the major principles of taxonomy were formed, refined and published first by Linneaus in his *Critica botanica*, followed by Adason, deCandole, John Lindley, and Sir Jospeh Hooker. These publications invited some criticism also as it was mostly based upon morphological characters and failed in establishing genetical relationship between taxa(s). Today association is based upon morphological characters in close association with the findings of anatomy, cytology and genetics. In the early times of this period natural system of classification was followed and was based upon understanding of the nature at that time. As time passed, new researches with better tools and equipment and with acceptance of 'the theory of evolution' provided a useful platform to the taxonomists to classify the known plants on the basis of their actual genetic and ancestral relationship. With these findings it was concluded that characters of the present day living plants through the process of evolution have evolved from simple structure(s) to more complex ones including their genetic organization over their ancestors. This system of classification was different from the early classification 'natural classification' and was called as the phylogenetic system and indicated much better genetic relationship between plants. Presently the taxonomist, use all old and new information from various fields of morphology, ecology, plant geography, physiology, embryology, genetics and cytology for plant classification. Charles. E Bessey (1915) after extensive taxonomical studies on different plants laid down some very important principles which were later called as 'Besseyian principles' in respect of the phylogeny. According to these principles

1. The life on earth has moved from simple to complex forms but due to regressive evolution it becomes simple on account of loss or degeneration of part or organ.
2. Evolution is irreversible.
3. The simple form of life currently present resemble more to their ancestors compared to the complex forms.
4. Terrestrial forms are more primitive in comparison to epiphytes, aquatics, saprophytes and parasites.
5. Usually perennials are more primitive than biennials and biennials more than annuals.
6. Woody forms is primitive than herbaceous.
7. In Angiosperms bisexual flowers are more common than unisexual.
8. Among flowering plants simple leaves are more common than compound leaves.
9. Occurrence of single flower is usual compared to flowers in inflorescence
10. Monoecious form is more ancient than dioecious form.
11. Flowers with free petal were formed much earlier than fused form of petals.

12. Symetric/Zygomorphic flowers are more developed than actinomorphic ones.
13. Epigynous and syncarpous forms are at more developed stage then hypogynous and apocarpous.
14. Apetaly condition is most advanced, flowers with many parts were formed earlier compared to flowers with few parts.
15. Non endospermic seeds or seeds with very little endosperm are more advanced than endospermic seeds.
16. Stamens in united condition is more advanced than free ot polyandrous condition.
17. Compound form of fruit is at more developed stage than simple fruits.

At early times several names were given to a single plant in different languages. A plant with many names usually creates confusion and these popular names are unsatisfactory when used all over the world. To over come these nomenclature confusions, an attempt was made in 1923 by a Joint American committee on Horticultural Nomenclature and published a list of Standardized Plant Names. This attempt could not succeed as very few people adopted these names who were not part of the government agencies,. Usually the botanical names have been given by the botanist to all the plants known in this universe and are same in all the languages and hence provided an international means of referring to them. Linnaeus in his wonderful work compiled in the form of *Species plantarum* (1753) used binomial system or binomial nomenclature This system was accepted and adopted by all botanists. Later in 1821, Stendel in Nomenclature botanicus gave a list of the Latin names of all the flowering plant that were available at that time. Around 1840 he published 'Index kewensis' which was the second edition.

Rules of Nomenclature

With the passage of time and better understanding of different form of plants and their relationship, better tools, efficient mobility, the number of plants known to man increased. It was now realized that to avoid confusion in names, some sort of uniform and generally acceptable principles had to be developed and adopted. Botanist through various discussion at international levels, and rectifying the shortcomings in the prevailing naming system gradually adopted a definite system of naming after several meeting which were called as International Congress. During such meetings number of rules were adopted and published which were called as International code of Botanical Nomenclature.

The 1956 edition of the code has given detailed instruction on the steps needed especially when a change appears necessary. The required change is first submitted to the different nomenclature committee which have been constituted by the International Botanical Congress. These committees then make their recommendations on the subject. During the International Botanical Congress every change that has been proposed is discussed at length before any decision is taken on the subject. Any proposal for a change receives due consideration before it is put to final vote for confirmation or rejection. During the last few International Botanical Congresses there has been only minor changes in the rearrangement of the code.

Major points of the International Code is that the Code is composed of three main points or parts, Principles, Rules and Recommendations alongwith few appendices. The most important component of code is the principles. It does not give detailed rules about nomenclature but show the main ideas that have guided the compilers of the code. The botanist must keep these sections of code in mind when he wants to publish a new taxon. The main point of this section is that a plant can have only one correct name. The second component of the code-Rules or articles provide detailed information on all the aspects that are related with the naming of the plants and lastly the recommendations deal with the practical aspect of the rules.

History and Evolution of Plant Classification

The subject related to the history and evolution of plant classification is of immense interest in the way that it not only tells about the persons related to it and their large contribution, but also that from their contribution various systems of plant classification have been evolved which are mainly based upon different biological facts and figures.

In plant kingdom angiosperms are the most diverse group and has given maximum problem in systematic botany to the taxonomists. Angiosperms is a very large group with wide variation and heterogeneous collections. When the thought of classification crept into the mind of botanists & after seeing the tremendous variability, at first seemed impossible to arrange such a material. Man in his quest for a settled life came across several plants but he started giving some names to the plants which were suitable to man as food. From this time the systematic botany or plant taxonomy took its origin. From this initial beginning to the present day, taxonomy has undergone vast changes. As time passed by the primitive man came across large number of plants, gradually he learnt more about their uses and started to give some sort of classification to them. The primitive man started to learn more about the plants which were of direct use to him as food. In this process leaves, shoots, roots and seeds, ones those were made into brews or poultices for treatment of various illness were given some names and classified them. By and by in the early stages of settled life or civilization man started to recognize various plants which were useful to him as food, fuel and also the ones which proved fatal to his near and dear ones. Such sort of early classification were primarily practical in nature and indicated their importance from economic point of view. These grouping were mainly based on characters like taste, smell etc. At this stage he lacked knowledge of morphological characters. After finding the usefulness of plant several ages passed by and gradually man starting cultivating them. Initially no field or land was prepared and seeds were thrown all over the land. Later he started soil preparation with the raw implements, he also started selection of better varieties and cultivated them.

The Indians of the Vedic period (2000 BC-800 BC) knew the art of agriculture and cultivated number of food crops. During this period anatomy and external features were studied by the scholars for better crop understanding especially of those plants which were commonly used. For better performance of both the crop and soil crop rotation was practiced. Vriksh ayurveda science of plants and plant life in the ancient India formed the early basis of botanical teaching and the medical studies especially of those plants that were used as poultices, pain reliever or for healing of wounds etc.

In the middle ages there was an upsurge of Islam among the Arabians. The Islamic scholars showed exceptional interest in medicinal plants and these scholars contributed immensely to the botanical science. During the sixteenth century the Yunani medical science was well established and number of books were published that contained description of numerous medicinal plants, various plant parts that were used as medicine.

Plants of one kind or the other are present in the universe at one or different places. They have wide adaptability and can resist extreme conditions very well. Most plant types are finely rooted in the soil, algae live in water and few of them can even grow in hot springs at a temperature of about $179^{0}F$, bacteria on dirt particles, and seen floating in air. Lichens show still wide adaptation and are easily seen growing on rocks under extreme hot, cold and drought conditions. Australian Eucalyptus are the largest growing plant and may attain a height of 400-500 feet. Besides these there are some bacteria which are very small in size. The job of the taxonomists is to arrange such a wide range of plant types into some system is extremely very difficult. To overcome these difficulties plant taxonomist have used three types of classification which are as follows.

Artificial : The main feature of this classification is that one or few important characters that are useful in identification of plants are used for the sake of convenience. Though very simple in form but the major drawback is that sometimes closely resembling plants are placed in different groups and the ones which in fact are different are placed in the same group eg. colour or shape of the leaf, fruit colour etc.

Natural : In natural system of classification of plants, emphasis is laid to include as many large number of important characters as possible and the plants are then classified in relation to their related characters. This system shows the plant status as it exists under natural conditions.

Phylogenetic : This system of classification is the most recent one and according to this system plants are classified in respect of their evolutionary and genetic relationships. Through this system it is possible to find out the ancestors or the derivative of any taxon. However at present through the existing knowledge a perfect phylogenetic classification is difficult to be constructed and all the phylogenetic system of today are formed by the combination of natural and phylogenetic evidences. The system of classification has been divided into four types and periods.

Period I : *Classification based on habit:* The classification based in habit was formulated by the Greek botanist and the herbalists as early as about 300BC and it continued upto middle of eighteenth century. Plant taxonomy in fact started with these early efforts of the Greeks. This classification was basically based on the habit of the plants, and the botanists were of the view point that the system showed natural closeness or affinities. Various workers associated with this system and their outstanding contribution to the classification is briefly described.

Theopharatus (370-285BC.) : He was known as the father of Botany and was a student of Plato and Aristotle. He divided plants into trees, shrubs, undershrubs and herbs on the basis of their form and texture, established differences between annual, biennial and perennial. In his 'Historia Plantarum' he has described and classified nearly 480 plants. This work was simple in nature and was easily understood by all and due to its simplicity nothing much outstanding was achieved for the next eighteen centuries.

Albertus Magnus(1193-1280) : He was born in southern Germany and in his Devegetabilis gave in detail the plant description. Practical methods of gardening and orcharding were described in detail. The work was widely distributed and read both by common people and as well as by university students and professors for about 200 years after its compilation. On the basis of stem structure he established differences between mono cotyledons and dicotyledons. These differences were possible with the help of so called crude lenses that were available at that time.

Otto Brunfels (1464-1534) : He established the perfects and imperfect group of plants on account of presence or absence of flowers. It is often said that modern systematic began with his work. Apart from plant classification he was also interested is medicinal values and domestic uses of plants. He gave a total account of the plants that were known at that time.

Jerome Bock (1498-1554) : He studied botany as a hobby was a German, started has career as a school teacher later become a minister and a physician by profession. After Otto Branfels he was referred to as the 'German father of Botany'. His outstanding contribution was that he succeeded in classifying the plant into herbs, shrubs and trees and tried to bring together the plants which showed affinities with each other.

Andrea Cesalpino(1519-1603) : He published his classification work pertaining to 1500 plants in 'De Plantis'. His classification was based on the characters of the seed and the embryo. He was an Italian botanist and a physician. Plants were classified according to habit (herbs-trees), characters of fruits and seeds. He established a herbarium of about 768 plants, is regarded to be one of the oldest herbariums and is still present.

Jean Bauhin(1541-1613) : Wrote 'Historia plantarum universalis' which was in fact published in 1650-57 after his death. The work comprised description of about 5000 plants with about 3500 sketches. This voluminous work was published by his son-in law who worked with him.

Gaspard Bauhin(1560-1624) : He travelled extensively all over the world especially Egypt and East Indies and collected large number of plants during his travels. Later he published Pinax in 1623 which comprised of 12 books. In his work he described nearly 6000 species on the basis of form and texture. He was the first to describe binomial nomenclature, but did not live to give it a solid form.

John Roy (1628-1705) : Published 'Historia plantarum' in three volumes from 1686 to 1704 on the basis of presence of embryo of one or two cotyledons and continued with the old systematic basis-plants habit which was divided into trees and herbs. He laid more emphasis on morphological and less on anatomical characters.

J.P. de Tournefort (1656-1708) : He is referred to as the founder of the modern concept of genera. He divided the flowering plants into trees and herbs, each was further divided into separate groups based on characters like flowers with petal or apetalous, simple or compound irregular or regular. This classification had flaws that it exhibited on differences between phanerogams and cryptogams or between monocot and dicots. By this time most of the botanist showed very less interest in the natural system however there was added interest amongst the botanist to learn more about the plants.

Period II Artificial systems based on numerical classifications

During this period the botanist in real terms did not follow the classification based on form what was in vogue since the time of Aristotle. The main idea behind this system of classification was the plant identification. In order to do so classification systems were modified but largely revolved around the main conclusions that were highlighted by Aristotle at very early stages of classification.

Carolus Linnaeus (1707-1778) : The most renowned and celebrated botanist of modern times. He was a Swedish a Physician professor of practical medicine a botanist and called as the 'father of taxonomic botany and zoology'. He had number of creditable papers in his name important ones being Genera plantarum, Flora Japponica, Species plantarum in which all genera and species known at that time were described. Bauhin's binomial system was followed and was established henceforth. The plant descriptions were very clear and accurate. Further he propsed sexual system of classification by which plants were divided into 24 classes on the basis of number, union length and some other stamen characters. By 1760 this system became well established because classification system was artificial, simple in plant identification, it continued for more than a century after his death. He established binomial system of nomenclature, established delimitations for species and brought precision to the art of description.

Period III System based of form relationship

Around 1800, it was realized that plants revealed greater levels of natural affinities than as shown by Linnaeus sexual system. With better tools and advancement in knowledge of worlds flora, large number of seeds, living plants and herbarium samples from various parts of the world were sent to different centres in Europe for their classification and nomenclature. The said system took all important characters and showed natural relationships was called an Natural System. This indicated the situation as it was present under the natural conditions. Important botanist/workers and their contribution in the said system of classification is briefly described.

Jean B.A.P.M de Lamarck (1994-1829) was a French biologist. He profounded the 'Theory of Lamarckism'. According to this hypothesis the changes that occur in the environment causes variation in the structure of the organisms and the resultant changes are then inherited by the off spring. The principles related to the natural classification were indicated by Lamarck in his work 'Flore francoise' which was published in 1778.

Bernard de Jussieu was a well known botanist. He modified the classification arrangements of Linnaeus, further made several improvements and soon Linnaeus natural system was known as that of Jussieu's. He divided the flowering plants into four groups on the basis of a) monocot or dicot b) presence or absence of petals c) ovary position d) union or distinctness of petals. Later his nephew Antoine Laurent de Jussieu (1748-1836) worked with him and started an era of natural system of classification after making several improvement in the earlier systems of classification. In 1789 published his Genera plantarum. This system was claimed to as the first real natural one. He divided the plants into 15 classes which included 100 orders.

A.P.de Candolle(1778-1841) : The 19th century was marked with great improvement in the systems of classification. de Candolle was a Swiss botanist. He improved Jussieu's system by adding few features of vascular bundles, and added 61 families compared to 100 of Jussieu's. His outstanding work was 'Theorie elementaire' in which he detailed his views towards the approaches in respect of plant classification. He pointed out that characters which were significant for life functions of plant were usually not important from systematic or classification point of view. He laid more stress on morphological rather than physiological characters while establishing relationship among the plants.

Robert Brown (1773-1858): By finding out the naked ovule in female flowers of Gymnosperm he established differences between Angiosperms and Gymnosperms. He made some outstanding observations on the flower and seed morphology. He followed the classification system of de Candolle. He critically studied various features of the ovule both prior to and after fertilization. On the basis of their outstanding work Robert Brown and de Candolle separated systematic botany from morphology.

John Lindley (1799-1865) : Published 'Introduction to the Natural orders of plants' which was the first detailed natural system of classification and was based on de Candolle's system of classification. It being in English the same was accepted and became very popular both in America and England.

Stephen Endlicher (1805-1849) : He bifurcated the plant kingdom into Thallophyta and Cormophyta. Under Thallophyta algar fungi and lichens were placed where as ferns mosses and seed plants were part of Cormophyta.

Bentham (1800-1884) and Hooker (1817-1911) were the renowned English systematists and described Spermatophyta in detail in their publication 'Genera plantarum'. In 1759 a botanical garden in Kew House was established. This garden comprised of collection of plants from all over the world. In 1840 John Lindley and some other renowned authorities in this field recommended that Kew may be developed into a National Botanical garden, which later became the most famous center of botanical research. Sir William Jackson Hooker (1785-1865) was Kew Garden's first director. Later his son Joseph Dalton Hooker (1877-1911) worked under his father. By 1855, junior Hooker was recognized as the most talented botanist of Europe. He travelled extensively all over the world ie Antartic, Mount Lebanon, Himalayas etc. His life was full of explorations and critical analysis of the results, George Bentham (1800-1884) while young came in contact with J.S.Hooker and published number of outstanding literature on plant floras. Genera plantanium was published in three volumes first appeared in 1862 and the third and last in 1883. In their work they described 97,205 species of known genera of seeded plants. Bentham and Hooker followed de Condolle's system of classification. The said system was first published in 1887 and later its scope was further expanded into a remarkable work of that time called as 'Die Naturilichen Pflanzen familien'. This method provided means for identification of all the known genera of plants from primitive algae to the most highly evolved plants (Compositae). The system was readily accepted and adopted by the botanists all over the world except United Kingdom where Bentham and Hooker system was followed. At present most of the herbaria in the world are arranged according to the classification scheme outlined by Engler and Prantl. The first part of Die Naturilichen Pflanzenfamilien though appeared in 1887 but the complete work in 23 volumes was published in 1915. Later in 1924 work on second edition was started, but due to death of Engler in 1930 and Second worldwar, despite these hurdles eight volume were completed and published.

Engler and Diels (1936) : After the death of Engler, Ludwig Diels (1874-1945) became the Director of the Berlin Botanical Gardens and was instrumental in the publication of the revised edition of Engler's work. Engler and Diels laid down the principles for a systematic arrangement of angiosperms. According to the system families were arranged according to the increasing complexity of the flower, fruit and seed development. The lower groups were recognized as having bract like members in each series, whereas the more advanced and evolved forms showed distinction in form and color in the two series. The union or fusion of petals (corolla) revealed highly evolved stage. Sympetalae hence was designated as a sub class which is different from the rest of the dicots forming the sub-class Archichlamydeae. These sub-classes markedly showed advancement from hypogyny to complete epigyny.

The plus points of Engler's system was that a large artificial group of Bentham and Hooker, the Monochlamydeae was abolished and its families with related forms with free petals were placed in large series the Archichlamydeae. Sympetalae a new group of dicot was made and it resembled to the Gamopetalae of Bentham and Hooker. According to them the monocots were more primitive than dicots and orchids were highly evolved compared to the grasses. Some other objectionable features in Engler's classification were that he accepted dichlamydeous flowers (perianth in two series) to be derived from monochlamydeous ones (perianth in single series), parietal placentation derived from axile, free central from parietal placentation and simple unisexual flowers to be most primitive precedence of monocots over dicots etc. The classification given by Engler laid more emphasis on the character of petal whether fused or free. The total dicots were split into three different groups is Polypetalae, Gamopetalae and Monochlamydeae. The Monocots were divided into seven series.

Period IV. Systems based on phylogeny

As time passed by the science developed number of theories on evolution were profounded. Taxonomists were looking out for a system of classification that would indicate relationships of plants with evolutionary view point. Ernst Heinrich Haeckel in 1866 coined the term phylogeny. According to phylogenetic system plants are classed keeping in view their evolutionary and genetic relationship, further the plants are classified according to their complexicity, starting with simple and proceeding to the most complex ones. Despite number of phylogenetic systems no one gives real/true relationship for the reason that substantial data or relevant information on evolutionary origin of plant kingdom is either lacking or is too less for the development of a perfect phylogenetic relationship. The phylogenetic systems that are in vogue presently are based on natural and phylogenetic evidences. With the advancement in research more and new facts are indicated hence there are large possibilities that this system would gradually improve. The phytogenetic taxonomy although is more than a century old but still it is regarded as an infant and much more remains to be done. Therefore the prevailing phylogenetic system is much like the natural system that was practiced in the earlier part of the 19th century. Hence presently phylogeny remains to be based upon natural classification and not classification based upon phylogeny. The workers who contributed towards this method of classification is briefly described.

Julius von Sachs (1832-1897) : Sachs was Professor of Botany at Warzburg and in taxonomy for the first time in 1868 proposed a phylogenetic system in place of the old systems of classification. The system being somewhat cumbersome and was not readily accepted. To simplify, Sachs sometime later proposed a classification of algae and fungi. With this classification he wanted to highlight the evolutionary relationship between these two groups of organism whereas some fungi groups may have been evolved from some groups of algae by the loss of chlorophyll.

Heinrich Gustav Adolf Engler (1844-1930) : For about thirty years he worked as professor of Botany at the Berlin University and later was Director of Berlin Botanical Gardens from 1889 to 1921. He published a phytogenetic system of classification which was quite similar to that of Eichler but differed from it in some aspects with his associate Karl Anton Eugen Prantl (1849-1893).

Richard von Wettstein (1802-1931) : He was an Austrian and followed the dicta or principles of Engler's classification but with some differences and was more close to phylogenetic relationships. Unisexual and naked flowers were regarded to be more primitive than bisexual ones with a distinct perianth. He was of the opinion that angiosperms have descended from gymnosperms and differed from Engler's view that dicots were more primitive than monocots.

Charles Edwin Bessey (1845-1915) : The first American who contributed immensely to the knowledge of plant relationship and based his classification on the phylogenetic lines. He put forth a number of his original ideas and his classification resembled to that of Bentham and Hooker's. He was of the view that seed plants had a polyphyletic origin and comprised of three distinct phyla and dealt with Anthophyta (angiosperms) only. Further all the three phyla were regarded to separately have been derived from the vascular cryptogams. The angiosperms were divided into two groups and three lines on the basis of historical or palaeobotanical ontogenetical and morphological studies of homologies. According to Bessey, Ranales were primitive angiosperms from which monocotyledones and dicotyledones originated. This classification was supported and liked all over the world.

Hans Hallier (1868-1932) : A German systematist developed a phylogenetic system of classification on the same principles as proposed by Bessey. He considered wider ontogenetical observations anatomical and palaeobotanical evidences than taken by Bessey. According to Hallier dicots were more primitive than monocots. He followed the principles/dictas of Bessey according to which the strobiloid type of flower is primitive, polycarpy and spiral arrangement of parts was considered primitive than syncarpy and cyclic arrangement, paid greater emphasis on ovule position and morphology.

Alfred Barton Rendle (1865-1930) : Published classification of flowering plant in two volumes. The plants were classified as per the system detailed by Engler and Prantl with some small differences. On the lines of Engler, monocots were recognized to be more primitive than the dicots. This system was said to be of convenience rather than based upon modern phylogenetic importance. In this system orders were arranged/grouped considering differentiation in the floral structure. The dicots were divided in three different groups.

Monochlamydeae: This group comprised of orders and families that had simple type of flowers some reduced forms having closeness with higher grades or representing lines of development from earlier extinct groups.

Dialypetalae: In this group orders were arranged in ascending order starting with Ranales in which flower parts are arranged spirally and ends with Umbelliflorae having cyclic arranged and reduced parts.

Sympetalae: Group comprises of orders that had higher grades of floral structure which had evolved from different dialypetalous groups.

The author indicated polyphyletic origins for the components of each of these three groups. Rendle's two volume work attracts great merit in respect of recording execeptional cases, discussions of affinities, historical review of earlier classification, clarity and completeness of descriptions.

Carl Skottsberg : Was born in Sweden at Goteberg in 1880 worked as Professor of Botany. Worked on the concept of Wettstein, modified Engler's system and later proposed a system of classification. According to him monocots were derived from some unknown primitive dicots, treated apocarpous and polycarpous conditions as primitive, syncarpous and monocarpellary as advanced. The apetalous families were said to be of polymorphic origin and their positions given by Engler, Wettstein and Pulle were redistributed. Divided the dicots into Choripetalae and Sympetalae and were further divided in different orders and families.

August A. Pulle. Was attached to Utrecht Botanical museum Netherland. In 1938 published a classification which was modification of Engler's system. Pulle did not accept that Spermatophyta was divided only into two subdivision ie Gymnospermae and Angiospermae, but consists of four subdivisions ie Pteridospermae, Gymnospermae, Chlamydospermae and Angiospermae (Monocotyledoneae and Dicotyledoneae). Classification of gymnosperms

(Pulle) is more phylogenetic in its components. Although angiosperms were classified mostly on the principles of Wettstein but was original in number of respects.

John Hutchinson, Curator of the museum of Botany of the Royal Botanical Gardens at Kew England. Developed the most recent classification of flowering plants based on phylogenetic lines. After extensive study of herbaria at Royal Botanical Garden at Kew and knowledge of African plants, published. Families of flowering plants in 1926, 1934 and 1959. His classification was based upon principles that were close to those of Bentham and Hooker and Bessey than that of Engler and Prantl. His basic phylogenetic principles were almost similar to those of Bessey but had some modifications. In 1948 he published a book British flowering plants, where he pointed out that apetalous families have partly been derived through Magnoliales and Ranales.

General Principles for the Classification of Flowering Plants

In 1959 Hutchinson revised his classification which was based on twenty four different principles that were comparable to Bessey's principles.

1. Evolution work both upwards and downwards former resulting in preservation and latter in reduction or suppression of characters former resulting in sympetalous and epigyny and latter towards apetalous of many flowers and unisexuality in flowering plants.
2. Evolution may or may not involve all plant organs at one time. Single organ or set of organs may be advancing while others may be retrograding or be stationary
3. Evolution largely has been consistent when particular progression or retrogression has set, it results in the end of phylum. Zygomorphy petals with reduced stamen number in Engler's hypogynous Metachlamydeae and greater tendency to perigyny and epigyny in Archichlamydeae and Metachlamydeae.

Principle in Respect of Plants Habit

4. In some groups trees and shrubs are more primitive than herbs
5. Tree and shrub plant habit is older than climbers in any family or genus, it being acquired through particular environment.
6. Perennials are older than biennials and from them annuals have emerged.
7. Epiphytes, saprophytes and parasites have originated in recent times than plants of normal habits. Acquatic plants have terrestrial ancestors.

Principles Relating to General Structure of Flowering Plants

8. Dicotyledons with united vascular bundles arranged in ring are more primitive than monocotyledons that have scattered bundles but necessarily monocots may not be directly derived from dicots.
9. Special leaf and floral leaf arrangement on stem is more primitive than opposite and whorled.
10. Simple leaves are more primitive than compound leaves.

Principles Pertaining to the Flowers and Fruits

11. Solitary flower nature is more primitive than inflorescence, latter's highest forms being umbel and capitulum.
12. Unisexual flowers are advanced form compared to bisexual, dioecious plants are more recent than monoecious.
13. Apetalous flowers are derived from petaliferous ones, former condition resulting from organ reduction.
14. Free petals (polypetaly) condition is more primitive than fused one (gamopetly).
15. Actinomorphy is more primitive than Zygomorphy.
16. Flowers parts in spirally imbricate are older than whorled and valvate.
17. Polymerous many parted flower condition is primitive compared to oligomerous mainly due to progressive sterilization of reproductive parts.
18. Perigyny and epigyny condition is derived from more primitive hypogyny.
19. Polycarpy (many carpels) conditions precede oligocarpy.
20. Free carpel (Apocarpy) is older than connate carpel (syncarpy) condition.
21. Indefinite stamens indicate greater primitiveness than with few stamens.
22. Separate stamens condition is older than connate one
23. Endospermic seeds with small embryo are said to be older than non-endospermic seeds with large embryo.
24. Aggregate fruits are highly developed forms than single fruits, capsule precedes drupe and berry.

Angiosperm is a very diverse group of plants comprising of nearly 250,000 species including the fruit crops. Angiosperms are very ancient and their first apperance in fossil form occurred in the Cretaceous period. Besides this old record few reports of the flowering plants in the fossilized form have also been reported from the Jurrasic period. The term angiosperm is derived from Greek word 'angeion' meaning receptale or vessel, and sperm, the seed. In brief the flowering plants with enclosed seeds are termed angiosperms. On the other hand gymnosperms (Gymno-naked, sperm- seed) bear naked seeds and both these forms evolved from spermatophyta i.e. the seed plants. Spermatophyta is said to have some 300 families 12,500 genera and about 300,000 species. With new technologies extensive surveys, more critical examination, new species from different parts of the world are being added to the existing ones very rapidly. Today angiosperms comprise of a dominant fast expanding plants of the world.

Aparts from huge variability present in the existing flowering plants, man through his efforts and care has collected large number of new varieties or strains especially which accounts for their economic or commercial significance. A numbers of factors are responsible for success of angiosperms, major being their ability to adopt to varied environmental conditions, and adaption of their flowers, fruits and seeds for dispersal over large areas.

The beauty and special features of these flowering plants is that they show wide variation in their form and mode of life. Some are very small acquatic plants, others are huge tree that may be more than 400 meters tall. Their life period may be limited just for few week contrary to some who are to live for over centuries. Their adaption is variable, some love to be on the seashore, others adapt to the scorching heat of the deserts, some are present on the mountain tops, some in the tropic forests and others on the frozen soils of the glaciers and snowy regions of the arctic and antartic regions. Most of the angiosperms are terrestrial plants, but a few have a secondarily adapted to water also. Further they are usually mesophytes, some are xerophytes, and few have also become epiphytes. They contain cholorophyll, the green pigment which help to produce food material in the presence of sunlight.

Angiosperms have provided mankind with a variety of economically useful plants. Plants of this group supplies enormous quantities of food which is required for the sustainability and maintenance of other organisms that live on the earth. Amongst the great variety of food cereals like wheat, barley, corn, rice, rye, legumes like soyabean, beans, peanuts, fruits, nuts and vegetables all have been extensively used by man and millions of individuals that form part of animal kingdom.

Apart from food these valuable flowering plants also supply fibres like cotton, jute, hemp, Indian and Manila hemp, linen for preparation of nets, ropes, strings and manufacture of clothings. Different parts of these plants have contributed immensely such as oils, fats, drugs, sugars, fibres, dyes, besides spices, coffee, paper, tannins, gums, rasins, alcohol, fuel, rubber, tea, coffee needs special mention. Perhaps the most significant role of these green and flowering plants is the purification of the atmosphere by taking in carbon dioxide and giving out oxygen during photosynthesis.

The varied types of flowering plants have taxonomically been classified into different orders and families etc. The term taxonomy has been derived from the Greek words (Taxis meaning arrangement and nomos means the law). Hence taxonomy is mainly concerned with the laws that governs the arrangement or classification, nomenclature and identification of the plants. This type helps in the orderly study of plants and their systematic knowledge about the differences and resemblance among and between different plants. The branch of botany where taxonomy is of the prime interest is known as systematic botany and when dealt with fruit crops it is called as systematic pomology. The aim of the systematic pomology is to indicate clearly and definitely the differences and resemblances by following the rules of the identification, nomenclature and identification in such a manner that their relationships with regard to their descent form a common ancestry is well described.

According to Hutchinson system of classificatin, angiosperms are divided into two 1) Dicotyledons 2) Monocotyledons, the dicotyledons on the petal character is further subdivided into Archichlamydeae-in which the petals are either free or absent and metachlamydeae, where the petals are united.

In the angiosperms dicotyledons group is very large and diverse and is characterized as:

Dicotyledons: Plants may be herbs, shrubs or trees, in perennial type of plant the stem has central pith with number of concentric wood layers, bark is separable that increase in thickness and size by means of cambium. Leaves with reticulate venation, floral parts pentamerous or in its multiple free or united. Ovules enclosed in the ovary, embryo with 2 cotyledons.

Monocotyledon: (Monocotyledoneae): Seeds with one cotyledon leaves with parallel veination, flowers hemicyclic trimerous carpels often free rarely united, vascular bundles scattered and closed, endosperm absent.

Brief description of orders that fall under both dicot and monocot groups, families important from horticulture or fruits point of view alongwith characteristic features of various orders in detailed below. Families important from fruit point of view are listed for different orders.

Order Sapindales: Families Sapindaceae and Anacardiaceae belong to this order.

Plants are mostly trees or shrubs, flowers are bisexual or unisexual, actinomorphic rarely zygomorphic, hypogynous peri or epigynous, disc glandular with which stamen and perianth are closely attached. Ovary superior 1-4 celled with 1-2 ovules per locule.

Order Myrtales (Myrtiflorae). Punicaceae and Myrtaceae are part of this order.

Plants are woody trees, shrubs or herbs. Leaves simple and opposite. Usually bisexual with 4-6 merous perianth differentiated into calyx and corolla. Pistil inferior, style simple one to many, stamens usually numerous. Ovary uni to multi locular. Ovules many on axile placentation, endosperm mainly absent.

Order Rosales: Has temperate fruit and crop families like Rosaceae and Saxifragaceae, are included in this order. Plants range from herb, shrubs or trees, leaves with stipules, may be simple or compound. Flowers bisexual, rarely unisexual, actinomorphic or zygomorphic, pentamerous hypo- peri to epigynous. Stamens often many and usually free, carpel few to many, free or united.

Malvales: Flowers bisexual, actionomorphic, cyclic, hypogynous with numerous indefinite stamens which often cohere in a bundle or bundles by filaments or at least having such tendency. Sepals often valvate united or rarely free. Herbs, shrubs or trees with fibrous stems, mucilage ducts and stipulate leaves. Tiliaceae, Bombacaeae families have fruit crops.

Order Ebenales: Horticultuarally important families included in this order are Ebenaceae and Sapotaceae.

Plants are trees or shrubs with regular, perfect actinomorphic, sympatalous, sometimes unisexual flowers, usually hypogynous. Stamens in one or more whorl, ovules one or few.

Order Loganiales/ Gentianales: Comprises of families Oleaceae and Apocynaceae.

The order comprises of plants which may be either herbs, shrubs or trees, leaves usually opposite, rarely compound. Flowers bisexual, perfect actinomorphic and hypogynous with 4-5 merous, perianth differertiated into calyx and corolla. Aestivation of petals contorted or twisted in buds. Stamens 2-5 or same number as corolla lobes and alternate with them and carpels, epipetalous. Carpels 2-4, ovary 2-4 celled.

Order Ranales: Comprises of about 8 families. Annonaceae and Lauraceae are fruit bearing families.

Plants are herbs or woody trees, flower bisexual, regular rarely zygomorphic, hypo or epigynous. Thalamus(torus) elongated may be convex, sometimes concave on which floral whorls are placed in cyclic, spirocyclic or acyclic fashion. Perainth petaloid free differertiated into sepals and petals, stamens numerous, carpels many to solitary, seeds endospermic.

Order Parietales: Passifloraceae and Caricaceae are the two families.

Plants are usually herbs or shrubs rarely trees, flowers bisexual or unisexual with pentamerous perianth differertiated into calyx and corolla, hypogynous, sometimes perigynous, stamens many or few, carpel 2- many syncarpous. Ovary superior ovules many on parietal placentation, endosperm absent. Leaves alternate, simple and often much lobed and divided.

Order Rhamnales/Celastrales: This order includes family Vitaceae and Rhamnaceae.

Plants are usually woody large trees, shrubs or climbers with decumbent habit, leaves alternate or opposite, simple rarely compound. Flowers bisexual or unisexual (abortion of sex organ). Actinomorphic, tetra or pentamerous peri or epigyny. Disc well developed. Perianth members free, valvate, sepals minute. Stamens in the one whorl epipetalous, endosperm present.

Order Solanales: Has only one family Solanaceae.

Plants are usually herb or shrub or climber with simple alternate or rarely opposite leaves. Flowers are typically actinomorphic, hypogynous sometimes zygomorphic, pentamerous perianth differentiated into calyx and corolla, calyx persistent, corolla various. Stamens epipetalous alternate to petals. Ovary bi, or tri carpillate, syncarpous, superior, with axile placentation.

Order Guttiferales: Has only one family Guttiferae important from horticulture point of view.

Plants usually woody shrubs or trees with intercellular secretory passage, flower usually bisexual, actinomorphic rarely zygomorphic, hypogynous, sepals and petals 5 each free, stamens generally numerous, free or variously united by filaments, ovary multilocular with axile placentation, seeds with or without endosperm.

Order Proteales: Has only one family Proteaceae.

A monotypic order has one family Proteaceae. Plants are woody trees or shrubs. Flowers generally bisexual or unisexual (by abortion of one of the essential members), 2-merous, actinomorphic or zygomorphic. Stamens opposite the perianth segment and usually adnate with them, carpel one, seed non endospermic.

Order Urticales: The order has only one family Moraceae.

Plants are herbs, shrubs or trees. Flowers are unisexual small, perianth members greenish 4-5 rarely more, when 4 then biseriate. Stamens few to several. Carpel 2, ovary superior unilocular, one ovuled.

Order Juglandales : This order has 3 nut crop families i.e. Juglandaceae, Fagaceae and Betulaceae.

Plants are woody trees or shrubs, deciduous monoecious with simple or compound leaves. Flowers unisexual apetalous subtended by bracts and bracteoles which are persistent forming an involucre round the fruit, which may be drupe, nut or nut like. Ovary usually inferior of two or more carpels. Ovules 1-2 or sometimes more.

Order Euphorbiales : In respect of fruit significance has only one family Euphorbiaceae.

Plants are trees, shrubs rarely herbs, flowers are almost always unisexual with reduction in the perianth whorl. Ovary syncarpous or monocarpous, ovules in the ovary or in each locules, solitary or in pairs. Endosperm copious fleshy, scanty or absent. Styles equal in number to the carpels. Perianth sepaloid, minute or absent. Stipules usually present.

Monocotyledon

Order Farinosae/ Commelinales : Has family Bromeliaceae which is important for pineapple fruit.

Plants are herb or rarely under shrubs. Leaf sheath usually closed. Flowers are bisexual, perianth usually differentiated into outer calyx and inner corolla ovary compound and superior. Endosperm is mealy, embryo is small.

Order Spathiflorae/ principes : The order includes two families Palmae and Araceae.

The plants are woody palms, flowers are trimerous, small on the compound thick spadices subtended by large herbaceous spathe unisexual or bisexual, ovary superior tricarpillate, fruit a berry or drupe, seed endospermic.

Order Scitamineae : Order has one family Musaceae which is important horticulturally.

Plants are usually large herbs often tree like in appearance, leaves are large with parallel veins, distichous, or in spirals basal leaf sheath open. Flowers zygomorphic, stamens 6, functional 1-5, ovary inferior, seed endospermic.

Before taking up detailed pomological description of family, species and variety it is essential to know the important characters which needs to be studied.

1. **Tree Characteristics**

   (a) Growth

   1) Height : spreading, upright, drooping, open, vase form, round, topped.
   2) Height-low, medium, tall
   3) Vigour-vigorous, medium, weak.

   (b) Trunk

   1) Nature of bark-tight, loose, rough, smooth
   2) Colour of bark-brown grey, dark, any other
   3) Pubescence-pubescent, glabrous
   4) Lenticels-size, shape, colour, intensity, other variation, other variation

2. **Twig Characters**

   (a) Branches

   1) Nature-spreading, erect
   2) Bark nature-tight, loose, smooth, rough
   3) Bark colour-brown, grey, greenish, dark or any other colour
   4) Lenticels-Size, shape, colour, intensity.

   (b) Growing shoots

   1) Nature-round, angular.
   2) Colour-yellowish green, reddish, other colour
   3) Abundance of thorns-short, medium, long.
   4) Pubescence-pubescent, glabrous.
   5) Lenticels-present, absent.
   6) Length of internodes-long, medium, small.

3. **Foliage Characters**

   (a) Leaves-

   1) Colour of new emerging leaves-yellowish green, copper colour.

2) Size-large, medium, small; broad, medium, narrow; thick, medium, thin.
3) Shape-lanceolate, oblanceolate, elliptical, broadly elliptical,ovate, obovate.
4) Nature-smooth, rough, tough, leathery.
5) Pose-upheld, out held, down held.
6) Folding and reflexion- folded upward, downward, wavy, wrinkled, flat.
7) Apex-acute, acuminate, cuspidate.
8) Base-obtuse, acute, cuneate.
9) Margin-dentate, serrate, crenate.
10) Upper surface-colour, pubescence.
11) Lower surface-colour, pubescence.
12) Mid-rib and veins-conspicuous, inconspicuous
13) Thickness of leaf-thin, medium, thick

(b) Petiole

1) Size-long, medium, short, thick, medium, thin.
2) Shape-round, angular.
3) Pubescence-pubescent, glabrous.
4) Colour-yellowish green, reddish tinged, any other colour
5) Glands-present or absent; if present, number, shape and colour.
6) Stipules-present, absent.

4. Floral Characters

(a) Blossoming

1) Flowering habit-on shoot or spurs.
2) Flowering-before, with or after the leaves.
3) The opening of the first flower
4) The duration of effective blossom
5) The opening of the last flower
6) Flower types-hermaphrodite, intermediate and male flower
7) Abundance of flowers- sparse, medium, profuse.

(b) Flowers

1) Arrangement-single or in clusters, number of flowers per bud and per node.

2) Type of inflorescence-raceme, racemose, cymose, panicle.
3) Size-width across.
4) Size of flower bud-small, medium, large.
5) Shape of the flower bud-ovoid, obovoid, oblong, ovoid-oblong, elliptical, sub cylindrical.
6) Colour of the flower bud-green purple, brown purple maroon.
7) Pedicles-size, colour, hardiness and number.
8) Sepal-hardiness colour and number.
9) Petal-colour and number.
10) Reflection of corolla-slight, moderate and strong.
11) Stamens-arrangement, number anther colour.
12) Styles-length, hairiness, colour.

5. Fruit Characters

(a) External characters

1) Nature of attachment- hangs well, drops readily.
2) Size-length and diameter, large, medium, small.
3) Weight and volume- large, medium, small.
4) Shape-round, oval, ovate, cordate, oblate, oblong.
5) Basal cavity-irregular, regular, deep, medium, shallow.
6) Apex-rounded, flattened, pointed.
7) Suture-deep, medium deep, only a line, indistinct.
8) Skin- thick, medium, thin, astringent, leathery, tough, free, semi-free, adherent.
9) Bloom-heavy, light, scanty.
10) Dots-size, intensity, colour.
11) Colour-red, purple, yellow.
12) Stalk-long, medium, short, thick, medium, thin, pubescent, glabrous.

(b) Internal characters

1) Colour of flesh-yellowish, reddish, white.
2) Texture of flesh-fine, melting, meaty.
3) Juiciness-abundant, medium, scanty.
4) Taste-sour, sub-acid, sweet.

5) Adherence to stone-clinging, semi-clinging, free.
6) Weight of the pulp and pulp/stone ratio.

6. Stone Characters

(a) Stone

1) Size-large, medium, small, broad, medium, small, plump, compressed.
2) Shape-round, oval, ovate, obovate, elliptical.
3) Surface-smooth, wrinkled.
4) Taste of kernel-sweet, bitter.

The "Dictionary of colour" by Maerz and Paul is used for recording the colour of various organs. The glossary of different terms used in description is given in after the chapter cherries.

1.3 Structure of Flower and Fruit

Prior to detailed information of various fruit crops, appropriate knowledge of flowers and fruit structure is very essential as fruit is the ultimate product of flower development in which embryo is present that initiates next generation. Fruit is generally defined as a ripened or mature ovary.

Pistil is the important part of flower and consists of three structures namely stigma, style and ovary. Stigma is the uppermost part of pistil and receives the pollen for the formation of embryo. At the time of pollination a shining fluid is secrerated, which help in pollen adherence and in its growth. The central portion is style and is derived from 'stylus' meaning pencil- like. The lower part of pistil is ovary and is the reproductive organ. The term ovary has been derived from Latin word 'ovum' meaning egg. In most of the cases the fruit is derived from pistil, but in some fruits the associated flower structures also take part or are ultimately fused with it to form the fruit. To exactly known the type of fruit one must also pay due attention to the position of gynoecium in relation to other fruit parts. In all there are three situations, which are usually prevalent in different fruit crops. Flower may be termed as hypogynous or ovary is said to be superior when the petal, stamens and calyx are attached to the receptacle below the ovary, e.g. orange, persimmon, grape, avocado etc. Perigynous- in this the ovary is said to be half inferior for the reason that the petals and stamens are attached to the receptacle which has grown up about the level of ovary but not attached to it. Examples of perigynous flower are apricot, plum, peach and cherry. The third type is epigynous. Petal and stamen are attached to the receptacle and has grown up about the ovary. Ovary in this case is said to be inferior, e.g. are banana, apple, pear and cranberry etc.

There are different kinds of fruits, which have been classified on the basis of structure, morphology and consistency based upon whether the fruit wall is succulent or dry, number of carpels and presence or absence of some accessory part of flower other than the carpel wall. The following fruit classification is very important for understanding the nature of various fruits.

A) Ovary or pistil- No. of carpel and accessory part in the fruit.

1) Ovary simple (derived from one carpel)

(a) Legume (pea, beans, peanut)

(b) Achene (strawberry fruitlet, fig fruitlet, buttercup)

(c) Berry (avocado, date, pineapple fruitlet)

(d) Drupe (plum, peach, cherry, olive)

2) Fruit formed from two or more carpels.

(a) Berry- grape, papaya, persimmon, tomato

(b) Capsule- okra, tobacco, cotton

(c) Silique- cabbage, reddish

3) Fruit formed from two or more carpels plus other accessory parts.

(a) Strobilus- hops

(b) Nut- filbert, chestnut, acorn

B) Fruit classification based on the texture of the pericarp/ mature ovary wall.

1) Dry- The whole of the ovary wall or pericarp becomes dry as the fruit attains maturity, fruit under this are further divided into two types;

(a) Dehiscent- at maturity fruit splits and seeds are exposed.

(b) Indehiscent- at maturity and there after the fruit does not split at all.

2) Fleshy fruit- in these types, the fruit seed coat / pericarp along with the associated flower structure usually becomes succulent.

3) Dry- fleshy fruit- some parts of the pericarp would become dry or may harden and other parts of ovary wall remains fleshy.

Classification of Fruit Types

1) *Dry fruits:* The dry fruits are usually of two types, indehiscent and dehiscent

   (a) Dehiscent fruit: The important dry dehiscent fruits are 1. Follicle 2. Legume 3. Capsule 4. Silique.

   Follicle: Such types of fruits usually develop from a single carpel, further there is one suture or line along which the dehiscence takes place and is located on its adaxial surface.

   Legume: Develops from a solitary carpel. Pea is an example of legume. The splitting or dehiscence takes place along the two sutures or lines that are formed in the pea pod.

   Capsule: Fruit usually develop from one or more carpels along with a part of axis is also involved. The dehiscence in capsule is usually of two types- loculicidal and septicidal. In the former type dehiscence lines directed to the point of union or non divergence of the adjacent carpel lengthwise in the adaxial bundle. Whereas in the later dehiscence lines are directed to the points of union or non divergence of the adjacent carpel.

   Silique: The fruit develops from two carpels that form a bilocular ovary with lengthwise septum. The two fruit valves separate longitudinally.

2) Indehiscent fruit : Fruits do not split open when fully ripe. Usually the indehiscent fruits are of four types namely of (a) Caryopsis (b) Achene (c) Schizocarp (d) Nut.

   (a) *Caryopsis*: Fruit develops from a solitary carpel and usually has only one seed. In the mature grain the seed coat remains very close to the pericarp and can not be separated because the integument remains somewhat disorganized.

   (b) *Achene*: The resulting fruit developes from one carpel and is usually one seeded and the pericarp can be removed from the seed coat when mature. In family Compositae the achene develops from two carels, whereas in Renanculaceae achene is derived from one carpel

   (c) *Schizocarpie*: The fruit develops from two or more carpels, each is independent from one another and contains one seed each.

   (d) *Nut*: Is a one seeded indehiscent fruit usually develops from a compound ovary in which the carpel wall may be hard in texture. True nuts are chestnut, hazelnut, and filbert. In filbert nut is derived from hypogynous flower and may be covered by a dry or a fleshy involucres. Is one seeded, but in most cases nut develops from two carpels.

(3) *Fleshy fruit:* The fleshy fruit usually comprises of berry, hesperidium and pepo.

(a) *Berry*: The berry fruit is formed from two or more carpels in which the ovary wall remains fleshy and succulent and produces a fleshy pericarp as the fruit matures e.g. grape, banana, and date.

(b) *Hesperidium*: All citrus fruits basically are berry but of a special type in which the dorsal wall of the carpels develop into a thick rind like structures, the outer portion of which is bumpy and thin- exocarp and inner layer thin to thick white in colour the mesocarp constitutes the albedo. The thin paper memberanes the endocarp encloses numerous placental hairs that form the edible part. Inner to vesicles the central portion of the fruit involves the axis as well as the carpel hence a special berry designated as hesperidium.

(c) *Pepo*: The fruit is derived from inferior ovary (epigynous flowers). The non-divergent receptacle tissue that encloses or surrounds the pericarp forms a rind like structure, which may be very hard (pumpkins and squashes). In pepo the edible portion is derived from carpel wall, usually placentae and enclosing receptacle.

(4) *Dry-fleshy*: Most of the horticultural fruits fall under the category as most of the fruits are in part dry and part fleshy. This group comprises of very diverse fruits that include drupe, pome, aggregate and multiple types.

Drupe fruits: Drupe fruits develop from a single carpel that has one seed but two ovules, one aborts as the fruit develops. The pericarp is usually differentiated into thin skin like structure called exocarp, mesocarp is generally fleshy and juicy and endocarp is hard and stony in texture (plum, peach, apricot, cherry etc.)

Pome: The pome type of fruit is one in which the several carpels are leathery or partially so and ranging to nearly stony in some forms. The receptacle becomes fleshy and constitutes the edible portion eg., apple pear, quince and loquat etc.

Compound fruits: The compound fruit comprises of both aggregate and multiple fruits.

Aggregate fruit: The aggregate fruit comprises of number of simple fruits, which develop from several simple pistils of a flower. Further the stem axis is not involved in the formation of edible part of the fruit. Raspberry in an aggregate fruit in which the fleshy edible drupes separate easily from the receptacle, whereas in blackberry there is organic union between drupes and receptacle and mature fruit is picked as whole. In strawberry the achene's are embedded in the fleshy receptacle, which forms the edible portion of the fruit. The achenes and the receptacle together form the aggregate fruit.

Multiple fruits: The fruits are derived from number of flowers and not from several pistils of a solitary flowers as in case of aggregate fruits. The number of flowers develop together resulting in multiple fruits. There are number of multiple fruit types that differ from one another in respect of the accessory parts that form the fruit structure and also in variable level of differentiation. The important multiple fruits are fig, mulberry, pineapple, hop etc.

Mulberry: The male and female flower in mulberry is formed in different inflorescences. The female flowers develop in a fruit which is multiple and covered / surrounded by a fleshy pericarp. On fruit development these structures are aggregated with fleshy receptacle and finally forms a multiple fruit.

Fig: The hollow fleshy receptacle of fig is called syconium. On the inner wall of the receptacle are present the flowers which may bear both male and female or they are borne on difference inflorescences. The flowers are reduced to only essential organs calyx and corolla are absent.

Hop: Horticulturally multiple fruit of hop is called as strobilus. In this fruit male and female flowers are usually borne on separate plants. The female flowers are clustered together to form heads and with growth bracts grow to form hops or "cones". The male flowers are usually small present in loose panicles and soon after anther dehiscence wither and falls off. The individual fruits are in fact the achenes.

Inflorescence of Fruit Crops

To a student of Systematic Pomology knowledge of inflorescence, various types of inflorescences is very essential to understand the bearing habits of different fruit crops. The distinct arrangement of flowers on the stem or axis is called as inflorescence. Fruit crop exhibit variation in respect of flower arrangement on the plant, bloom order and flower number per unit of the structure. Flowers may be borne singly or in clusters, the position may be either terminal or lateral in leaf axis. If borne terminally the growth ceases and hence termed as determinate in nature, or indeterminate if growth does not ceases, but continue to grow after flowering has taken place. As inflorescences show variation they are classified on the basis of distinct features that they possess. In general there are three types of inflorescences namely solitary flowers, racemose flowers and cymose flowers.

1) *Solitary flowers:* When only one flower is borne the resultant situation is termed as solitary flower(s). The flower may be borne at the terminal point of the shoot, laterally on the leaf axils or may be sometimes borne on the special structures like stalk or stem.

2) *Racemose inflorescence:* A number of different types of inflorescences are observed under this inflorescence type like raceme, catkin, and corymbs. The special feature is that growth is indeterminate. The flower-bearing axis continues to grow and the flowers are produced in the sub terminal growing region. The old flowers are produced at basal portion and younger ones growing in terminal. The flower opening and blooming proceeds from base to apex. Brief description of different types of inflorescences is described below.

   (a) *Raceme*: The flowers are borne laterally on pedicels whose length is almost equal, examples are wild cherry, currant etc.

   (b) *Catkins*: In catkins the flowers are usually without petals- apetalous, unisexual and are borne on a spike or raceme, which has a thin slender rachis. The male catkin generally falls off soon after anther dehiscence is completed. Hazel, mulberry, walnut, are some very common fruit examples.

   (c) *Umbel*: In this type of inflorescence the main axis is short and flower bearing pedicels arise from one point and the pedicel lengths are almost equal that spreads like an umbrella. No fruit crop has this type of inflorescence.

   (d) *Corymbs*: The axis of inflorescence is elongated and the pedicels, which are lateral in position arise from different points on the axis. The pedicels are of unequal length. The oldest flower has the longest pedicels and the younger one, which is located towards the tip, has smallest pedicel length. All flowers attain one horizontal plane. The flowering-anthesis proceeds upwards. Pear is an example of corymb inflorescence.

3) *Cymose inflorescence:* In cymose inflorescence the flower bud is produced at the terminal or growing point as a result further growth cannot commence from this terminal. New flower buds are differentiated below the apex on the lateral shoot. The younger flower clusters are produced from the apex to the base. Such a growth is termed as determinate growth

and is opposite to what takes place in racemose inflorescence are not so common compared to racemose. Further mixed type of inflorescences complicate cymose inflorescences.

(a) *Cyme*: The inflorescence in cyme is determinate in nature, apical or central flower opens first. A number of fruits show this type of inflorescences like persimmon, strawberry, quince, red and black raspberry.

(b) *Fascicle*: The characteristic feature of fascicle cyme is that peduncle is very short or reduced, pedicels are long and of almost equal lengths and the flowers are very closely crowded together. Example plum, cherry.

Key to Genera(s) of Fruit Crops

Bombacaceae

A. Leaves simple, feather winged, entire cotyledons plane leafy or fleshy—*Durio.*

Tiliaceae

A. Calyx compressed of distinct sepals

B. Petals pitted at the base, inserted around the base of less elevated torus

C. Fruit unarmed, glabrous or tomentose—*Grewia.*

Malpighiaceae

A. Fruit a fleshy 3 stoned drupe—*Malpighia.*

Guttiferae

A. Style very short or 0, ovules solitary in each locule of the ovary

B. Sepal 4 —- *Garcinia.*

Annonaceae

A. Fruit is an aggregation of many carpels closely crowded into a spheroid or ovoid mass, ovules solitary

B. Carpels fused together with the receptacle (torus) into a fleshy (often edible) syncarpium

C. Corolla gamopetalous 3 lobed or 3 spurred, almost closed with only a minute opening, above the stamens and pistils- *Rollinia.*

CC. Corolla polypetalous, 6 in 2 series, inner series sometimes or even wanting, outer petals valvate -*Annona.*

Inner petals with margins not involute—-*Asimina.*

Dilleniaceae

A. Stamens and carpels infinite. Winter buds enclosed in a swollen base of the petioles - *Actinidia.*

Rhamnaceae

A. Calyx lobes deciduous.

B. Disc lining the shallow calyx tube nearly or quite free from the ovary, fruit drupaceous mostly fleshy and often edible with a single 1- 4 celled stone enclosing as many seeds or 1 seeded by abortion, seed coat membranous

C. Fruit a fleshy drupe plants prickly, leaves 3 nerved - *Zizyphus.*

Vitaceae

A. Stamens free, climbing shrub or herb

B. Petals expanding, flowers in cyme, bark close, pith white

BB. Petals cast off from the base while cohering by their tips, hypogynous disk or 5 nectariferous glands alternate with the stamens, flowers in panicles, berries usually edible, leaves rarely compound, never pinnate — — — *vites.*

Anacardiaceae

A. Leaves simple

B. Stamens 8-10 (all or some fertile), style accentric stigma a mere dot - *Anacardium.*

BB.Stamens 1-5, style lateral, stigma simple– –*Mangifera.*

AA. leaves pinnate or compossed of 3 leaflets

C. Ovules suspended by a basilar funiculus

D. Styles 3— — —*Pistacia.*

Caricaceae : Has only one genus *Carica.*

Punicaceae: Has only one genus *Punica.*

Passifloraceae

A. Hypanthium short, petals 4-5, rarely 0, stamen 4-5 -------- *Passiflora.*

Rosaceae

A. Ovary inferior, carpels 2-5, more or less connate and adnate to the cup shaped receptacle, whole developing into a fleshy fruit (pome) tree or shrub with free stipules -------- *Pomeae.*

B. Fruit not follicular, indehiscent, or carpels growing into druplets

C. Pistil borne on flat hemispherical on convex receptacle subtended by cup shaped portion of the receptacle (Hypanthium) usually many

D. Ovule 1, carpels many, calyx usually with bractlets, alternates with the lobes – *Potentilleae.*

E. Carpels become druplets, ovules two, but seeds solitary – – *Rubeae.*

EE. Carpels, rarely two, fruit drupaceous, calyx usually deciduous, flowers symmetrical style subterminal, ovules pendulous, radicals superior— —-*Pruneae.*

Various Genera under each tribe

A. Sepals 5

B. Carpels solitary

C. Style terminal, leaves usually serrate, pith of branches solid– *Prunus.*

Tribe : *Potentilleae*

A. Style deciduous

B. Receptacle in fruit much enlarged, coloured

C. Flower white, receptacle white — — *Fragaria.*

BB. Receptacle not fleshy even in fruit

CC. Petals white or yellow, obtuse or emerginate

D. Pistil numerous— —*Potentilla.*

Rubeae

A. Druplets pulpy —-*Rubus.*

Pomeae

A. Carpels bony at maturity, faint hence with 1-5 cells/stone

B. Pistils with two fertile ovules, leaves entire or crenate

AA. Carpels with leathery or papery walls at maturity fruit hence 1-5 celled, each cell with 1or 2 rarely many seeds

C. Flowers in compound corymbs

D. Styles1-5, distinct or crenate, carpels partly free ----- *Malus.*

DD. Styles 5, distinct, carpels wholly connate, fruit pear shaped, rather large, yellow. Leaves evergreen with excurrent veins— — *Eriobotrya.*

AAA. Carpels 4 to many seeded, styles free, leaves entire —- *Cydonia.*

E. ovary 3-5 celled, flowers in umbels, leaves deciduous—- *Pyrus.*

Oleaceae

A. Fruit fleshy and indehiscent, a drupe or rarely a berry, not lobed, ovules twin laterally affixed near the apex, seeds solitary, suspended or pendulous with endosperm, radicle superior

AA. Fruit is a drupe, endocarp hard thick or thin, inflorescence axillary, rarely terminal—-*Olea.*

Apocynaceae

A. Anther cells not appendaged at base

B. Fruit is a berry, indehiscent, ovary 2 celled, cells 1-4 ovuled

C. Ovules laterally affixed, cyme terminal, few flowered, spines axillary - *Carissa.*

Solanaceae

A. Stamens all perfect not didynamous, normally 5

B. Fruit is a berry or atleast indehiscent

C. Corolla limbs subequally plicate or divided into valvate or in duplicate lobes

E. Anthers longer than filaments, connivent connate in cylinder or cone, acuminate at base or dehiscent at 2 apical pores—*Cyphomendra.*

F. Fruiting calyx inflated or bladdery, calyx cut shortly or to middle -*Physalis.*

Lauraceae

A. Anthers extrorsely locellate, valves dehiscent upwards

B. The whole perianth persisting under the fruit, appressed or slightly spreading, perianth sometimes deciduous from the base – *Persea.*

Proteaceae

A. Fruit follicular, capsular or rarely indehiscent and sub drupaceous, flowers usually in pairs along the rachis with only 1 bract for each pair

B. Ovules 2, collateral, pendulous, orthotropous

C. Flowers, racemose or fascicled, involucre non or inconspicuous, bracts deciduous

D. Fruit scarcely or tardily dehiscent, pericarp thick fleshy or hard, seeds with thick often unequal cotyledons

E. Perianth straight—- *Macadamia.*

Moraceae

A. Anthers reversed on the bud with inflexed filaments

B. The male flowers, spicate, racemose, capitate, female globose capitate

BB. The flowers of either sex spicates, spikes short and dense or long and lax— —*Morus.*

C. Plant trees or shrubs, flowers usually on a fleshy receptacle, globose or ovoid, clearly enclosing the numerous flowers, but with small mouth which is bracteate introrsely, the mouth is enclosed in fruit– – *Ficus.*

CC. The flower cluster unisexual, with or without 3-4 bracts at the base, in heads, spikes, rarely in racemes or the female one flowered

D. stamen 1– – *Artocarpous.*

Juglandacea

A. The staminate flowers in pendulous catkins, pistillate flowers spicate with a small fleshy erect point

B. Husk at length splitting into segments, nut smooth or angled—- *Carya.*

BB. Husk is indehiscent, nut wrinkled or sculptured—*Juglans.*

Betulaceae

A. Staminate flowers with 4 perianth—segments or by abortion fewer

AA. Staminate flowers with no perianth (hazel tribe)

B. Nut large enclosed by a leafy involucre, staminate flowers with 2 bractlets, pistillate flowers, capitate— — —*Corylus.*

Fagaceae

A. Ovary of pistillate flowers 6 celled, spikes of either sex eract and strict, fruiting involucre or burr densely covered with strong pickles— — — *Castanea.*

Musaceae

A. Calyx tubular, later split-spathaceous— — —*Musa.*

Bromeliaceae

A. Fruit a berry, indehiscent ovary inferior, seeds not winged nor plumbed

B. Pollen grains entire not provided with pores or a longitudinal membranous fold

C. Petals furnished with 2 ligules inside

AA. Berries connate among themselves and also to the bract and axis— — *Ananas.*

Palmaceae

A. Leaf segments infolded in vernation, spadices interfoliaceous

B. Flowers dioecious

C. Leaves pinnasect, segments acuminate, spathe solitary ovary of 3 distinct carpels, only one maturing, seed deeply grooved ventrally— — —*Phoenix.*

AA. Leaf segments folded back in vernation

CC. Seeds adherent to the endocarp, hilum diffused, embryo opposite pore, spacies interfoliaceous, flowers usually monoecious in the same spadix, the lower ones in 3 with middle one pistillate— — — —*Cocos.*

CCC. Raphae ventral, embryo dorsal— — —-*Areca.*

Myrtaceae

A. Fruit a berry or rarely an indehiscent drupe leaves opposite, punctate

B. Stamens straightish in the bud seeds with endosperm– – –*Feijoa*

C. Stamens inflexed or involute in the bud seeds with endosperm

D. Calyx limbs closed in buds deeply divided at anthesis– – – –*Psidium.*

E. Embryo thick and fleshy– – –*Eugenia.*

Dilleniaceae

A. Stamens and carpels numerous, winter buds enclosed in the swollen base of the petioles - *Actinidia.*

❏❏❏

Section - I
TROPICAL AND SUB-TROPICAL FRUITS

1
Mango

Genus : *Mangifera*
Species : *indica L.*
Family : *Anacardiaceae*
Chromosome No. : 2n = 40

In the early Sanskrit literature Amra, mango was known to Indian at very early times. Mango occupies an important place in the Hindu mythology as well as in various religious observances. Various archaeological discoveries also mention the presence of Mango in India. Alexander the Great in 327 BC during his invasion found mango cultivation in the Indus Valley. The Buddhist and Chinese pilgrims in their travel notes have also indicated mango cultivation. Amir Khusru a famous Turkoman poet has quoted 'Mango as the pride of garden, the choicest fruit of Hindustan'. During the Mogul Emperors, mango occupied an important place in fruit culture. The *Ain-e-Akbari* an encyclopaedia work written during the rule of Akbar, gives detail account of mango, varietal characteristics, quality etc. In praise of mango, British authors lavished huge praise and designated it as King of Fruits and that the Goa mango is perhaps the largest and most delicious, wholesome and best tasted of any fruit in the world.

The family Anacardiaceae to which mango belong also includes cashew (*Anacardium occidentale*) which is widely cultivated in the tropics for its edible fruit and pistachio nut (*Pistacia vera*) grown in temperate regions for its delicious nut.

Further if it is apple to the North America, mango is much more than that to lakhs of people living in tropics and subtropics where mango has been growing from time immemorial. Presently mango is found growing in abundance naturally from Kumaon to Burma in the tropical Himalayan region at an elevation of 1000-3000 feet, especially in the peninsular India where it is cultivated extensively. All credit must be given to the Portuguese who first introduced this important and useful fruit from India first to Africa and from there to South America, Brazil being the first country in tropical America were it was established. Today it is found in almost every tropical part of the world and at some places its cultivation has even spread into the subtropics. The name 'manga' is derived from Mango a Portuguese word which itself is an adaptation of Tamil "Man-Kay" or 'Man-Gay'.

The mango tree is usually evergreen and shows variation both in plant growth and height. The tree is generally erect, upright broad dome shaped crown and some form are low growing.

Origin and Distribution

Mango probably has originated in the Indo-Burma region and grows wild in the forests of India, more so in the hilly areas of north east. It is considered to be an allopolyploid in origin, further differentiated by gene mutations and hybridization. Records show its cultivation at an early times (4000 yrs. or more). The Indians introduced it to Malaya and other east Asian countries in 4th and 5th century BC. Persians took it to east Africa in the 10th century AD. Later the Portuguese introduced mangoes to west Africa, New World and in the start of 18th century it reached Brazil. From Brazil it reached West Indies.

Uses

Mango is generally used in three distinct ways, i.e. unripe, ripe and processed like chutneys, pickles, dried slices, canned slices, in syrup, juices,

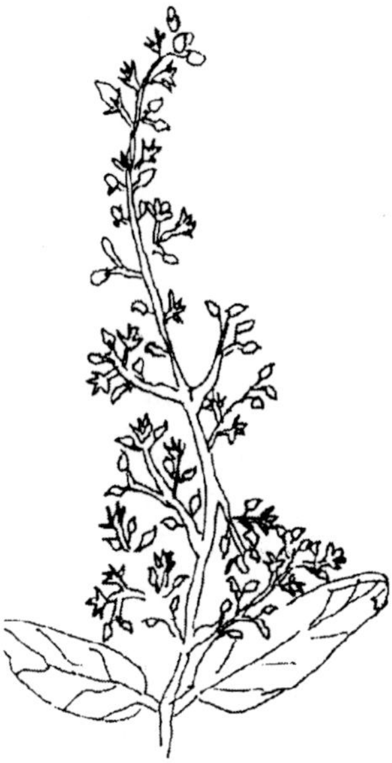

Flowering shoot

Fruiting shoot

Flower types

Stamen

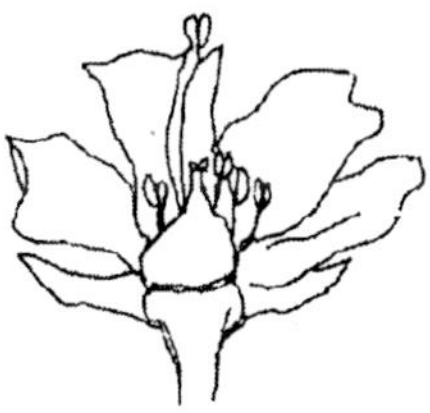

Male flower

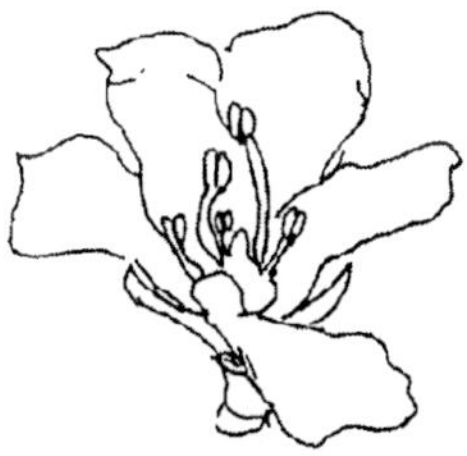

VS Flower

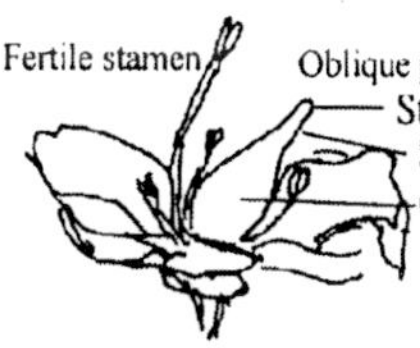

Perfect flower

Fruit

Seed

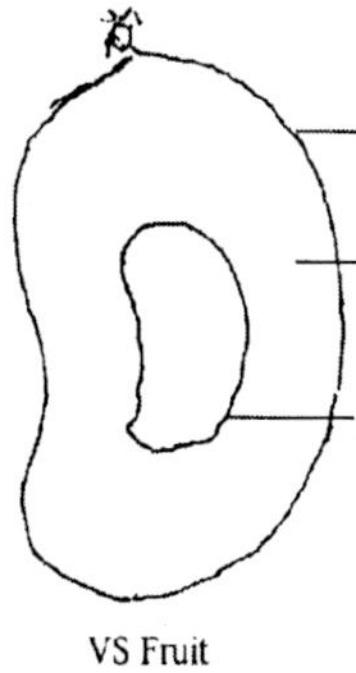

VS Fruit

paste or puree. Meat and fish dishes are flavoured with green fruit. In India, kernel serves as an important famine food after boiling to remove its astringency, as bye product used as feed for cattle and poultry. Young leaves are eaten fresh or cooked as vegetable. Extracts of unripe fruit, bark, leaves and stems show antibiotic activity. Wood is quite hard, strong and easy to work for construction and other wood work. Wood must be treated with preservatives.

Description

Mango plants are large trees (10 to 40 m), evergreen, erect, branched with dense round top, long lived, accompanied by well established both tap (6m. or long) and fibrous root system. Branches and twigs somewhat stout, trunk greyish brown, rough with prominent fissure. Leaves usually produced in flushes, simple, arranged spirally, exstipulate, coriaceous, shiny, smooth, opening leaves are usually red/purplish tinged, later turning dark green, short (1-2 cm)and thick petioled (some times may be 5-10 cm long) flattened on the upper surface and prominent pulvinus at base. Leaf shape elliptic or oblong-ovate to oblong-lanceolate, apex acuminate or acute, entire margin, surface undulating, midrib thick and prominent on lower surface, with 20-30 pairs of lateral veins. Inflorescence appear terminally on last season growth, fruiting terminal is a large branched panicle (20-60 cm), red or purple tinged and pubescent with one to six thousand flowers borne per panicle, erect, rigid, well branched rachis pink or purple, may also be in various shades of green. Polygamous i.e. male and hermaphrodite (1-36%) flower in same inflorescence. Flowers of two types perfect (bisexual) and male. Individual flower small, 5 to 8mm across, sub-sessile, pentamerous (may range 4-7 rarely), calyx 5, yellowish green, free, concave, ovate lanceolate inserted at the base of fleshy disc. Petals generally 5 (may range 4-7), elliptic lanceolate to obovate lanceolate, creamish, with 3-5 dark yellow ridges on inner surface, almost twice as long as calyx, later becomes pinkish, disc prominent fleshy placed between stamens and petals. Stamens generally 5 (3-7 rarely), inserted on outer margin of disc, 1 or 2 are fertile, others are staminodes, anthers pink, at anthesis turns purple red. In hermaphrodite flowers stamens same as above, ovary sessile, one celled, one ovuled, set on disc with lateral style, stigma simple about the same length as that of fertile stamen. Fruit is fleshy, laterally compressed, drupe, variable in size, shape, (ovoid-oblong) and colour (green, yellow, red). The exocarp (skin) coriaceous, thick, smooth, gland dotted, mesocarp the edible portion is variable in texture thickness and flavour (turpentine), soft, sweet and juicy, fibreless to fibrous, endocarp bony in texture, thick smooth, fibrous which may be free from mesocarp or extending into it. The pistillate/calyx end has conical projection or beak, above which there is a sinus/ typical depression or curvature, stalk end may be elevated, depressed or intermediate. Seed bony, thick, variable in size and fibre content. Seeds may be mono-embryonic (zygotic embryo) or poly-embryonic with 2-12 embryos, apomictic embryos result from epidermal cells of the nucellus, whereby zygotic embryos may or may not be suppressed.

The horticultural description of some important *Mangifera* species are summarized below :

M. foetida Lour. Widely present in Malay Archipelago. Plants are large trees, leaves elliptic-oblong to obovate about 10-12 inch long, apex acute or rounded, base acute. Petiole thick, panicle large, bright red in colour, thick. Flowers sub-sessile, calyx ovate-obtuse, corolla pink or red linear lanceolate, reflexed. Stamens 6, 1 long is the perfect, 5 are small and imperfect. Ovary smooth green oblique and shiny. Fruit flesh yellow, stone very fibrous, quality inferior.

M. zeylanica Hook. f. found plenty in Sri Lanka at an elevation of 3000 feet. The plant resembles very closely to *M. indica*. The plant is shiny, leaves 2-3 inch long oblong-obovate to elliptic-lanceolate, apex obtuse, panicle stout individual flower with thin pedicel, calyx rounded, corolla elliptic oblong, disc large, stamen 1 fertile 6-8 staminodes, ovary somewhat hairy.

M. odorata Griff. Specie is indigenous to Malacca and fruit lacks turpentine taste common with other species. Entire tree shiny, leaves elliptic-lanceolate to oblong measure 6-12 inch in length, apex acute or acuminate, petiole nearly two inch long thick and green, flower fragrant, calyx ovate-oblong, corolla greenish to blood red linear oblong. Perfect stamens two, staminodes present, thin ovary shiny, fruit yellowish green compressed, stone also compressed and fibrous.

M. caesia Jack. Found in abundance in Malacca and Malayan islands. Tree large growing. Branches thick, leaves obovate to elliptic very large 6-16 inch long slightly shiny, petiole small 1 inch long, panicle branched and thick, pedicles small and thick, calyx broad ovate and pubescent, corolla linear erect purplish, disk somewhat lobed, stamen perfect, imperfect are like small teeth, style thin. Fruit oblong-obovate.

M. caloneura Kurz. Medium sized tree, leaves oblong or oblong lanceolate thick at base, reticulated on both surfaces with 16-20 pairs of nerves, petiole upto 1 inch long, apex pointed. Panicle spreading, tomentose, flowers sessile and clustered, petals 5 with 3 ridges, fertile stamens 1-2, style lateral, ovary rough. Fruit small in size smooth yellow with sweet and acid taste.

M. longipes Griff. Plant is a tree, branches spreading, leaves oblong or elliptic- lanceolate, apex acuminate with 10-14 pairs of nerves faintly reticulate above, conspicuous below. Panicle long, flowers pedicelled, calyx ovate, petals white veins yellow, narrow and recurved, disc conical and 5 segmented. Stamens 5 and one is perfect other are imperfect with long filaments. Ovary usually smooth, style long thin and lateral.

M. pentandra Hook f. The Malay name 'Mam Ploni' means ripened artificially. Medium to large tree, leaves are oblong or oblong lanceolate, apex pointed (like that in *M. indica*), reticulated on both the surfaces, panicle tomentose and spreading, flowers subsessile and yellow, sepals oblong, petals yellowish white with yellow brown ridges. Stamens 5 and all perfect and of unequal length, style terminal and ovate and is suspended laterally.

M. sclerophylla Hook. f. A woody strong tree, branches usually very thick, angled and shiny. Leaves shiny elliptic in form, coriaceous, base round with narrow apex, veins faint and in about 10 pairs, smooth on both the surfaces, midrib conspicuous, petiole thick about 1 inch long. Panicle pubescent, flowers sessile small in size, petals about the size of sepals. Calyx pubescent and ovate, corolla ovate oblong with 3 short ridges, 1 fertile stamen, style lateral with small disk. Fruit nearly round.

M. Griffithii Hook f. Plant is a tree, leaves usually small 3-5 by nearly 2 inch in size, oblong or oblong-obovate base narrow, veins in 10 pairs, petiole thick. Inflorescence-panicle comprises of number of numerous sub erect compound pubescent racemes, not much longer than the leaves. Individual generally very small and subsessile, calyx pubescent densely broad ovate, corolla round ovate with 2 ridges, disc with 5 to 6 lobes, stamens short only one is fertile.

M. sylvatica Roxb. A common tree in jungles of Sikkim and Khasi hills. Plant are like common mango, leaves lanceolate to oblong lanceolate, 10-12 by 2-3 inch in size, smooth membraneous with 16-20 veins which are thin. Panicle glabrous slender and branched flowers loose, pedicelled sepals 5, petals 5, with 3 prominent ridges, stamen none is perfect, style lateral ovary smooth.

Cultivars

In India most of the cultivars have been originated as chance seedling either from natural crossing or from gene mutation and superior types have been maintained through vegetative propagation. Besides region wise cultivars have been recommended for cultivation. Important cultivars with their distinct traits are summarized below.

Dashehari : In north India, it is the most important, famous and commercially grown cultivar. Produces heavy crops but is alternate in bearing. Mid season variety-matures by June end. Fruits medium sized, elliptic-oblong with attractive greenish yellow colour smooth shiny surface, flesh yellow fibreless, sweet juicy excellent in taste and flavour suitable for canning.

Langra : A mid season cultivar with tendency of biennial bearing but heavy cropper. In quality it is sometimes rated even better than Dashehari. Fruits are larger in size round to oblong oval in shape with very small beak, light green (lime green) to yellowish green smooth leathery surface, flesh yellow with more fibre, has very good combination of sugar and acid, flavour pleasant aromatic in taste.

Chausa : Vigour in growth, is one of the most sweetest cultivar, very low in acidity or may be completely lacking when fully ripe. Bearing is late 15-20 years. Fruits are large in size, oblong in shape surface smooth and shiny yellow. Flesh yellow, soft, juicy, flavour pleasant, quality very good. The cultivar suffers from mango malformation and alternate bearing.

Bombay Green: Susceptible to malformation and biennial bearing. Matures very early in north India-mid June. Fruits are green in colour when ripe medium in size somewhat irregular in shape-oblique ovate, bears moderately, taste good, quality and keeping quality not very good.

Alphonso : Cultivar is rated very highly both in India and abroad, an export cultivar. Has performed well on the west coast of Maharashtra (Ratnagiri) specific in its climatic requirements, mid season, medium cropper, but fairly regular in bearing. Fruits are glossy attractive yellowish with bright pinkish blush at the basal end, large oval-ovate oblique (ventral shoulder is prominent). Has an excellent blend of sugar and acid, superb in taste, flesh yellow pleasantly flavoured with very less fibre has excellent table and processing qualities.

Pairi : Though bears quality and heavy fruits but it also suffers from biennial bearing. An excellent variety adapted to west and southern parts of India. Fruits are medium sized, ovate, apex (beak) significant with very slight sinus, surface smooth shiny yellowish green with very brilliant red tinge on the shoulders.

Bangalora : A very popular mid season cultivar of south India. Quality wise though not very good, but on account of its regular cropping and good processing traits, it is preferred by processing industry. Bears regularly and heavily, fruits are large with neck at base, apex with conspicuous beak and sinus, attractive in appearance with bright apricot yellow colour overlaid with red blush, skin thick flesh yellow to orange, firm fibreless and somewhat dry with pleasant flavour and sweet acid blend, storage quality very good.

Neelum : Very regular and high yielding cultivar grown commercially in south India, has high adaptability, storage ability fairly good, quality wise better than Bangalora. Fruits medium in size ovate-oblique in form somewhat rounded, beak and sinus prominent. Surface orange-yellow at ripening, smooth shiny flesh firm fibres few, pleasant flavour, taste good with acidic blend.

❑❑❑

2 Banana

Genus : *Musa*
Species : *paradisiaca L.*
Family : *Musaceae*
Chromosome No. : 2n =22,33,44

The genus Musa has been named after Antonio Musa, who was the official physician to Octovius Augustus the first emperor of Rome, 63-14 BC. Banana (Musa spp.) was perhaps the first among the fruit plants to be cultivated and there is hardly an difference of opinion that banana was one of the first food on which human relied upon. Banana was cultivated in India at very early times which is evident from the fact that Alexander the Great found banana cultivation in Indus river valley. In tropics, banana constitutes an important food crop for man and is a staple food of Buganda (Uganda), Wahaya (Bukoba) and Wachagya (Tanzania).

Banana is the most nourishing amongst all the fruits crops, is rich source of energy, cheapest and plentiful source of essential nutrients including minerals and vitamins, besides has number of medicinal properties.

Banana is a very useful fruit plant and all parts are used in one or other forms. Nearly half of the bananas produced are eaten fresh and the remaining half as vegetable after cooking. In Africa beer in made from banana. A number of products like slice, flour, powder, soft drink, chips, jam, hydration of core, tissue paper, and paper board from pseudo-stem, as well as baby food are made from banana.

The native place of edible banana is considered to be the hot tropical regions of south east Asia, and India is said to be one of the centres of banana origin. Banana belongs to the family Musaceae, which has two important genera viz Musa and Ensete with about 50 species. The genus Ensete comprises of 6-7 species, of which *E. ventricosa* fruits in Ethiopia are used as food crop. The genera is said to have originated in Asia and then later spread to Africa continent. The other genera Musa (x=10,11 rarely 7 or 9) with some 40 species have their origin in the Assam-Burma, Thailand area. Being plants of lowland tropics their basic requirement for growth is high temperature, humidity and light intensity and fails to sustain drought and poor drainage conditions.

The genus Musa is further divided into five section i. Eumusa ii. Rhodochlamys iii. Calimusa iv. Australimusa v. Incertae sedis. In the botanical nomenclature a number of Latin names were in common use, more frequently used names were *Musa paradisiaca, M. cavendishii,* and *M. sapientum.* Presently genome nomenclature is used for identification of cultivars which have been derived from the two wild species *Musa acuminata* and *Musa balbisiana.* The basic chromosome number is n=11, and the edible bananas may have 22,33, 44 chromosomes and are respectively referred to as diploid, triploid or tetraploid.

The worlds best and commercial bananas belongs to the *acuminata* group AAA, while resistance to drought and diseases is highly associated with the *balbisiana,* genome BBB, whereas the hybrids of AB, AAB, ABB (diploid and triploid level), show great genetic variability in respect of their adaptability to monsoon areas that experience prolonged dry season. The pure AA or AAA are more suited for cultivation in areas where assured water either through rainfall or irrigation is available.

Origin and Distribution

Bananas are tropical by origin and generally show intolerance to frost. They are distributed throughout the warmer regions of the world and rarely

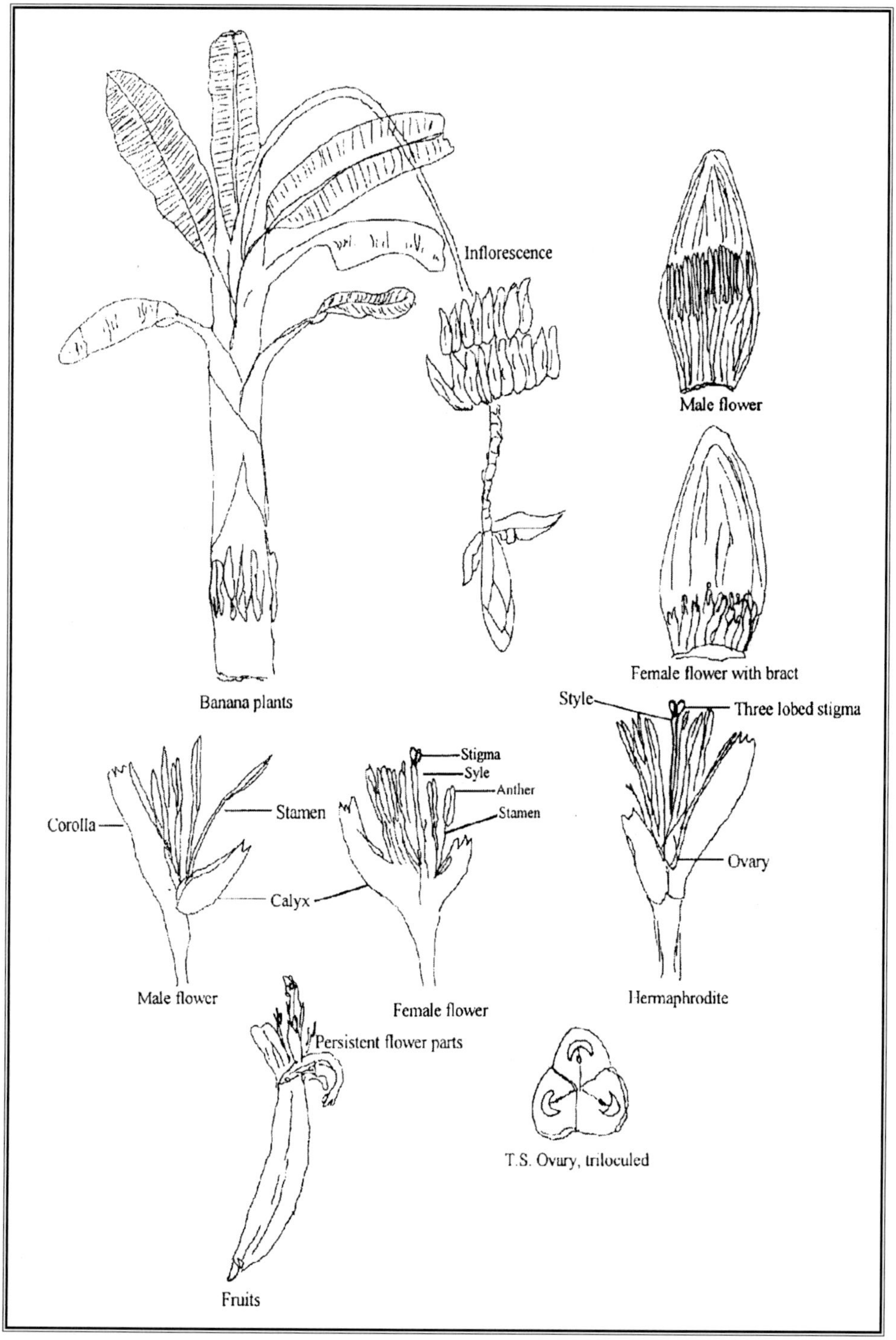
Inflorescence
Male flower
Female flower with bract
Banana plants
Style
Three lobed stigma
Stigma
Syle
Anther
Stamen
Corolla
Stamen
Calyx
Ovary
Male flower
Female flower
Hermaphrodite
Persistent flower parts
T.S. Ovary, triloculed
Fruits

beyond 40^0 N and S latitudes. The wild bananas are diploid and distributed in south east Asia and pacific, one related Genus Ensete (2n=2x=18) is distributed in tropical Africa and south east Asia. Banana evolution is very old , the crop was established in west Africa, much before the arrival of Europeans. From here the crop reached New World in the late fifteenth century. In all likely hood the crop entered Africa from Malaysia through Madagascar or the east coast by the migrants and travellers mainly from Indonesia.

Uses

Fruit is the main product, used either raw or cooked, processed into starch, chips, puree, beer, vinegar or dehydrated. Leaves are used to polish floor, as pot line in which rice is cooked, as wrapper for different foods, pseudo-stem yields fibre. Vegetative parts and reject fruit is used as fodder. Young unfolding leaves medically serve as a cool dressing for unflashed or blistered skin. Fully ripe fruit is a laxative, flour from banana is used to overcome dyspepsia with flatulence and acidity.

Description

Banana consists of large to gigantic herbs rarely trees, with rhizomes and pseudo-stem of leaf sheath, leaves large arranged spirally, distichous with large stem clasping sheath. The new leaves are formed from meristem near ground level push up through pseudo-stem in a tightly rolled condition. Leaves are arranged in a loose rosette and comprises the sheath, a petiole and a blade. The sheaths are almost circular, packed tightly into non woody pseudo-stems which functions as plant trunk. Leaf lamina large, oblong edges red tinged, midrib thick stout, large number of pinnately parallel veins which end at margins, greyish waxy deposits on midrib and along the veins. The inflorescence is terminal complex spike on peduncle in the centre of the pseudo-stem which emerges from the centre of leaf crown, flowers in nodal clusters, each cluster in axil of large spathaceous bract which are deciduous, crowded and spirally arranged, broad ovate/lanceolate dark purple or brownish outside and pale purple or crimson inside. Flowers zygomorphic usually bisexual, rarely unisexual on account of abortion or sterility of either carpels or stamens, ovary inferior in large mixed spatheceous inflorescence. Flowers placed in the axils of the bracts are about 12-20 in number. Lower flowers are pistillate, whereas terminals are staminate. Perianth comprising of 6 members arranged in 2 whorls, differentiated into sepals and petals, sepals usually two toothed, somewhat thick white or light pink tinged, petal thin, 3 toothed, large pinkish, connate or some times free large in size with well developed ovaries. Stamens

6, free, of which 5 are fertile, 1 staminode, rarely all 6 are fertile, filaments filiform, free anthers two lobed. Anthers two celled, linear and long. Carpels 3, connate in a trilocular inferior ovary, each locule with numerous ovules in axile placentation, epicarp, mesocarp slightly fibrous and endocarp, fleshy, gland present in the ovary. Style simple stiff and long, stigma capitate sticky, 3 lobed, fruit fleshy soft berry. Seed triangular or grooved with thick hard testa, embryo straight.

Description of Genus *Musa* L. (X = 10, 11 rarely 7 or 9)

Musa is a plant of lowland tropics comprises of about 40 species, perennial stooling or rhizomatous herbs native of south eastern Asia and Pacific. Centre of diversity and of origin is centred in areas of Assam- Burma- Thailand. It requires high temperatures, light intensities and humidity, susceptible to drought, poor drainage. Tightly clasping leaf sheath form the pseudo-stem, swollen at base, suckers freely. The flowers and bracts are placed independently on peduncle, falling by abscission. At the base of the spike functional female ovaries are present in basal hands, flowers here are usually female only, the male or neutral flowers are placed on distal hands at the distal end of spike, bract oval reddish purple or violet, pollen grain very finely granular, seeds 5-7mm in diameter.

The genus *Musa* is divided into following 5 sections :

a) Eumusa
b) Rhodochlamys
c) Callimusa
d) Australimusa and
e) Incertae sedis

a) Eumusa : (x=11, 2n=22, wild spp.22, 33 and 44)

Very old diversified and widely distributed banana section, has 13-15 species, found in southern India, Japan and Samoa. Important characters are pseudo-stem more than 3m. in height, inflorescence pendent or semi pendent, bracts dull coloured, flowers numerous, in two series in each bract. Includes seedless bananas. *M. basjoo* Sieb., yields fibre which is made into textile in Japan.

b) Rhodochlamys : (X=11, 2n=22)

Relatively small section, has 5-7 species extending from India to Indonasia, some what related to Eumusa through *M. flaviflora* Simmonds. Pseudo-stem less than 3m high, inflorescence erect, less flowers per bract, generally in single whorl, parthenocarpy absent. *M. ornata* Roxb. and *M. velutina* Wendl. & Drude are ornamental type.

c) Callimusa : (X=10, 2n= 20)

Native of Indo China, Malaya and Borneo, small section with 5-6 species, diploid, seeded type, pseudo-stem small inflorescence erect, bracts purple. *M. coccinea* Andr. is ornamental type.

d) Australimusa : (X=10, 2n= 20)

Specie extending from Queensland to Philippines, recently differentiated Musa section. Inflorescence pendent semi pendent or erect, includes spp. *M. textilis* abaca or Manila hemp and Fe'i banana, referred to as *M. fehi* Bert. ex Vielli, later related to *M. maclayi,* wild in New Guinea and the Solomon Island. From here it was taken by Polynesians eastward to Tahiti and attained maximum significance. The fruits are eaten after cooking. Inflorescence erect, has red sap, female sterile, parthenocarpic, skin orange coloured, flesh yellow.

e) Incertae sedis

Some spp. in this section are of doubtful affinity. Species included are *M.ingens* Simmonds (x=7, 2n=14) found in New Guinea, pseudo-stem 10m tall.

Uses

Most important amongst tropical fruits, second to grapes in production. Ripe bananas are eaten fresh, roasted or fried which are sugary and easily digestible, on cooking yields starchy foods, form single biggest item in international fruit trade, staple in Africa, fresh unfermented juice is drunk. Sweet meats from dried slices of ripe fruits, chips from mature unripe fruits are prepared. Banana flour (unripe) and powder (ripe) used in confectionary, male buds (with bract) are eaten as vegetables, pseudo-stem, peduncles, fruit peels fed to cattle, leaves as plates, umbrella and wrapping material, sheaths, petioles and leaves as tying material, for thatching, as mulch material, sap as indelible ink, fibre from pseudo-stem.

Pomological description of some important species is described as under.

Musa ensete Gmel (*Ensete edule,* Horan). Abyssinian Banana. One of the widely cultivated oldest and largest specie amongst the decorative bananas. Tall upright plant, stem high about 15-20 feet, base swollen, stolons are not produced. Leaves are very long 20 feet by 3 feet oblong, midrib red tinged, spike erect, bracts heavily imbricated about a foot long dark reddish brown. Individual flower white nearly 2 inch long, about 20 or more in a cluster, calyx 3, petal 3 free, both are lobed. Fruit small 2-3 inch long, seeds few 1-4 but large black and shiny. Produces good fibre also.

M. Fehi Vieill. (Syn. *M. Fei, M. Seemanii, M. Uranoscopos*). Cultivated extensively in Ecuador. Plant stem erect and tall 15-20 feet with violet sap, produces stolons. Leaves firm and larger than *M paradisiaca*, veins conspicuous, base obliquely rounded, petiole 1-1½ feet long. Spike upright long somewhat curved at base, flowers in clusters of 6-8, calyx 5 lobed equally splitting down to the base. Petals free, short. Fruits many in a bunch somewhat angled 5-6 inch long 1 inch across almost straight. Skin thick yellow pulp medium firm, seeds small and dull black in colour.

M. textilis Nee. (*M. mindanensis, M. silvestris, M. troglodytarum textoria, M. Abaca*). Abaca Manila hemp. Widely cultivated in Philippines. Plant tall, 20 feet or more, stolons produced freely, leaves usually oblong base deltoid bright green above, shiny below with prominent brown spots, petiole small 1 feet, spike drooping male flower deciduous, calyx curved with horn like hooks. Fruits three angled, small somewhat curved 2-3 inch long 1 inch across, seeds many black in colour and inedible.

M. Basjoo Sieb and Zucc. (*M. japonica*, Hort). Japanese Banana. Grown extensively in Japan for fibre. Plant medium tall, stolons produced freely, stem about 6-10 feet high, leaves bright green thin oblong, petiole short and stout, spike about 1-2 feet long dense, bract dull brown, oblong with about 15 flowers, female flower in cluster of 3-4, male in clusters of 8-12, bracts persistent, calyx white, with about 45 short teeth at apex. Fruit 3 angled, oblong, seeds few.

M. paradisiaca Linn. Adam's Fig, Cooking Banana, Plantain. Native of India, tall upright growing. Plant 25-30 feet, stem 2-2.5 feet, stolons produced freely, leaves, thin, bright green, oblong 5-8 feet, base rounded, spike 4-5 feet long, drooping, bracts dull violet, shiny, persistent, lanceolate to oblong lanceolate in form. Male flowers mostly persistent with about 12 flowers per cluster white or pale yellow, 1½ inch long, calyx at apex 5 toothed, corolla free oval, half the length of calyx. Fruit yellow to yellowish green ½-1 feet long, cylindrical.

M. sanguinea (Hook F.). Plants are showy and decorative, native of India (Assam). Stem slender 4-5 feet high, leaves bright green thin and oblong, petiole slender, spike earlier erect, later drooping. Bracts blood red, persistent lanceolate, male flowers in few clusters, calyx bright yellow in colour 1½ inch long, corolla, free as long as calyx. Fruit oblong 3 angled pulpy pale green with red variegation, 2 inch long, seed small and black.

M sapientum, Kuntze. (*M. sapientum,* Linn). Common Banana. Grown in plenty in India. Is regarded to be a subspecies of *M. paradisiaca* and has been the progenitor of most of the banana commercial varieties. For most of the characters it resembles *M. paradisiaca,* but differs in that male flowers are deciduous, 3 angled about 3-8 inch long and 1½ -2 inch across. This specie has number of botanical varieties and their short description is as follows.

M. sapientum, var. Dacca, Baker (*M. Dacca*, Horan) Dacca Banana. Native of India. Fruit flavour good, tender, yellow in colour, skin very thick, tip and base bright green in colour, petiole base red, leaves pale green, shiny beneath.

*M. sapientum v*ar. cinerea Blanco. Letondal or Chotda Banana. Introduced from India to other places. Fruits are small thin skinned, rounded which splits at maturity and turns black, flesh greyish and delicious. Perfect seeds not found.

– – – – – – – var. rubra Red Jamaica or Spanish Banana. Almost all plant parts stem, petiole, leaf and fruit are red or red tinged, very decorative.

– – – – var. Champa Baker. Hart's Choice, Lady Finger or Golden Early Banana. All plant parts are red except fruit, flower pale straw in colour. Fruit 6 inch long, skin thin and soft, flesh with delicate flavour, ripens very fast.

– – – – var. magna Blanco Chotda Tondaque. Grown extensively in Philippines and cloth is made from its fiber but is inferior in quality compared to that of Abaca. Fruit very long 1 feet by 3 inch across, several angled. After cooking becomes more palatable.

– – – – var. seminifera Baker (*M. seminifera*, Lour.). Found wild in India, Malayan and Philippine islands. Fruits are small not edible, because of numerous seed, surface greenish or yellowish, is the common seed bearing wild form of banana.

Cultivars

The number of banana cultivars are variable, but to one estimate there are about 250-300 cultivated cultivars, Description of some cultivars grown in India are described below.

Poovan (AAB) : Has number of synonyms Champa, Lalvelchi, Chini Champa. In India it is considered to be an important table cultivar mostly grown close to the coastal areas. Plant is slender, tall vigorous and hardy, midrib of petiole is marked with pink colouration. A good bunch may weigh upto 15 kg with 220 fingers but average bunch weight is around 15 kg. Fruit small to medium in size usually yellow, flesh firm, sweet with pleasant sourish or sub acidic taste, keeping quality good. Resistant to panama wilt and fairly to bunchy top disease.

Harichal (AAB) : syn. Sapri Peddapacharti (large green bananas) Bombay Green. Mostly grown in Maharasthra, Andhra Pradesh, and Tamil Nadu. Plant is semi tall 3 ½ -4 mts. tall growing, regarded to be a bud sport of Dwarf Cavendish. Requires more humid conditions. Fruits are large, skin greenish to dull yellow and thick fingers more straight, bunch less compact, has good keeping quality, average bunch weight is about 20 kg with about 160 fingers.

Kanchkela (ABB) : syn. Monthan, Ponthan, Bontha, Bainsa etc. A very important commercial and culinary banana grown extensively all over banana growing areas of India. Plant is strong tall hardy and grows very well under low moisture regimes or irrigated place. Bunch on an average weighs about 15-17 kg.

Dwarf Cavandish (AAA) : syn. Basrai, Kabuli, Pacha Vazhai Hirvi etc. Very popular and commercially grown in Maharashtra prefers drier climates. Plants are dwarf, 2 m. tall, fruits/fingers are large somewhat curved, skin greenish yellow or dull yellow, soft and sweet in taste, ripening is late and it does retain some of its green colour even or ripening. Keeping quality not good, bunch large about 25 kg and with 140-160 fingers, has tendency to drop fingers when fully ripe resistant to panama wilt, susceptible to leaf spot and bunchy top.

Martaman (AAB) : syn. Rasthali, Mutheli, Malbhog, Rasabale, Sonkel. Is rated as one of the most preferable cultivar of West Bengal. Characteristic features is that plant is tall growing and has distinct brownish specks/blotches on the bright yellowish green stem, petiole and leaf sheath margin reddish in colour. Fingers are of medium size like that of Poovan, skin thin to medium thick, yellow in colour sweet acidic in taste with pleasant flavour, flesh firm, keeping quality therefore good. Bunch about 15-18 kg with 130-140 fingers. Susceptible to Panama wilt and finger drop are limiting factors.

Lalkela (AAA) : syn. Sevazhai, Anupam, Rathambala, Red banana etc. Almost all plant parts being red or red tinged, hence the name red banana. The pseudo-stem leaf margin petiole, midrib and fingers all have marked purple red or reddish pigmentation. The fingers are large, stout, straight and with pleasant flavour but do not match in quality to other bananas. Bunch is heavy and weighs about 18-20 kg.

Hill Banana (AAB) : syn. Virupakshi. Very popular in the hills of Tamil Nadu, usually are perennial in nature and quality wise are also good, prefers cool conditions. Finger are medium in size angular, thick, skin, dull yellow, pulp white quite slimy or moist with buttery consistency, sweet in taste. Bunch is small weighing about 10-12 kg has some 90-100 fingers.

❑❑❑

3
Citrus

Genus : *Citrus*
Species : *Sinensis aurantium, limon, medica grandis* L. *paradisi Macf. reticulata Blanco*
Family : *Rutaceae*
Chromosome No. : 2n = 18

Citrus is one of the most important and popular fruit of the world which is widely cultivated in almost all of the tropical and subtropical parts of the world. Amongst the subtropical fruits in respect of average it occupies the third position. In India too it is next to mango and banana in commercial perpective. Citrus fruits on account of extensive adaptability are successfully cultivated in tropical, subtropical and even in the marginal temperate regions where the condition are somewhat favourable with its attractive appearance (foliage) distinct flavour and above all its excellent fruit quantity, citrus is liked by one and all. Fruits are not only refreshing and delicious but are also very rich in vitamin C and is good source of A and B, with fruit sugar acids, minerals like Ca, P, Fe. They are essential and play significant role in health promoting constituents in human diet, significance of citrus fruits in diet is very well established. Limes and lemons lack in fruit sugars but are rich source of vitamins, minerals as the oranges are. They are rich in acids therefore are not consumed fresh but are used for flavouring salads, dishes, vegetables etc. used for pickle making. Albedo yields pectin which is employed to make jellies, marmalades. Oils are extracted from the inter portion of the rind and used for flavouring. Citrus acid and citrate of limes are commercially made from lime and lemon juice, which is useful as a natural cosmetic and also as hair rinse, skin lotion and as mouth wash.

Since very early times citrus fruits were in cultivation in Indo-China and the sub Himalayan region of India, which also happens to be its place of origin. In the beginning of the Christian Era, citrus fruits were introduced to different parts of the world especially the New World. It goes to the credit of United States that this introduced fruit was developed to its maximum level and later developed citrus industry that covered all aspects of citrus starting from cultivation to harvest and finally its utilization as juice, processing, marmalades etc.

Oranges amongst the citrus fruits are the most common which is evident from the fact that large areas are put under it. Most of the citrus fruits are eaten as fresh i.e. sweet oranges, mandarins, pummelo and grapefruit. On the other hand limes and lemons are too acidic and are not used as fresh but are used to flavour salads, vegetables, dishes, meat and to make pickles.

The early history and also the origin of citrus fruits despite being in cultivation over a long period of time is still very confusing primarily due to very limited information. Varied views are expressed over its origin. Some say that citrus fruits appeared about 30 million years ago. According to Chinese written records its presence is said to be as early as 2200 BC, whereas citrus fruits do not find any place in list of fruit crops which have been in cultivation for nearly 4000 years.

For very long period of time citrus origin had been a matter of confusion and controversy, but on account of consistent efforts by various citrus workers, todays most of the taxonomist are in total agreement of citrus origin being in Himalayan region and south China. Two species of subgenus Eucitrus - C. *indica* and - C. *Assamensis*, and three species of subgenus Papeda - C. *latipes*, C. *ichangensis* and C. *macroptera* are native of Assam region. Likewise C. *medica*, citron is said to be native of north west of India, as it is found growing in plenty in Assam, Nilgiris and lower Himalayas. Lemon is sometimes cited to be indigenous to India, but its actual home is Malaya region.

The sweet oranges are considered to be native of southern China or Cochin China and then spread to India. About its present form there is again dispute some say it developed in India and then spread to other parts, whereas others believes that sweet oranges developed to its present form in China, later reached India and then moved westwards.

In the later part of the fifteenth century, the citrus fruits were introduced to America by the voyages carried by Colombus and the Spaniards, on their return journeys to their native places, and by the end of the same century most of the citrus species had reached and established in most of the tropical and subtropical part of the Eastern hemisphere.

All citrus fruits, belong to the family Rutaceae which is very large and diverse family. Members exhibits wide adaptability, most are tropical and subtropical in nature, but some members grow successful in warm temperate regions also.

The citrus classification revolves around two classical noteworthy works of Swingle (U.S.A.) and Tanaka (Japan). According to Swingle's classification genus Citrus is divided into two subgenera, 16 species and 8 botanical forms.

Subgenus 1. Eucitrus (10 species) which are given below.

1. Citron — C. *medica*
2. Lemon — C. *limon*
3. Lime — C. *aurantifolio*
4. Sour Orange — C. *aurantium*
5. Sweet Orange — C. *sinensis*
6. Mandarin — C. *reticulate*
7. Pummelo — C. *grandis/maxima*
8. Grape fruit — C. *paradise*
9. Indian wild orange — C. *indica*
10. Tachibana — C. *tachibana*

Subgenus 2. Papeda (6 Species)

1. C. *latipes*
2. C. *ichangensis*
3. C. *macroptera*
4. C. *hystrix*
5. C. *celebica*
6. C. *micrantha*

The other worth while classification was given by Tanaka (Japan) who happened to be Swingle's student. After his return back to Japan, and extensive survey and studies, gave his own classification which was very elaborative and was of the opinion that a species should represent a definite unit, which is quite different from the other unit. His liberal approach was quite different from the others and resulted in large number of species. He grouped genus citrus into two subgenera namely Archicitrus and Metacitrus. Tanaka's classification is as follows.

Genus Citrus (2 subgenera, 144 species)

Subgenera 1. Archicitrus comprised of 5 sections, 98 species.

Section

1. Papeda - 12 species
2. Limonellus - 16 species
3. Aruntium - 28 species
4. Citrophorum - 21 species
5. Cephalocitrus - 21 species

Subgenera 2. Metacitrus consist of 3 sections, 46 species.

Section

1. Osmocitrus : Has 9 species
2. Acrumen : A big section and has 36 species
3. Pseudo fortunella : Has only one specie.

On the basis of their general appearance and other characteristic like the fruit shape, colour, smoothness or bumpiness, vesicle colour number, juiciness or dryness taste etc the citrus fruits of prime commercial significance have been categorised into four group, the oranges, mandarins, the pummelos, grapefruits, and the common acid members. The last group comprises the citron, lemons and limes. Besides these fruits, all of which belong to the genus Citrus there are kumquats which belong to Fortunella and a very distinct related trifoliate orange (*Poncirus trifoliate)*. Fortunella is important for its fruit whereas *Poncirus* and its hybrids are significant as rootstock.

Origin and Distribution

The cultivated species of citrus are believed to have originated in tropical and sub tropical parts of south east Asia and old records also show that citrus occurs naturally from India and southern China to northern Australia. The earliest records on cultivation comes from China 2200 BC i.e. at remote times.

Most probably citrus has originated in the drier monsoon areas rather than the tropical rain forests, because plants show dormancy and water storing hairs (pulp vesicles) which help seed to develop under such conditions. Natural hybridization takes place between cultivars and species without much difficulties, thereby resulted in wide range variation in the form of complex hybrids. Further man selected ones, taking into consideration edible juice, desirable flavour, juice storing vesicles etc.

Citrus aurantifolia (Christm.) Swing 2n =18. Lime, sour lime, common lime.

This citrus species has its origin in northern India, parts of Burma adjoining India, northern Malaysia. According to Swingle lime is native to the East Indian Archipelago and is seen growing wild in northern India and may have originated here and later spread to other parts of tropics.

C. aurantifolia Lime

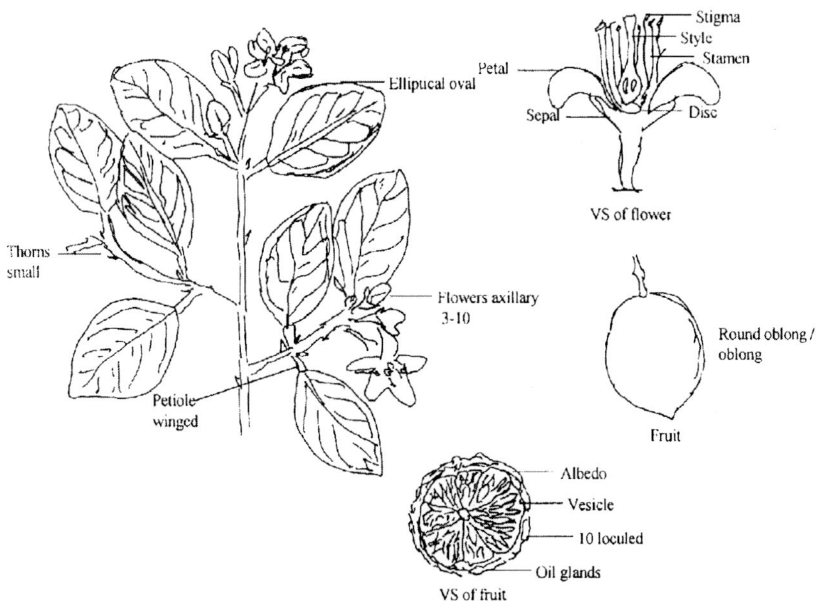

Description

Lime is very distinct specie, hybridizes easily with other citrus species, but is not related closely to them. Plants are short trees (5 m tall) much branched, irregular in outline, heavily marked with short sharp spines. Leaves are simple, alternate, small (4-8 x 2-5 cm), ovate-elliptic to sometimes oblong-ovate, margins crenulate, wing narrow. Inflorescence short occur axillary usually 1-7 flowered, sometimes 10 flowered. Flowers are white, small flowering over long period, calyx cup shaped 4-6 lobed, petals white, 4-5 (12 x 4mm), stamens 20-25, ovary 9-12 loculed, style distinct. Fruit is a berry, small, greenish yellow, oval, ovoid or globose in shape about 3-6 cm in diameter, with distinct apical papillae at calyx end, rind / skin or peel thin with numerous glands, adhering tightly to very acid yellow green vesicles, which are juicy and fragrant. Seed, plump, small, oval, ovoid cotyledons white, polyembryonic, are more tender than other citrus species.

Horticulturally limes are divided into acid and sweet limes. Apart from the most commonly grown 'Mexican' or 'West Indian' or 'key' 'lime', Tahiti or Persian lime (2n = 3x = 27), a triploid type and hybrid of lime and citron (*C. medica)* are also important. This lime is more hardy and is more adapted to subtropical conditions. The fruits are globose, longer, high in acid, almost seedless, less fragrant than the common lime. The sweet lime is a hybrid of lime sweet, lemon or sweet citron. It is used as a root stock.

Citrus aurantium L. (2n =18). Sour or Seville Orange.

Cochin China, in south eastern Asia is regarded to be the place of origin of sour orange. The specie later moved eastwards. In the 11th century it reached Europe, almost five centuries ahead of introduction of the sweet orange. Amongst the *Citrus* spp., sour orange is often regarded to be the one of the first citrus which was introduced to the New World. Today it is cultivated all over tropical and sub tropical parts of the world.

Plants are relatively tall (10 m), thorn small thin and slender. Leaves alternate, simple, dark green, shiny, ovate to elliptic, large (10 x 7 cm) broadest in middle, round to short pointed apex, margins crenulate, petiole 2-3 cm long, broadly winged. Flowers are borne axillary, flower large, white, very fragrant (5-10% staminate), stamens 20-25, ovary 10- 12 loculed. Fruits usually bright orange-red, very aromatic, globose, rind/peel thick, rough (surface bumpy), pulp very sour to bitter, core generally hollow, seeds borne are many, polyembryonic.

Citrus limon (L.) Burf. (2n = 18). Lemon.

Lemon has originated in south eastern Asia, but its exact place of origin is yet unknown as some say it is in north Burma, others say east of Himalayas, south China or north western India. The Arabs knew the specie in the 10th century, from here reached Europe in the 12th and 13th centuries. Later the crop spread all over tropics and subtropic parts of the world. Less important than limes in the tropics but more successful in temperate countries. After treating in dry weather fruit can be stored for 6-8 months.

C. aurantium

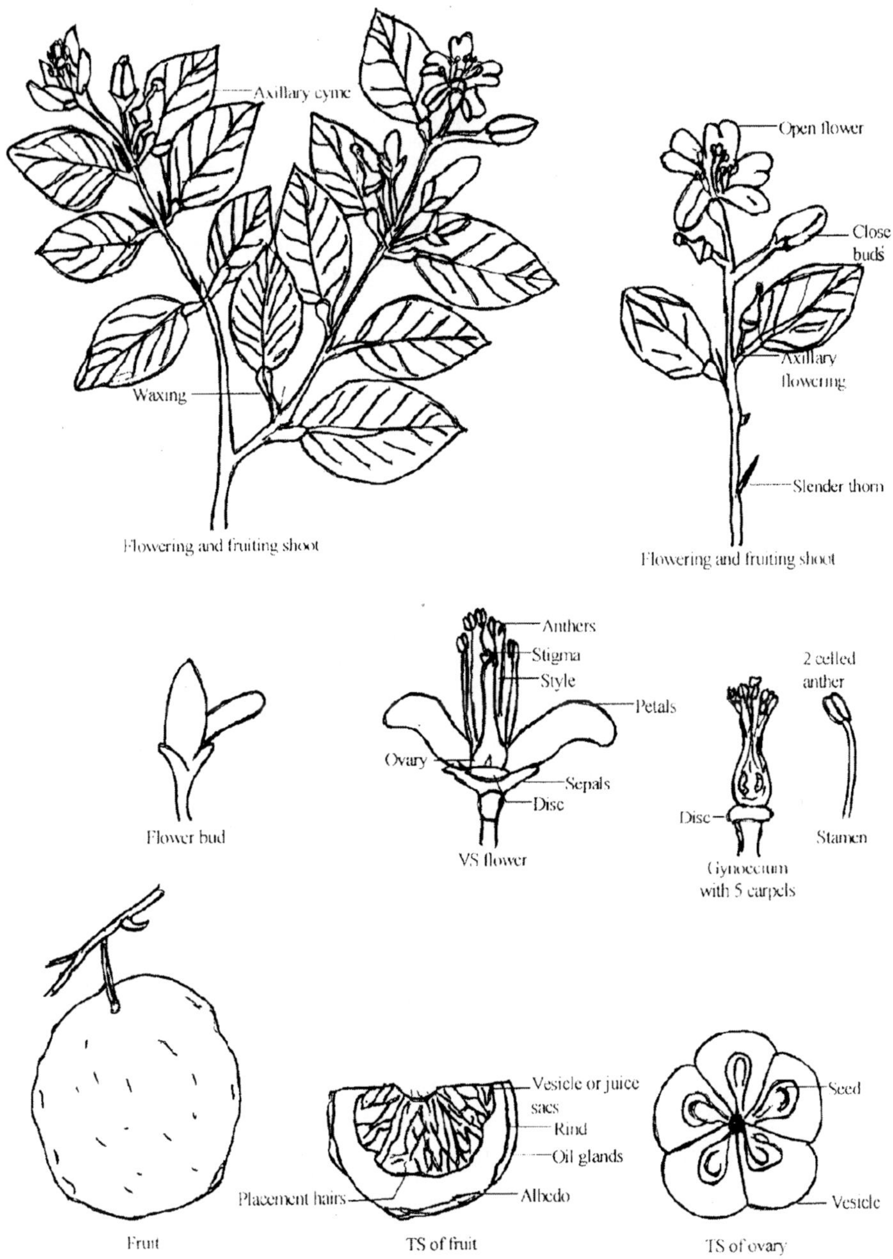
Axillary cyme
Waxing
Flowering and fruiting shoot
Open flower
Close buds
Axillary flowering
Slender thorn
Flowering and fruiting shoot
Flower bud
Anthers
Stigma
Style
Petals
Ovary
Sepals
Disc
VS flower
2 celled anther
Disc
Gynoecium with 5 carpels
Stamen
Fruit
Vesicle or juice sacs
Rind
Oil glands
Placement hairs
Albedo
TS of fruit
Seed
Vesicle
TS of ovary

Being very sour, lemons rarely are eaten fresh, used to prepare squashes, lemonade, in confectionaries and culinary purposes like garnish flavouring, yields lemon oil, citric acid and pectins, candied peel from rind is good source of vitamin P, successful rootstock for sweet orange, mandarins and grapefruits.

C. limon (Lemon)

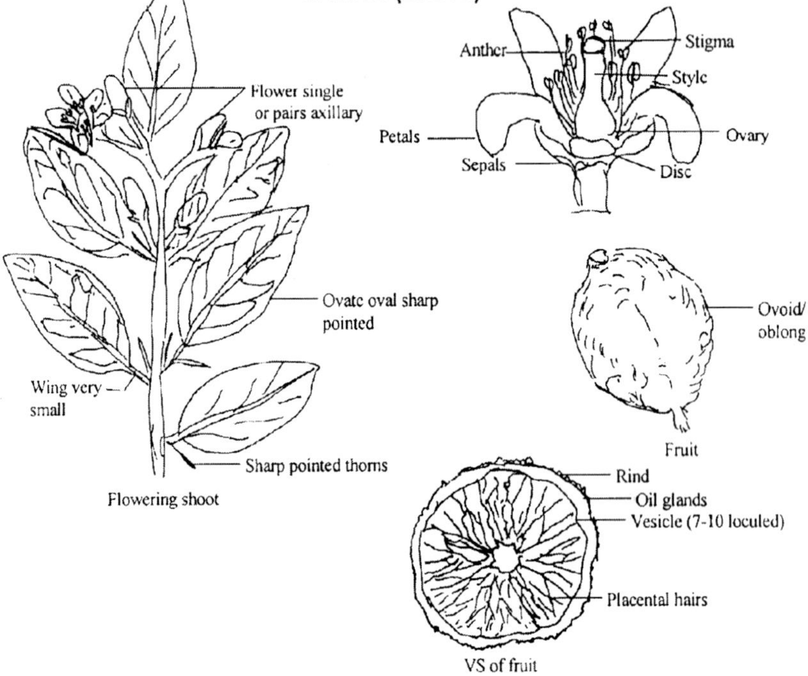

Lemon plants are small trees (3-6 m tall), thorns stiff, stout. Leaves medium (10 x 6 cm), simple, alternate, light green, margin slightly serrated, petiole short, narrowly winged with distinct articulation with petiole. Flowering occurs in leaf axils, flowers solitary or in clusters 3-5 cm in diameter, produced in all seasons, calyx 4-5 lobed, petals 4-5, pink in bud, white above, and purplish below, stamens 20-40, style single, ovary 8-10 loculed. Fruit oval-oblong (5-10 cm long) with distinct nipple at stylar end, yellowish green (light yellow) at ripening, rind thick, densely dotted with glands, adhering tightly to sour tasty pale yellow pulp/vesicles, seed plump ovoid with greenish cotyledons, polyembryonic. Lisbon, Eureka and Villafranca are the most common and widely grown cultivars. Shows complete nucellar embryony, stock therefore is uniform.

Citrus medica L. (2n = 18) Citron.

Citron is considered to be native of India, but others say that probable region of its origin lies in south western Asia, where it has been in cultivation over a very long period. The crop spread westwards and later reached Persia. Citron is often ranked as the first species to be cultivated in the western world and in China. The specie later reached most of the tropical regions of the world, but is cultivated on a very limited scale. Commercial cultivation is however confined to some islands in the Mediterranean region especially Italy, Greece, France etc. Citron is used as a potential source of candied peel, mostly used in confectionary and to flavour cakes. Almost all plant parts i.e. leaves, shoots, roots, flowers, fruits, seeds have been used by the Chinese as medicine and as an ornamental. Medicinal preparation are known and made for the treatment of headache, asthma, stomach-ache, physiological disorders etc.

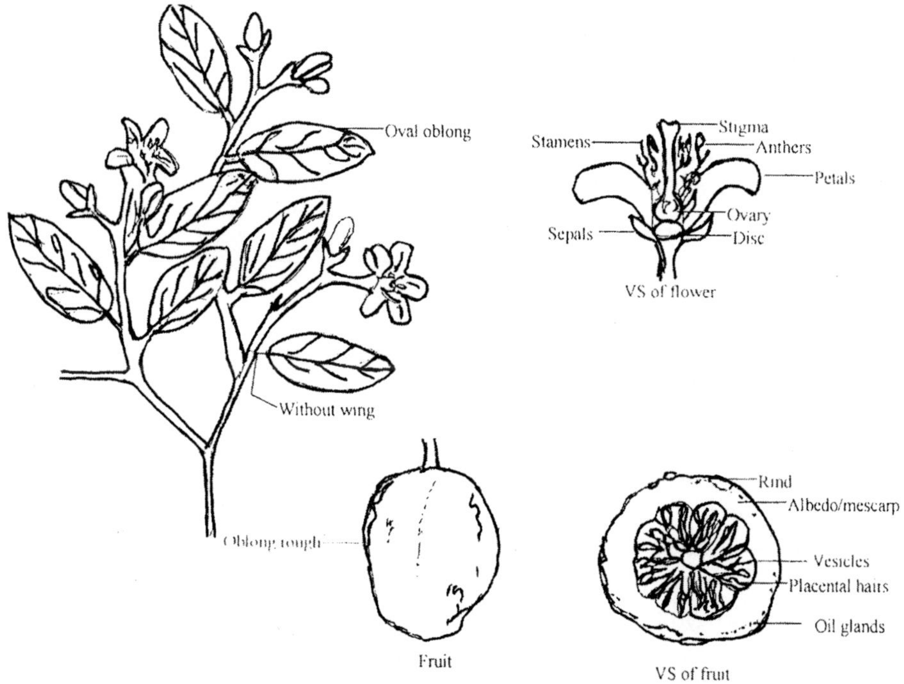

Citron plants are shrub or small tree (3 m tall) wood somewhat soft, current grown angular, later becoming terete (circular) purple tinged when young, spine single, stout. Leaves, alternate shiny ovate-lanceolate to elliptic ovate, medium large (8-20 x 3-9 cm), apex pointed or rounded, rounded at base, petiole short, almost wingless and not articulated with lamina. Flowers occur axillary over an extended period in few flowered racemes, flowers are large 3-4 cm across, hermaphrodite and staminate, large proportion being staminate, sepal 5 lobed, petals 5, pink tinged outside, stamens 30-40, style single, thick ovary usually large with10-13 locules. Fruit ovoid to oblong-ovoid, large sized 10-20 cm long, yellowish coloured, rind/peel very thick, surface rough very bumpy, segments small, pulp or vesicles greenish, sour in taste. Seeds small, white, polyembryonic.

Major citron cultivars are divided into two groups eg.

a) *Acid cultivars :* Floral buds and shoots are pinkish, pulp acidic, inner seed coat and chalazal spot dark coloured.

   Known cultivars are

   Diamante – Important/commercially grown in Italy.

   Etrog - Main cultivar of Israel.

b) *Non acid cultivars :* Floral buds and shoots not coloured, pulp non-acidic, inner seed coat and chalazal spot are pale yellow.

   Corsican is a well established of Corsica (France) and in California. Another botanical variety of Citron *Citrus medica* var sarcodactylis is well known in Japan, China, India and Indo China, is cultivated on a large scale, is often referred as fingered citron, because fruit split into number of segments (like finger) with very less pulp. Valued for rich fragrance, being dwarf type planted as an ornamental in the south and far East.

Citrus sinensis (L.) Osbeck. Sweet Orange 2n = 18, triploid and tetraploid occur

C.sinensis is regarded to be native of southern China or Cochin China. From here sweet orange reached Europe in latter half of 15th century. With the introduction of better / improved types from China large orangeries were established in Europe, and in another 15-20 years it was well established in Mexico and Florida. Presently sweet orange is grown all over tropic and sub tropics except the places which are very wet or located at high altitudes.

Is most important of all citrus species, generally eaten fresh, in recent times more than 90% of produce is employed for juice making, squashes, for marmalade and flavouring purposes, pectin from rind / peel, essential oils ie neroli oil (flower), orange oil (peel), petit grain oil (leaves), is extracted pulp and molasses serve as cattle feed.

Sweet orange plants are evergreen trees (6-12 m tall), crown usually rounded, young twigs angular, often with stout spines (on young seedlings). Leaves are dark green shiny, ovate-elliptic (5-15 x 2-8 cm), rounded at base, apex short pointed, margins sometimes slightly serrated / crenate, petiole small medium (1-2.5 cm), narrowly winged articulated. Flowers axillary, borne singly or in few flowered racemes, 2-3 cm in diameter, pentamerous, bisexual and fragrant, calyx 5 lobed, corolla usually white, stamens 20-25, style single slender with globose stigma, ovary with 10-14 loculi. Fruit sub globose 4-12 cm in diameter, peel / rind about half centimetre thick, adhering tightly to juicy sub acidic vesicles, yellow to orange red in colour (often remains green in tropics), central axis solid. Seed nil to many, obovoid, white inside, polyembryonic. Seedling of sweet orange grow upright, and are very spiny.

***C. sinensis* Sweet orange**

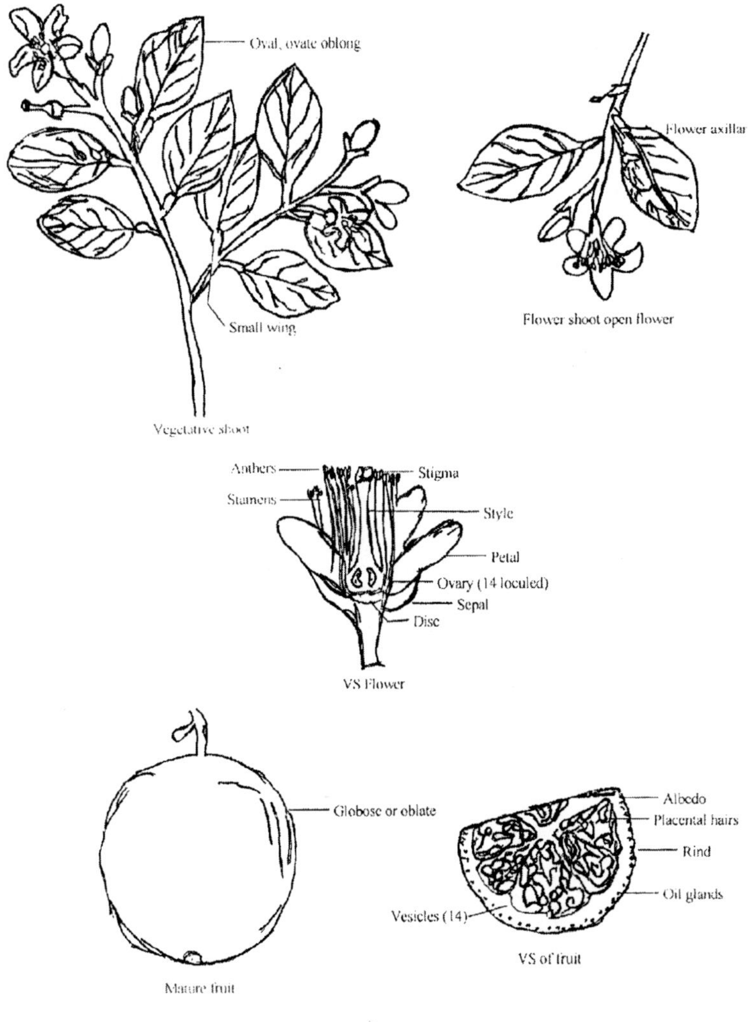

Citrus reticulata Blanco. Mandarin 2n = 18

Mandarins are very often referred to as 'fancy fruit', and this citrus species is one of the three most highly differentiated species of the genus. The mandarins are confined more to south east Asia, including the Malaysia Archipelago. According to some other writers its origin lies in Indo China region Various mandarin groups, Satsuma mandarins originated in Japan, Mediterranean mandarins in Italy and King mandarins originated in Indonesia, common mandarins in Philippines, after the specie was introduced into Europe and in United States in the middle of 19th century. Mandarins now are widely grown all over tropical and sub tropical parts of the world on account of its very distinct flavour, and loose peel / rind, taste ranging from very sweet to sour.

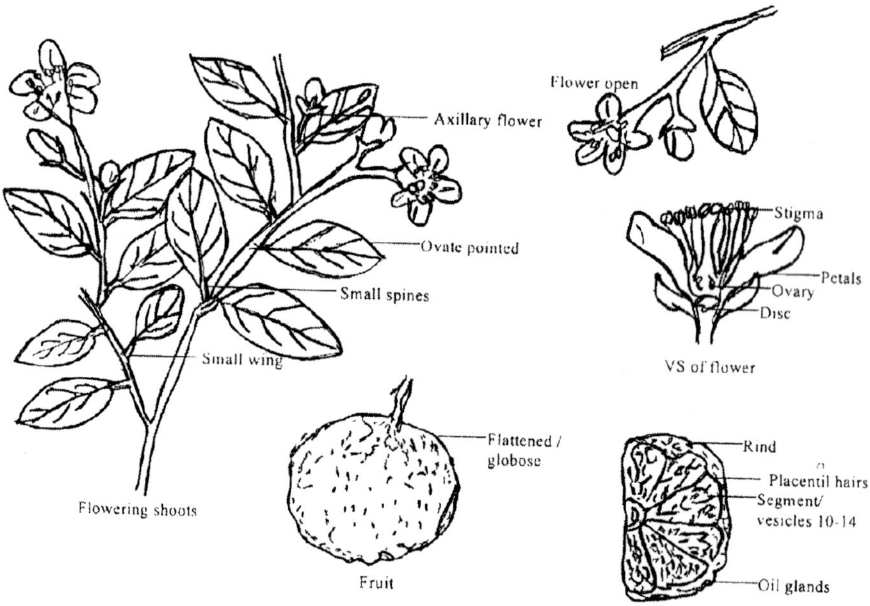

Usually they are eaten fresh, juice extracted from fruits, segments are canned, essential oils and pectin derived from rind. Mandarin plants are relatively small (2-8 m tall), spiny, twigs are slender. Leaves small dark green, shining, green above and pale below, narrow at both apices, narrowly or broadly lanceolate or elliptic with acute base and tip, margin crenate, petiole usually narrowly winged. Flower occur singly or in small clusters in the leaf axils, flowers small (1.5-2.5 cm diameter), white fragrant, pentamerous, calyx 5, corolla 5, white, stamens 20, style single, ovary with 10-15 loculi. Fruit is a depressed globose or sub globose berry, 5-8 cm in diameter, yellow or orange-red when ripe, rind thin, loose separating easily from segments, which are sweet juicy, orange in colour. Seed small, pointed at one end, embryo green, polyembryonic. In some countries mandarins and tangerine are used indiscriminately, however mandarins are used for yellow fruited cultivars and tangerine for deep orange types.

Citrus grandis (L.) *Osbeck*. Pummelo, Shaddock *2 n= 18*

Pummelo is considered to be native of Thailand and Malaysia and it is here that best cultivar, environmental riches and orcharding skills, are present. From here it spread to Indo- China, southern China, southern most parts of Japan and westward to India, and later to Mediterranean and tropical America. Captain Shaddock, commander of an East India ship introduced Pummelo into Barbados in the 17th century. Is not important in temperate regions.

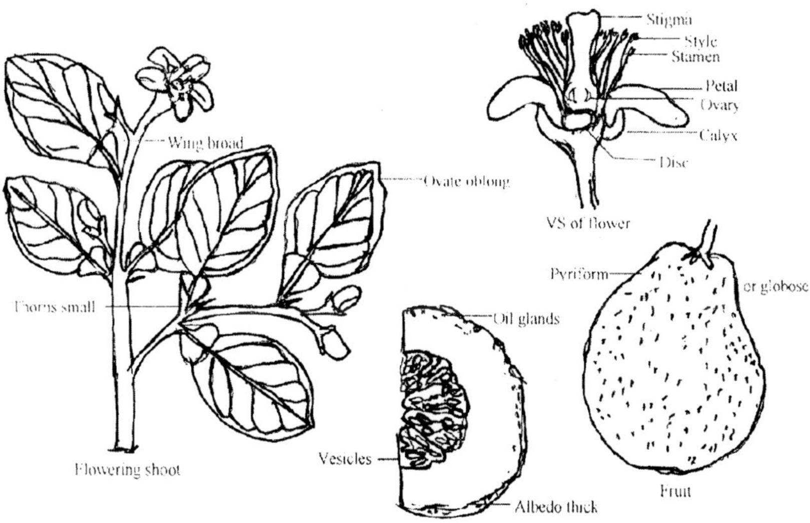

In East the fruit is highly esteemed as a dessert fruit, fresh juicy pulp vesicles are eaten fresh. The thick white inner part of rind (albedo) is candied after removal of exocarp containing oil glands, all plant part have medicinal values for treatments of fevers, coughs, gastric disorders.

Trees are medium tall (5-15 m), spreading and spiny low branching, young twigs pubescent (many persists for year or more) spine prominent large and long (5 cm). Leaves simple alternate, dark green shiny, large (5-20 x 2-12 cm), ovate to broad, elliptic, base rounded to subcordate margin entire to slightly / shallowly crenate, apex obtusely acute, sometimes may be slightly notched, under surface pubescent along midrib and veins, petiole broadly winged (7 cm broadest amongst all citrus spp.). Flowers are borne axillary, solitary or in clusters of few flowers. Flower white fragrant large measuring 3-7 cm in diameter, pentamerous, sepal 5 lobed, petal 5 cream coloured, stamens 20-25, style one, ovary 11-16 loculed. Fruit very large (largest of all citrus spp.) globose, subglobose or pyriform in shape, 10-30 cm in diameter, yellowish when ripe, rind very thick densely dotted with glands, pulp vesicles also very large, pale yellow or pink well filled with sweet juice (in inferior types vesicles tend to be dry). Seeds are large yellowish and ridged, monoembryonic.

Citrus paradisi Macf. Grapefruit 2n = 18

The origin of grapefruit is not very well established. Since the crop is closely related to pummelo, it is said to have originated in Barbados (West Indies), as old records refer to 'forbidden fruit' or small shaddock in Barbados in 1750 and in Jamaica in 1789. It has become famous as a breakfast fruit. Commercially crop is grown in Florida, California, Texas and Arizona, besides exports are made from Israel, South Africa, West Indies and Brazil. Grapefruit is regarded to be either an interspecific hybrid of pummelo and sweet orange or a hybrid or a mutant of pummelo.

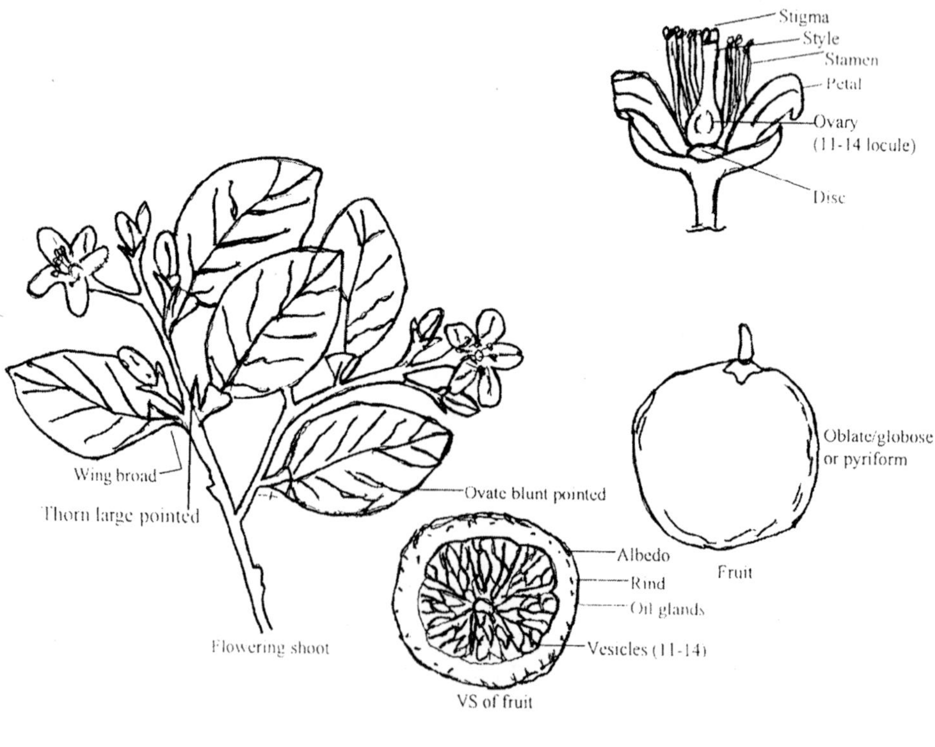

Grapefruit is distinctly recognised as breakfast fruit because of typical flavour and mild bitterness of the juice, also canned, pulp and molasses used as cattle feed.

Grapefruit plants are large tree (10-15 m tall), spreading, evergreen tree round topped, foliage dense, young twigs angular and sparsly pubescent. Leaves, green evergreen, smaller than pummelo and larger than sweet orange, pale green when young, petioles broadly winged (less than pummelo), oblanceolate obovate (7-15 x 4-8 cm) often crenulate. Flowering occurs axillary, solitary or in small clusters, about 4-5 cm across, pentamerous, calyx 5 lobed, corolla 5 white fragrant,stamens 20-25, style single, ovary 12-14 loculed. Fruits, large, globose borne in clusters, 8-15 cm in diameter, greenish-yellow coloured, rind thinner, vesicles small (compared to pummelo), adhering to rind. Seed white, smooth, cotyledons white, polyembryonic.

Poncirus trifoliate (L.) Raf. Trifoliate Orange

Plants are small trees with upright growth, densely branched. Young growth smooth angular, dark green thorny, old shoots circular. Thorns very stiff, stout, sharp, 1½-2 inches long flattened at base. Leaves trifoliate, deciduous, obovate, central leaflet notched, margin crenate. Flower buds produced singly or in pairs in leaf axils covered by scales and are formed in summer, in spring appear before leaves or on naked shoot. Flowers, short pedicelled or nearly sessile. Calyx 5, small oval (1-1½ cm long), corolla 5, white large (3-4cm across) obovate clawed at base, Stamens 20-25, filaments usually distinctly 2 celled. Pistil club shaped, oblong, pubescent, style short ovary 6-7 parted. Fruit surface rough, surface bumpy covered with pubescence, fruit stalk/stem short and stout, fruit hangs well if not picked. Fine gummy pulp acidic, seeds, many plump ovate brown in colour.

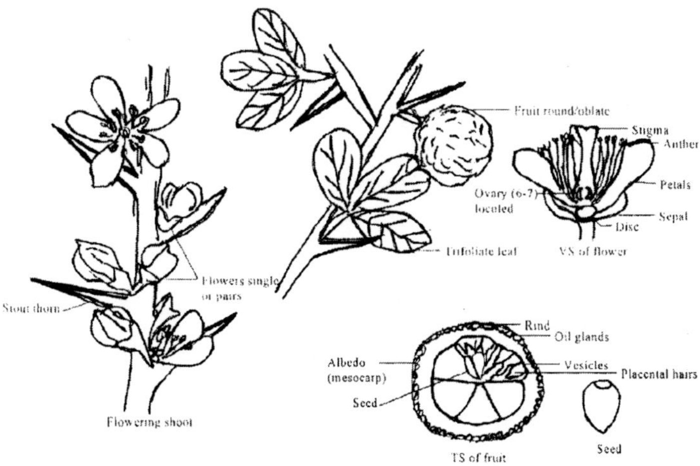

Fortunella margarita Fortunella/Kumquat

Kumquats are part of genus Fortunella, and in Chinese language the word kumquat means 'gold orange'. They are indigenous to south-eastern China and have been given differed names. As time passed it was separated from genus Citrus and were assigned as ornamental, and usually during Christmas time the fruits along with twigs and leaves are extensively used for decorative purposes. Fruits are usually eaten fresh without removing the rind as the spicy aromatic flavour of rind and acid pulp makes a very delightful combination. Fruits are extensively used for making marmalades, jellies preserves candied fruits, as it contains very high amounts of pectin.

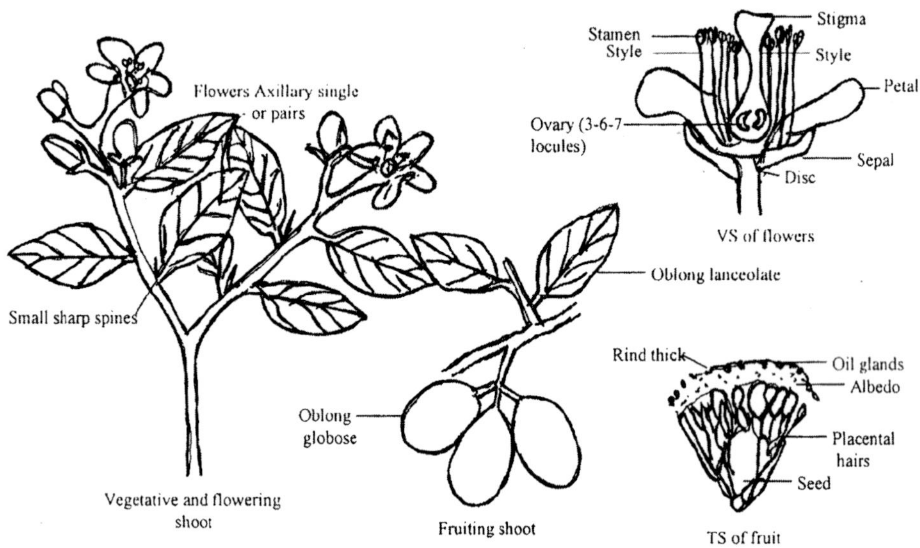

This distinct group of citrus has been named after Robert Fortune the well known English horticulture worker.

Plants are shrubs or small trees, branches are dense. Early growth of both branches and branchlets angular, becomes circular later. Spines axillary smaller than Poncirus present along with the bud. Leaves usually unifoliate petiole narrowly winged short pointed evergreen dark green above and lighter below, margins, entire to slightly crenate, oil glands thickly dotted on under surface. Flowers appear in pairs or in clusters, axillary on new or on last years growth, white coloured, smaller than other citrus types. Flower stalk with 1-2 small bracts, calyx 4-5 yellowish green, corolla 4-5, white oblong lanceolate, stamens 16-20 filaments in a ring form at base, free at top, length irregular, anthers attached at filament tip. Ovary generally 3-6 celled, sometimes more, 2 ovules per cell, stigma usually small. Fruit smaller compared to citrus, globose, or oblong, rind relatively thick, sweet and edible, juice vesicles small. Seed small, cotyledons generally greenish.

Other species of Fortunella are

F. crassifolia Swingle. Meiwa kumquat. Plants are highly branched shrub, spines none or short only. Fruits are more rounded with 6-7 cells. Leaves are thick and trough like pale/green below, tapering at tip petioles narrowly winged.

F. margarita Swingle. Oval or Nagami Kumquat. Plants are shrub or small trees, usually thornless, twigs angled when young slender and thin. Leaves lanceolate tip round base cuneate margins crenate, dark green above, pale below, with glands. Flowers appear in leaf axils solitary or in few flowered clusters with small pedicels, ovary 4-5 celled with 2 ovules in each cell. Fruit oval or oblong with glands embedded in the thick fleshy rind.

F. japonica Swingle. Round or Marumi Kumquat. Plants are branched shrubs with non to very small spines, has broad and blunt small pointed leaves, light green in colour, Fruit round small orange coloured no sign of rudimentary style, segments 5-6, seeds small. Very handsome of all citrus fruits because of dwarf plants, twigs much branched small and brightly coloured orange fruits.

F. hindsii Swingle (Hongkong wild Kumquat) Is indigenous to the hilly areas of Hongkong, China and the near by places. Plant is a spiny shrub or a small tree. Leaves oval-elliptic, tapers at both the apices, dark green above, pale beneath, petiole winged. Flower small and broad. Fruit orange red in colour, small, sub globose with few juice vesicles, pulp scanty, seeds plumb, oval to ovate in shape, cotyledons green.

Family Rutaceae apart form exhibiting tremendous variability both within 'various species and between different generas - Citrus, Poncirus and Fortunella also has some close relatives namely Bael and Wood apple. The brief description of wood apple is given below and bael is dealt in a seperate chapter.

Feronia limonia (Linn) Swingle. Wood apple. Is native of India, Indo-China and Sri Lanka. Plants are deciduous tree, very spiny, leaves with 3-7 pairs of leaflets, odd-pinnate, usually obovate in shape, apex blunt, margin entire, petioles generally small, spines axillary, long and straight. Flowers appear in terminal or axillary panicles, pentamerous, stamens 10-12, filaments each having many cells, stigma cylindrical. Fruits oblate-globose, rind hard and woody in nature, pulp pinkish in colour, acidic in taste, seeds numerous embedded in pulp. Very useful for making.

The varietal description of various cirtus fruit species which have been in cultivation over a period of time are briefly summerized below.

The Oranges : The major members of the orange group are the sweet and bitter orange though quite similar but also have specific differences. The sweet oranges are of four types. Common Orange (Blond orange of Mediterranean) has many varieties.

Blood orange (or pigmented orange. Is of two type a) light blood orange. b) deep blood orange.

Acidless orange. Acid is low, flavour insipid

Navel orange. Very important and Washington Navel variety represents this group.

Bitter or sour orange is represented by

a) Sour orange used mainly as a rootstock or to prepare marmalade
b) Sour orange that are grown as ornamental from which neroli oil is obtained
c) Bitter sweet orange, resemble common bitter orange but have low acid.

A. The Sweet Orange

The sweet orange is the most dominant in citriculture scenario in contrast to mandarins which are liked more in the Orient (China, Japan). No other citrus fruit is liked and used all over the world as this group is. The sweet oranges fall into four groups and each group has number of very popular cultivars which are cultivated commercially all over the world. Salient features of some well known cultivars are detailed below.

Jaffa (Florida Jaffa): Is a clone of Palestine beledi seedling group. Tree upright, medium large quick growing foliage dense moderate in production, alternate bearer. Fruit obovate-globose or ellipsoid, ripening mid season, basal end raised with radiating furrow, orange coloured medium sized, rind with moderate thickness pitted and pebbled. Flesh pale orange, tender juicy with pleasant flavour, seeds few. Has good shipping quality but does not hang well on tree.

Joppa : Resembles Jaffa for number of characters and is different in that it starts bearing early and prolifically, branches are stiff and thornless, branchlets are stout. Fruit medium in size globose to oblong, orange coloured, rind medium thin faintly pebbled. Flesh light orange, soft juicy with rich aroma. Mid season cultivar, seeds few.

Foster : Tree medium in growth semi spreading foliage dense. Fruit medium large roundish oblong in form, orange in colour, apex depressed base slightly collared. Rind thin pitted oil glands large, flesh orange in color texture coarse, vesicles or sacs large, very juicy, a fine blend of acidity and sweetness, flavour rich quality excellent pith rather open, seeds large oval, maturity mid to late season.

Hamlin (Norris): Originated as chance seedling and named after the owner of the orchard AG Hamlin. Tolerates cold better than most oranges, productive. Tree medium large with moderate vigour. Fruit small-medium globose-oblate, collar radially furrowed, areolar ring very faint orange in colour at maturity. Rind thin surface smooth finely pitted, flesh juicy tender orange coloured, pleasant sweet flavour, low in acid. Matures very early in the season, seed few to none.

Pineappple: Chance seedling very sensitive to frost, very productive excellent for processing. Tree medium large with moderate vigour. Fruit spherical-obovate medium sized, base somewhat depressed radially furrowed areolar ring faint, bright orange. Rind finely pitted and pebbled medium thick, flesh light orange, soft juicy sweet highly flavoured moderately seedy. A mid season ripening cultivar. Has two limb sport which are seedless. Seedless Pine apple and Blaquemines.

Parson (Parson Brown): Originated as a chance seedling. Known for its early maturity is relatively more seedy. Trees are large vigorous and productive. Fruit globose medium large orange in colour, radial furrows at base areole very faint. Rind pitted and pebbled moderately, flesh orange, firm juicy highly flavoured. Moderately seedy.

Valencia (Valencia Late, Hart's Tardiff, Hart Late): Has wide adaptation with alternate bearing, tendency to be heavy cropper. Tree large upright vigorous. Fruit orange in color, but turns green under specific conditions. Oblong-spherical medium large, areole ring faint Rind quite thick leathery tough smooth and pebbled, flesh orange, very juicy sweet but with some acidic taste, flavour good, processing quality excellent. Matures late. Storage and transportation qualities are very good hangs well on trees. Seeds few to none.

B. Pigmented or Blood Oranges

This group of oranges differs from the common sweet orange in that the fruit generally has pink or red coloration on the rind, in the flesh and juice and also has distinct flavour. Cultivars of this group have originated mostly in Malta and Sicily in the Mediterranean basin. Promological description of some cultivars which are grown commercially are presented below.

Ruby (Ruby Blood) : This cultivar was introduced from Mediterranean region. Plants are compact not very large moderate in vigour and production. Ripens in mid season. Fruit globose-oblong medium sized areolar spot faint, furrow circular intense red/reddish flush, rind not very thick, pitted and pebbled, flesh orange coloured and red under favourable conditions soft juicy with rich flavour, seeds few.

Spanish Sanguinelli (Syn. Sanguinelli, Sanguinella Negra) : Originated as a limb sport of Doblefina. Tree foliage light green, spineless, productive small medium in size. Fruit large very attractive with persistent style somewhat asymmetrical. Rind and flesh intense red coloured and no other cultivar matches it in colouration, surface smooth, texture fine, flesh firm juicy sweet, flavour excellent. Late to mid season in maturity, shipping and storage qualities very good. Seeded.

Torocco (Tarocco dal Muso, Taracco di Francofonte) : An Italian cultivar. Tree medium in size, irregular in bearing but produces moderately. Fruits are quite large variable in shape globose-broad-obovate prominently furrowed, collar yellowish orange blushed with red. Flesh somewhat thick adheres tightly, pebbles moderate. Flesh orange to red pigmented firm, juicy rich in flavour, seeds few or none. Shipping and storage quality good has mid season maturity, does not retain quality if left for long on trees. Seeds few to none.

Doblefina (Oval Sangre, Blood Oval) : Tree small poor in growth branches spreading crown open, foliage light green. Precocious and heavy bearer. Fruit small - medium, oval to oblong, yellowish orange, blushed with attractive rose colour. Rind medium thick leathery, smooth, firm adherence tight, peels with difficulty. Flesh firm, pink tinged, juicy, taste pleasant flavour not very strong. Shipping, storage quality good but does not hang well on tree, almost seedless.

Blood Red (Blood Red Malta) : Origin unknown probably came from Mediterranean basin. Grown widely and commercially in India and Pakistan. Cultivar is seedy light blood orang coloration of both rind and flesh, flavour very characteristic and rich. Good colouration is generally attained in the submountain region.

Maltaise Sanguine (Portugaise) : Origin of this variety is uncertain but the Egyptian variety Baladi Blood introduced from Malta is said to be Maltaise Sanguine. Likewise Maltase Blood and Bloodred Malta seems to be a clone of this cultivar. Trees are very productive medium large moderate in vigour. Fruit oblong globose medium in size, blood colour well developed both on rind and flesh. Rind soft easy to peel pebbled moderately. Flesh soft tender melting, taste and flavour excellent seeds few, storage and shipping quality poor.

C. Navel Oranges

Very distinct features of Navel oranges is that the navel is produced regularly by the fruits, are seedless for the reason that both viable pollen and ovules are either lacking or are rare. These oranges were established in Mediterranean region a number of centuries back. The navel orange cultivars of great significance are briefly described as below.

Washington (Riverside, Bahia, Baia or Baiana) : Is considered to be a limb sport of a variety 'Selecta' in Bahia, Brazil. Tree medium in size and vigour, crown round topped, anthers are without pollen. Fruit large spherical obovoid or ellipsoid early in maturity, apex projected with broad nipple base collared, navel medium large surface smooth pebbled deep orange. Rind medium thick tender flesh deep orange, firm less juicy, rich in flavour and taste. Processing quality poor, ships and stores well. Seedless.

Navelencia : In growth characters is less vigorous than Washington. Fruit lacks colour of Washington, paler rind more smooth, navel protudance prominent mostly closed. Flesh light in colour (as rind) firm more juicy, flavour good more than Thomson. Fruit matures earlier than Washington and after Thomson, and hangs well on tree, almost seedless.

Thompson : Limb sport of Washington. Tree compact semi dwarf type less in vigor than Washington. Fruit medium large, shape like that of parent but with long and deep basal furrows, apex with large open navel. Rind and flesh less coloured, rind smooth glossy pitted finely, flesh firm more juicy, taste and flavour good. Ripens early, seedless.

D. Sugar or Acidless Oranges

Is a small group that comprises of very low acid varieties, resembles sweet lemons, limes or limettas which are low in acidity, sweet in flavour, taste insipid and presence of cream coloured chalazal spot on the seed which is usually darker in acid type more so in sweet oranges. The Indian variety Mosambi is said to belong to this group but the dark brown color of chalazal spot does not support this view. Real acidless orange cultivars are very few and typical examples are Shamouti Meski obtained from Shamouti and Maltaise Meski from Maltaise Blonde (both common orange).

Maltaise Meski : Originated in Tunisia, a non acid orange cultivar. For all plant, foliage, flower and fruit characters it is like the parent cultivar Maltaise Blonde.

Shamouti Maski or Shamouti Moghrabi : Is a Labenese cultivar which is indistinguishable from Shamouti (Palestine Jaffa) baring that it is acidless (insipid in taste) and more seedy in nature.

In India, Mosambi, Sathgudi and Shamouti are invariably placed under sweet oranges as the acid content is very low. Brief description of these cultivars (common oranges) is given below.

Mosambi (Mosambique) : A very popular variety and grown commercially in India, early in maturity. Tree characters as those of Blonde oranges. Fruit medium large oblate-globose to obovoid, areolar ring shallow, light yellow to pale orange. Rind quite thick, stripes faint with longitudinal ridges and grooves. Flesh light yellow juicy, acid low, taste insipid, seeds few.

Sathgudi : Origin unknown a popular cultivar grown extensively in India. Tree vigorous produces moderately. Fruit medium-large globose, rind orange coloured medium thick tough leathery pitted and pebbled, peels easily central axis solid and thick. Flesh orange, juicy flavour fair sweet in taste, acid very low, mid season cultivar, moderately seedy.

Shamouti (Palestine Jaffa, Jaffa, Chamouti) : Tree upright moderate in growth and vigor, branches thick and thornless, petioles with narrow wing. Fruit medium large oval ellipsoid, base flat collar narrow apex round, areolar ring slight orange coloured. Rind thick leathery smooth pitted, oil glands faint, flesh light orange, firm juicy fragrant and sweet in taste, peels easily. Shipping and storage very good. Mid season cultivar. Seedless or nearly so.

Marrs (Marrs Early) : Precocious and heavy bearer matures very early. Tree moderate in vigour bears in clusters, fruit round to oblate medium large orange in colour well coloured under favourable conditions, smooth finely pitted. Flesh orange well coloured juicy, sweet in taste, acid very low. Fruit hangs well on tree and maintain the quality. For good quality and high juice picking should be delayed. Seeds moderate in number.

Mandarins

A very large, varied/distinct citrus group important in the Orient that has some prime and highly acclaimed citrus fruits. Mandarins resemble closely to oranges and are usually referred to as loose skinned oranges a confusing usage. Mandarins exhibit wide variation in respect of fruit size, colour, flavour, maturity time, adherence of rind. Most important cultivars have small-medium fruits (Dancy, Satsumas, Nagpur, Clementine etc). while large fruited mandarins are King and Ortanique. Very small fruited mandarins include Cleopatra. All are of horticultural interest as fruit, ornamental or rootstocks.

Mandarins have been divided into number of natural groups and each has several cultivars of importance. Important groups along with their representative cultivar are described below.

1 Satsuma mandarin (C. *unshiu,* Marcovitch.)

2 King mandarin (C. *nobilis,* Loureiro.)

3 Manditerranean mandrin (C. *deliciosa,* Tenore.)

4 Common mandarins (C. *reticulata* Blanco.).

***Satsuma Mandarin* (C. *unshiu* Marcovitch)** : Is very popular and cultivated extensively in Japan, called as Unshu mikan. Most cold tolerant of citrus fruits and ripens early than any other oranges as its heat requirement is very low. Brief description is given below.

Tree cold hardy slow growing moderate in growth spreading branches drooping almost thornless. Leaves long lanceolate, dark green base and apex tapering petiole long thin with wing. Fruit small medium oblate subglobse orange coloured areole small faint navel pressed. Rind smooth leathery thin, oil glands prominent, easy to separate. Flesh orange coloured soft melting subacid in taste, flavour rich. Segments 10-12, axis hollow, carpillary membranes loose. Very early to mid season in ripening. Fruits should be picked quickly when mature otherwise quality deteriorates, stores well.

Further five group of unshu mikans are well recognized which are as follow.

1. Wase (Early). Includes all early ripening clones.
2. Zairai (Native or indigenous). Includes old clones whose origin and parentage is not known.
3. Owari. Group includes cultivars that became very popular in Owari Province.
4. Ikeda. Fruits are small subglobose seedless moderate in quality matures late.
5. Ikiriki. Tree not productive, fruit seedless, quality excellent.

Important Satsuma mandarin cultivars grown are described briefly.

Owari (Satsuma) : An old Japanese cultivar origin unknown. Tree very productive slow growing branches spreading and drooping. Fruit oblate-subglobose neck small, rind thin smooth leathery, at maturity roughness and looseness both increases, flesh orange tender melting subacid in taste rich in flavour, segments 10-12, axis hollow separable. Seedless. Early to mid season in maturity.

Silverhill (Silverhill Owari) : Cold resistant and productive. Tree upright very vigorous in growth. Fruit more flat than any Satsuma mandarin, medium in size. Rind quite thin, smooth, orange, leathery, flesh orange, soft juicy very sweet acid low, excellent in quality. Storage quality good matures early in season.

***King Mandarin* (*C. nobilis*, Leureiro)** : The name is derived from an American variety. 'King' Characters are intermediate between orange and mandarin. Cultivar 'King' is representative of this group, is therefore is a natural tangor. Salient horticultural feature are given below.

King (King of Siam) : Tree cold hardy than many mandarins, very productive upright crown open moderate in vigour, branches few thick thornless to moderately thorny, foliage dark green, petiole narrowly winged. Fruit oblate or depressed globose (with short neck) base furrowed, apex depressed, areole faint. Rind thick (compared to mandarins) adherence medium, peels easily, deep yellowish orange to orange, smooth to rough. Flesh deep orange, segments 12-14, axis hollow and large, soft juicy with rich flavour. Seeds few to many. Maturity late to very late in season. Hangs well on tree.

Mediterranean Mandarin (*C. deliciosa*, Tenore)

Derived from species *C. deliciosa*, famous mandarin of Mediterranean basin. Particular features of this group is that it differs from other mandrin in that seeds are plumb and spherical, leaves are small narrow lanceolate and aroma of oil glands and juice is very aromatically flavoured. Description is detailed below.

Mediterranean : (Willowleaf, Mediterranean Common). Trees are cold hardy but bears irregularly, small slow growing, crown broad, branches drooping thin thornless, leaves narrow lanceolate. Fruit oblate medium sized with few lobes, neck distinctly furrowed, areole absent, small navel present at apex. Rind thin loosely attached yellowish orange, shiny smooth. Flesh light orange soft juicy aromatic, flavour sweet, segments 10-12, loose axis hollow. Matures early to midseason. Seeds many small plumb, cotyledons light green. With ripening rind separate rapidly lacks storage ability.

Common Mandarins (*C. reticulata Blanco)*

This group is varied type and shows wide varation in plant and fruit characters. Cultivar number is also much more than seen in other mandarin groups. Important cultivars grown commercially are described briefly.

Clementine (Algerian) : Originated in N Africa and is an accidental hybrid, early in maturity. Tree cold resistant, bears regularly and consistently, medium in growth, crown round, branchlets thin almost thornless, foliage dense, leaves variable in size narrow lanceolate in shape. Fruit variable in shape oblate-globose oblong or pyriform base necked or collared, apex depressed or with small navel. Rind firm and adhering, peels easily, deep orange to reddish orange, smooth glossy, oil glands prominent hence pebbled. Flesh deep orange with 8-12 segments soft melting, juicy, sweet subacid in taste, flavour aromatic. Seeds few to some, cotyledons green monoembryonic.

Dancy : Tree characters nearly resembles Clementine but has tendency towards alternate bearing, veination not so prominent. Fruit oblate-broad obovoid or pyriform, neck sometimes distinct, apex broadly depressed. Rind thin leathery loose, peels off readily, smooth shiny surface bumpy because of oil glands. Flesh deep orange, melting tender juicy, flavour pleasant slightly more acidic, segment about 12 separates easily, axis large and hollow. Stores quite well. Seed characters are like that of Clementine, seeds few to some, polyembryonic, cotyledons green.

Kinnow : Cold resistant, has strong alternate bearing tendency. Tree tall upright branchlets long, slender almost thornless, foliage dense, leaves broad lanceolate. Fruit oblate with both base and apex depressed, medium sized. Rind thin adhering (for mandarin) peelable, yellowish orange, smooth shiny leathery tough, very smooth pitted, flesh deep yellowish orange segments 9-10 axis solid - semi hollow, separable very juicy highly flavoured aromatic, seeds numerous pale greenish yellow, polyembryonic. Hangs well on tree.

Nagpur : (Ponkan, Warnucro). Citrus fruit of most commercial significance in India which is the famour Pokan of south China. Cold resistance is moderate, alternate bearer, upright in growth quite vigorous early-mid season in ripening. Fruit globose or oblate, large (for mandarin) base with small neck has distinct furrows, apex depressed sometimes with navel. Rind adheres loosely, orange coloured quite thick, smooth pebbled due to sunken oil glands. Flesh orange, tender, juicy melting flavour pleasant aromatic. Segments 10, separates easily, axis large hollow. Seeds few small plumb, cotyledon green.

Wilking: Is a sister of Kinnow both resulting from King x Willow leaf. Maturity mid season. Tree characters almost like that of kinnow, bears many small fruits in the 'on' year and no fruit in the 'off' year. Fruits with base and apex flattened, oblate, small-medium in size Rind orange coloured medium thin peelable shiny pebbled, flesh deep orange, very juicy rich in flavour, aromatic segments 9-12 firm separate easily, axis semi - hollow. Seeds many, cotyledons greenish yellow to yellow. Puffing seen in fruits.

Page : An early ripening cultivar obtained from a cross between Minneola tangelo x Clementine mandarin. Tree has moderate vigor, branches are upright, almost thornless productive. Fruit matures early, oblate-subglobose, apex and base rounded. Rind somewhat thin leathery, separate easily surface smooth peebled reddish - orange in colour. Flesh deep orange, juicy tender sweet rich in flavour, segment 10, axis solid somewhat open. Seeds many, cotyledons nearly white.

Ortanique : Tree moderate in growth, branches spreading and drooping thornless slender, foliage dense petioles winged. Mid season to late in maturity. Fruit variable broad obovoid oblate to subglobose, apex depressed or with small navel, base round or tapering short neck or collared areole distinct Rind thin adherence tight but peelable, yellowish orange glossy finely pitted. Flesh orange juicy flavour rich and characteristic, segments 10-12 solid to partly open, seeds moderate 10, plumb, polyembryonic.

Pummelo

In all likelihood is indigenous to the Malayan and East Indian archipelago. In respect of fruit characters the pummelos fall into two major groups the pigmented and the non pigmented types. The pigmentation in the former types is caused by carotenoid lycopene and varies from pink to deep red some of which are very attractive with very high in flavour. The non pigmented (common pummelo) are very variable in number of characters, have moderate to high acid contents and mostly seedy. Description of some known pummelo cultivars is described briefly.

Non Pigmented Varieties

Banpeiyu : The variety is considered to have originated in Malayan region and is very popular in Japan and Taiwan. Tree large very rapid growing, spreading, leaves large wing broad, all new growth pubescent. Fruit one of the largest spherical to subglobose, light yellow in colour. Rind thick tightly adhering to vesicles, surface smooth. Flesh pale yellow, soft and juicy, segments many (15-18) central axis solid and large, carpellary membranes thin and tough. Mid season ripening cultivar, flavour excellent and pleasant taste of acid and sugar, storage quality very good, seedy.

Hirado (Hirado Buntan) : Originated as a chance seedling at Nagasaki Prefecture, Japan and rated very highly, more resistant to cold. Early - medium in ripening, stores well, Trees are medium - large vigours in growth, leaves large thick, wing broad. Fruit large bright yellow, glossy, oblate both ends slightly depressed, surface smooth, rind medium thick, flesh light green yellow soft semi dry, flavour pleasant with balanced blend of acid and sugar with some bitterness, segments many, carpellary membranes thin and tough. Seedy.

Kao Phuang : One of the most famous cultivar grown in Thailand. Mid to late in ripening, fruit hangs well on tree for long. Tree upright very vigorous. Fruit large lemon – yellow, surface shiny smooth, broad pyriform, neck distinct at stalk end, depressed at apex. Rind semi thick adherence medium with the vesicles which are firm large and separates easily, juicy, flavour good, carpellary membranes semi thick, axis small and compact. Seedy.

Pigmented Varieties

In its place of origin-Orient a number of pink and red fleshed types are present some of which are pomologically described below.

Chandler : Is a hybrid cultivar obtained from a cross between Siamese Sweet x Siamese Pink. Flesh in pink and in other characters it is intermediate between the parents. Fruits mature early in the season with very good storage potential Fruits are medium sized, round flat to globose in shape light yellow with slight pubescence. Rind smooth medium thick, vesicles moderate in juice flesh firm later turns soft, flavour pleasant subacidic in taste, seeds many.

Ogami (Syn. Egami) : This cultivar has performed very well in Florida and other parts of the world. Fruit quite large broadly round flat (oblate), rind smooth shiny and moderate in thickness pink to deep pink in colour, flesh deep pink, pigmentation extends far into the albedo (mesocarp), seedy.

Siamese Pink (Siam) : Considered to be one of the finest cultivar in respect of its strong flavour, matures late. Tree very large fast growing, branches spreading, shoots, twigs, petioles, slightly pubescent. Fruit large broadly obovate-short pyriform depressed at apex light yellow but also with pink colouration. Rind adheres tightly medium thick, smooth and glossy. Segments many, carpellary membranes quite thick and tough but open at axis when mature. Flesh pinkish very juicy coarsely grained, flavour subacid with some bitterness, resembles that of grape fruit. Almost seedless.

Siamese Sweet : Commonly called as sweet pummelo practically non acid. Plants are dwarf type, branches drooping, leaves distinctly round pointed. All new growth markedly and densely pubescent. Fruits are broad oblate to obovoid. Vesicles separate easily, are large crisp, lack in juice, sweet in taste.

Grapefruit

The fruit in also called as forbidden fruit or small shaddock, obtained from pummelo, and the name has been derived from the fact that it bears in clusters, and flower resembles that of grape. Like pummelo, grapefruits are also of two types - the common and pigmented grapefruits, and the varieties obviously differs in number of characters i.e. maturity time, seedless or seedy, flavour etc. Very important cultivar grown are described briefly.

Common Grape Fruit Varieties

Duncan: Plants are large, very vigorous and productive early to medium in ripening and also among the few which are cold hardy. Fruit large broad obovate, oblate - globose, pale to light yellow in colour, areolar ring slight and the furrows at base are small and spreading. Rind surface even/smooth, glossy, medium in thickness. Flesh light yellow/buff, soft very juicy, flavour very characteristics strong and pleasant, seedy.

Marsh : (Syn. White Marsh, Marsh seedless). Cultivars most significant value is its seedlessness character and its late ripening. Tree large, fast growing, branches are spreading, very productive, adapted to very hot climates. Fruit medium sized, spherical or oblate, light yellow, areole ring almost absent. Rind surface even, smooth semi thick, flesh light yellow, soft very juicy, with pleasant flavour, but less than that in seedy cultivars. Has very good storage quality and fruit hangs for long on the tree. Seeds few to none.

Pigmented Grapefruit Cultivars

Foster (Foster Pink) : Cultivar is said to be the first pigmented grape fruit variety that ripens in mid season, originated as a limb sport of cv. Walters. Plants are large, very productive and vigorous. Fruit medium large in size, round flat to spherical, light yellow overlaid by pink blush, which is also seen in the albedo (mesocarp), areole inconspicuous, furrows at base are small and diverging. Rind smooth, medium thick, flesh light yellow, pink under favourable conditions, soft, flavour good. Very seedy.

Redblush (Syn. Ruby, Red Marsh, Red Seedless) : Ruby originated as a limb sport of Thompson. Tree characters are that of Thompson or Marsh. Fruit resembles Thompson for most of the characters except intense pigmentation of the flesh, albedo/ mesocarp also pink with bright red blush on the fruit surface. Hangs well on the tree as the parent cultivar.

Thompson (Pink Marsh) : Originated as a limb sport on a tree of cv. Marsh, being seedless it became very popular in a short period of time. Plants are very productive large with vigorous growth. Fruit medium sized, spherical -

oblate, areole indistinct light yellow, rind relatively thick, smooth and tough. Flesh colour dark yellow, under desirable condition develops pink colour (not juice), albedo not coloured, flesh soft juicy, flavour good, almost seedless or with few seeds. Has excellent storage and shipping quality, fruit hangs well for long period of time. Seeds few to none.

Citron this citrus fruit is most likely to have originated in the north eastern India and the areas near by. As the fruit is inedible, and tree lacks ornamental value, one ponders why the Romans and Greeks liked it so much. This was the first citrus fruit which was known to the Romans, further the fruit fragrance is soft, penetrating and lasting. Salient features of some citron cultivars grown extensively are described.

Diamante (Cedro Liscio) : A very important and commercial cultivar of Italy, origin unknown. Trees are small in size, spreading, crown, open, thorny with some large stout spines. New growth, buds and flowers are distinctly purple pigmented. Fruit usually large, long oval to ellipsoid, basal cavity furrowed, apex pointed - nipple like, lemon yellow in color. Rind very thick, fleshy surface lobed or ribbed, smooth. Flesh crisp, dry, acidic in taste, seeds many.

Etrog (Atrog, Ethrog) : New growth, flowers and flower buds all are purple pigmented. Tree small, productive, moderate in growth. Leaf apex obtuse. Fruit small to medium, ellipsoid, neck distinct, nipple at apex sharp, style persistent, lemon yellow. Rind fleshy, thick, surface bumpy/ribbed, flesh firm and crisp, acidic in taste, low in juice content.

Lemon The characteristic of lemon plant, flower and fruit are very well known, while describing the varietal characteristics, mention must be made with reference to nipple or apical mammilla, very tight attachment of the fragrant rind and the high level of acidity of the vesicles or flesh which are pale yellow in color. Varietal description of commercially important lemon cultivar is given below.

Eureka : Trees are with sparse foliage, almost thornless spreading, crown open medium in growth, fruits around the year, very precocious and productive, susceptible to cold. Fruit small medium, oblong elliptical, neck short, at base small collar is present but usually with a nipple, yellow in color. Rind semi thick pitted, oil glands sunken. Segments about 10 in number adhering tightly, central axis solid and small, flesh greenish yellow, tender, very juicy finely grained very acidic in taste. Fruiting more in winter, spring and early summer. Seeds few to none.

Lisbon : Originated in Portugal. Tree with dense foliage, thorny, upright, large in size and grows vigorously and very productive. Fruit medium, elliptical oblong, base tapering neck not distinct, apex ends in sharp pointed nipple which has irregular areolar furrow, yellow in color. Rind quite thick adherence tight, pitted surface smooth and less ribbed than Eureka. Segments about 10, core solid, small, flesh greenish yellow, tender, juicy, finely grained, very acidic in taste, fruits more in winter and early spring, seeds few to none.

Villafranca : Originated in Sicily. Commercially not very important compared to Eureka and Lisbon. Tree characters resembles to that of Lisbon, but the plants are more erect or upright in growth, foliage less and with few thorns. In fruit characters it is similar to Eureka but fruits more during winter like Lisbon.

Nepali oblong (Assam, Nimber or Pat Nebu) : An everbearing lemon cultivar. Tree vigorous branches wide spreading, crown open and flat almost spineless. Fruit medium large, oblong ovate to long elliptic, nipple broad, base round. Rind relatively thick, greenish yellow, glossy, smooth, segments about 11, axis usually hollow. Flesh fine grained, greenish yellow, juicy, not too acidic, seed none to few.

Meyer : Flowers throughout the year, but more so in spring. Tree small medium, spreading, hardy, productive, almost thornless. New growth and flower with purple pigmentation. Produces more in winter. Fruit medium short elliptical or oblong, ribs slight, neck very small, base and apex round, slightly furrowed nipple broad. Rind thin, adhere tightly, surface smooth yellowish orange to orange, segments 10, flesh light orange-yellow, soft, very juicy with typical lemon flavour, seeds few.

Limes

Like the citron and lemon, the limes in all likelihood have originated in north eastern India. Limes are generally of two forms - Small fruited acid limes (*C. aurantifolia*), West Indian lime and the large fruited acid limes (*C. latifolia*). In its natural habitat a number of forms are recognized which differ markedly in size, form, shape, spine, and seediness character etc. The West Indian or Mexican lime is the kagzi nimbu and also has number of vernacular names. Detail of cultivar(s) is given below.

Small Fruited Acid Limes

West Indian (Mexican sour lime, Key) : Small fruited. New growth, flower buds and flowers are purple tinged very susceptible to cold. Trees are medium in size, spreading, bushy, with many slender short branches, densely marked with small slender spines. Leaves broad lanceolate, petioles winged, foliage dense, light green, flowers throughout the year more so in spring and late summer. Fruit small, obovate, round or short elliptical, base also round but with small furrowed nipple. Rind thin, surface fine, smooth, adhere tightly, greenish yellow. Polyembryonic with few seeds, segments 10-12, aroma characteristic, flesh greenish yellow, fine, soft juicy, very acidic. Seeds moderate.

Large Fruited Acid Limes

Tahiti (Persian) : Cold resistant, same as that of lemons. Flowers throughout the year more in spring. Purple pigmentation present on shoot and flowers. Tree vigorous, very spreading, almost thornless, foliage dense green, leaves broad lanceolate, petiole winged. Fruit small medium, variable in shape oval, oblong, obovate, short elliptical base round with short neck and faint furrows apex round, areolar area forms a small nipple. Rind thin lemon yellow, adherence tight, flesh greenish yellow, tender juicy acidic. Segments usually 10, axis small, solid, almost seedless.

Sweet lime (*Citrus limettioides*) : In north-eastern India to which it is native it is said that soh synteng of Assam is the acid form of this fruit. Like acid limes, the Indian sweet lime is the mitha nimbu and number of its forms differing in fruitfulness, fruit shape, size and with or without nipple etc are easily recognized. Detail of important cultivar is given as

Indian (Palestine): New growth white (free of colouration). Tree has irregular growth habit, large, spreading, branches thick, thorny, foliage dense, leaves rolled, long oval cupped, petiole wing margined. Fruit variable in shape oblong short elliptic, subglobose, base and apex round, apex with low flat nipple and circular furrow. Rind very thin surface smooth, oil glands distinct greenish to orange yellow. Segments 10, axis semi hollow, flesh light yellow, juicy, free of acid with slight bitterness. Seeds few highly polyembryonic. Sweet lime is of value in India, Near East, Egypt and Latin America and is said to have medicinal properties especially in prevention and treatment of fevers and liver ailments.

Kumquats

Kumquats resemble the other citrus fruits and more to Calamondin and also to some other small mandarins, but differs from citrus in presence of few ovary locules (3-5), usually two ovules per locule, being very small, edible and sweet in taste, rind some what pulpy, and flower buds are some what angular.

Pomological description of important Kumquats is briefly summarized below:

Nagami or Oval Kumquat (F. margarita [Lour] Swing) : Popularly called as Naga or Nagami Kinkan of Japan. The characteristic features are that fruits are oval, oblong or obovate in shape, fewer number of segments usually 4-5, rind deep orange, both flesh and rind are richly flavoured. Tree and leaves are larger in size.

Meiwa – Large Round Kumquat (F. crassifolia Swing.) : Is less cold hardy compared to Nagami, but is considered to be best from fresh eating point of view. This specie is the Ninpo, Neiha or Meiwa Kinkan of Japan. A natural hybrid of oval and round kumquats. Characteristic features are that fruit is large in size, round to short oblong in shape, segments nearly seven. Rind is very thick, flavour sweet with few to no seeds.

Marumi or Round Kumquat (F. Japonica [Thumb.] Swing.) : Maru or Marumi Kinkan of Japan. Tree low in vigour with some thorns. Leaves are small, apex not sharp pointed. Fruit small round to slightly oblate - obovate. Rind thin and sweet in taste, segments vary from 4-7.

Lemon Like Fruits

There are some distinct fruits in which lemon characteristics are evident, however the differences are to such a magnitude which warrants their separate characterization and classification. In this group the most important are the karna, the galgal or hill lemon and Jambhiri or rough lemon and all of them are widely grown in India. Meyor lemon and the limettas are also lemon like fruits, which are described below.

Galgal (C. pseudolimon Tan.) : An ancient Indian citrus whose origin is unknown and is commonly called as hill or kumaon lemon, grown extensively in the submontane areas along the foothills of Himalayas and in Punjab as a substitute for lime or lemon. Tree upright, semi spreading, irregular with open crown, branches are thick stout armed with numerous thick large spines. Leaves large (like sweet lime). Flowers also large and purple tinged so is new growth. Fruit quite large, oblong ellipsoid, apex with pointed nipple, base with neck which is furrowed. Rind pale yellow to golden yellow adhere tightly, smooth or rough, medium thick. Flesh pale yellow, texture coarse, semi juicy, sourish and bitter in taste segments 10, axis large and hollow, seeds many and bigger in size.

Karna (*C. Karna* Raf) : Karna Khatta, Karna Nimbu, Khatta Nimbu. A very old Indian citrus fruit of unknown origin, moderately polyembryonic. Considered to be a natural hybrid between rough lemon and sour orange, as the characters exhibited resemble the two species. Widely employed as a rootstock in northern India, second only to rough lemon. Tree medium large, upright, branches spreading armed with spines, leaves large, petiole winged. New growth and flower (large) both purple pigmented. Flowers and fruits only once a year. Fruit medium large, shape variably usually round to oval nipple prominent at apex, may be absent too. Rind quite thick adhere tightly, golden yellow to deep orange, smooth or ribbed. Flesh orange, texture coarse, semidry, acidic in taste, flavour sour orange like. Seedy, cotyledons white. Soh - Sarkar (Assas) belongs to this specie.

Rough lemon (C. jambhiri Lush.) : Jatti Khatti, Lemon, Citronelle. The species is regarded to be native of north-eastern India, where even today it grows wild. It is presumed that Portuguese while returning back home introduced it in the southeast Africa. Later towards the end of the fifteenth or early sixteenth century it was brought to Europe from where it reached New World. Rough lemon though used as a substitute to lemon but its use as a rootstock is world wide recognized. Tree large vigorous in growth, not very resistant to cold, upright, branches spreading armed with many small thorns. Leaves light green, small to medium in size blunt pointed. New growth and flowers are purple pigmented. Flowers, small usually produced around the year more so in spring and summer. Fruits are medium sized, shape variable, usually oblate to elliptic oblong. Apex with a broad nipple, areole furrowed and irregular in form, base with a distinct neck or collar which is irregularly furrowed. Rind lemon yellow to brownish orange in color, medium thick, surface rough, bumpy (sunken oil glands) deeply pitted or ribbed, separates readily. Flesh pale yellow to pale orange, acidic in taste, juice moderate, segments 10, hollow and large. Seeds many, small highly polyembryonic, cotyledons light green in colour.

Tangors - Mandarin Like Fruits

They are the hybrids of the mandarins and oranges, and the common natural tangors are 'King', 'Clementine mandarin' 'Temple orange'. Campeona, Ellendale Ortanique and Murcott orange are of hybrid parentage. Of all these 'Temple' is the most important. Salient horticultural traits of some well known tangors are described.

Temple. Very cold hardly than the oranges or mandarins, medium to late in maturity. Tree characters are like that of mandarin, productive, tree spreading bushy, vigor medium, slightly thorny. Fruit broad obovate to subglobose, neck if present is short, furrowed and rough, apex with small navel. Rind quite thick, adherence moderate, peels readily, segments 10-12, axis solid. Flesh orange, tender, juice moderate with spicy rich flavour. Seedy.

Dweet : Evolved from a cross between Mediterranean Sweet Orange and Dancy Tangerine. Late in maturity. Tree character intermediate of the two parents. Fruits do not hang well on tree. Fruit reddish-orange in colour, globose - oblate, neck distinct, medium to large, surface pebbled. Rind adherence tight and peels very poorly, also puffs, flesh orange, firm very juicy, flavour rich, seeds numerous.

Mency : Originated from a reciprocal cross of Dweet. Fruits ripen early does not hang well on tree, susceptible to sunburn, small in size, reddish orange, somewhat oblate and necked, surface pebbled due to oil glands. Rind adherence not tight and peels easily. Flesh orange, juicy with acidic flavour. Seeds are many.

Tangelos

Hybrids of mandarin, grapefruit and pummelo are designated as tangelos. They exhibit characters of both the parents. Tangelo cultivars of present significance are detailed below.

Clement : Is a hybrid between Duncan grape fruit and Clementine mandarin. Tree and foliage characters intermediate, productive. Fruit subglobose to oblong, light orange yellow in colour, medium-large. Rind quite thick, pebbled, peels easily. Flesh dull yellow soft, moderately sweet. Early to medium in ripening.

Minneola : Hybrid of Duncan grapefruit and Dancy tangerine. Less cold hardy than Orlando, requires cross pollination for regular and high yields. Maturity midseason. Trees are vigorous and productive. Fruit oblate-obovate, large neck distinct. Rind deep reddish orange, smooth, medium thick, adherence moderate surface pitted. Flesh orange, juicy tender, flavour rich aromatic, somewhat tart.

Orlando: A hybrid of Duncan grape fruit x Dancy tangerine. Early in maturity. Tree like that of Minneola, more cold hardly, leaves cup-shaped, less in vigour. Fruit broad oblate to subglobose without neck, medium large in size. Rind, thin orange in colour, adherence quite tight, does not peel easily, pebbled. Flesh orange, very juicy tender, somewhat sweet seedy.

Seminole: Parentage is same as that of Orlando and Minneola. More fruitful. Late in maturity trees are productive, medium-large, leaves small to medium, rounded and cupped. Fruit broad oblate deep reddish orange in colour. Rind adhere moderately, thin, pebbled, peelable, axis usually hollow. Flesh dark orange, juicy soft, somewhat acidic in taste, seedy.

Mandarin Like Fruits

Besides tangors and tangelos, there are some other fruits in citrus (Orient) which are like mandarins. Important from horticulture point are Calamondin, Rangpur and Iyo.

Calamondin (C. madurensis Loureiro.). Has very less value as a fruit but hangs well on tree used widely as an ornamental. Very cold hardy. Trees are columnar, upright, very productive, almost thornless, leaves small, broad oval. Fruit very small, oblate to spherical, apex flat. Rind orange to orange red smooth, pitted, thin peels easily. Flesh orange in colour, soft, tender, juicy and acidic in taste. Segments 9, axis hollow, small. Seeds few polyembryonic plumb, small. Cotyledons green.

Rangpur (*Citrus limonia* Osbeck). Very cold hardy : Tree productive, vigorous, branches spreading and drooping armed with small thorns. New growth, flower buds and petals are deep purple tinged. Fruit small medium hangs well on tree, shape variable depressed globose, round to broad obovate. Apex with short nipple, base collared or small neck, furrowed. Rind yellowish to reddish orange, loose, thin, smooth faintly pitted. Flesh orange in color, soft juicy, very acidic in taste. Seeds many, small, polyembryonic, cotyledons green.

Some Wild and Semi Wild Species

As north-eastern region of India has been recognized as one of the major centre of citrus origin, a number of species are said to be native to this place.

Number of well established and distinct species have already been characterized, still a number of species both in wild and semi wild forms which are of less economic value as a fruit are found growing in abundance which in future may be used to improve some traits. Pomological traits of some species are described briefly.

C. indica. Indian wild Orange. Specie is found growing in many parts of Assam, Nagaland, Meghalaya and other north-eastern parts of India in wild form. Plant not tall growing, bush - like, medium in growth, branches spiny, leaves oblong lanceolate, petiole winged. Fruit small broad obovoid or sub-pyriform, appear singly on terminal twigs, about 2 cm in diameter, pedicel very small. Rind very thin, red in colour, segments few, orange red in colour, inedible, vesicles spindle shaped, very soft, pulp slimy, acidic in taste with unpleasant flavour. Seeds are very large, smooth, monoembryonic and occupy major portion of fruit.

C. assamensis. Adajamir. A very distinct specie known for its peculiar aroma that resembles to ginger or eucalyptus smell that emits from the crushed leaf or from the fruit. A new specie which was indigenous to Assam region. Tree medium sized, foliage medium dense with thick leathery glossy leaves. Fruit medium small, spherical to round in shape, surface smooth, very acidic juice hence has very limited use as fresh fruit.

The other species *C. latipes, C. ichangensis* and *C. macroptera* are usually placed under subgenus Papeda which has number of true wild species of citrus. Important characters that separates it from citrus are that pulp vesicles of all these species have many globules of very acrid oil. Petiole very large 1 ½ to 3 times longer than broad and broadly winged, flowers small usually less than 2 cm across, stamens free, but if united in bundles then flowers are large.

C. macroptera. Malanesian Papeda. Amongst all the species of Papeda group, this species is most promising as a rootstock. Widely distributed in Indo-China, Thailand, Philippines, New Guinea, New Caledonia and Polynesia. Small tree (15 feet), branches are angular, green (Orange). All plant parts are with amber coloured glands, spine axillary straight. Leaves, broad lanceolate (2-5 cm), petiole winged, margin broadly crenate. Flowers are like that of orange. Fruit globose, pale yellow, small (2.5 in dia.), smooth, glands small, segments 10-12 with 1-2 seed/vesicle. Juice scanty, highly acidic.

C. latipes : Khasi Papeda. Cold hardy. Native of northeastern Khasi hills (India), and northern Burma. A small spiny shrub or a small tree (10-15 feet), twigs angular with sharp stout spines (1.2-2.5 cm), reduced or absent on flowering twigs. Leaf ovate-acuminate, petiole large and broadly winged. Flowers four merous, borne in small axillary racemes with 5-7 flowers. Fruit medium sized, globular, rind somewhat thick, leathery, segments 9, quite large pulp vesicles few, spindle shaped, well developed. Seeds many 5-7 in each vesicle, round in shape.

C. ichangensis (Ichang Papeda) : Native of south western and west-central China. This specie in the most cold hardy of all evergreen types. Differs from other Papeda in that it has large flowers, stamens are connate. Leaves are very long with large winged petiole, seeds are very thick, monoembryonic. Small shrub or a tree (12-15 feet), twigs angular, spines stout. Leaf blade ovate acuminate, apex emerginate, base rounded. Flower large (2.5-3 cm), pentamerous ovary with 7-9 locules. Fruit small (3-4 cm dia.) glossy, peel rough, bumpy, medium thick, segments 7-9, each with many seeds large and thick, blunt at both ends, monoembryonic.

Citrus is highly heterogeneous group and various species hybridize freely with one another in nature. As a result of this several intergeneric and intrageneric hybrids significant from horticultural point of view have been obtained and they are described as below.

Intergeneric

1. Hybrids of Poncirus

Citrange: This group has parentage of trifoliate orange (*P. trifoliate*) and sweet orange (*C. sinensis*). Plant and fruit characters are intermediate i.e. unifoliate or trifoliate leaves, fruits may be yellow or orange with thin rind juicy and flavoured. Important cultivars are Troyer, Carrizo, Morton, Coleman etc.

Citrangedin : *A trispecific* hybrid of Citrange with the Calamondin

Citrangequat : *A trigeneric* hybrid between three genera - Poncirus, Citrus and Fortunella

Cicitrange : A back cross hybrid between Citrange and Ponicrus

Citrangor : A back cross hybrid of Citrange and *C. sinensis*

Citrumelo : A hybrid between *Poncirus* and *C. paradisi*

Citrandarin : A hybrid between *Poncirus* x *C. reticulata*

Citremon : A hybrid between *Poncirus* x *C. limon*

Citradia : A hybrid between *Poncirus* x *C. aurantium*

Citrumquat : A very difficult hybrid to breed between *Ponicrus* and *Fortunella japonica* or *F. margarita.*

Hybrids of Fortunella

Limequat : Cross between *C. aurantifolia x. F. japonica*

Orangequat : Cross between *C. reticulata* (cv. Satsuma) x *F. japonica* x *F. margarita* cv. Meiwa

Parocimequat : Double cross between *F. japonica* x *C aurantifolia* cv. Mexican) x *F. hindsii.*

Intrageneric

i) *Tangor* : Cross between Mandarin (*C. sinensis*) and Sweet orange (*C. reticulata*). Important cultivars are Temple, Clementine, Monreal etc., are monoembryonic

ii) *Tangelo* : Cross between Mandarin (*C. sinensis*) x grape fruit (*C. paradisi*). Important cultivars are Orlando, Minneola, Seminole etc.

iii) *Lemonime* : Cross between lemon (*C. limon*) and lime (*C. aurantifolia*). Representative cultivars is Parrine.

a) Lemonnage : (*C. limon* x C *sinensis*)

b) Lemandarin : (*C. limon* x *C. reticulata*).

❑❑❑

4
Papaya

Genus : *Carica*
Species : *papaya* L.
Family : *Caricaceae*
Chromosome No. : 2n = 18, 36

The papaya plant is a huge/giant herbaceous plant and in general appearance looks more like a palm although there exists no botanical relationship between them. Papaya is said to be native of tropical America where it is a common fruit however the exact place of its origin is not well determined. Presently the plant is widely distributed and is abundant in India, Sri Lanka and many other countries of the Malaya Archipelago, Hawaii. Papaya fruit as well as all plant parts contain a milky sap called papain. This enzyme resembles animal pepsin in its digestive action and has become an important article of commerce, is utilized in the clarification

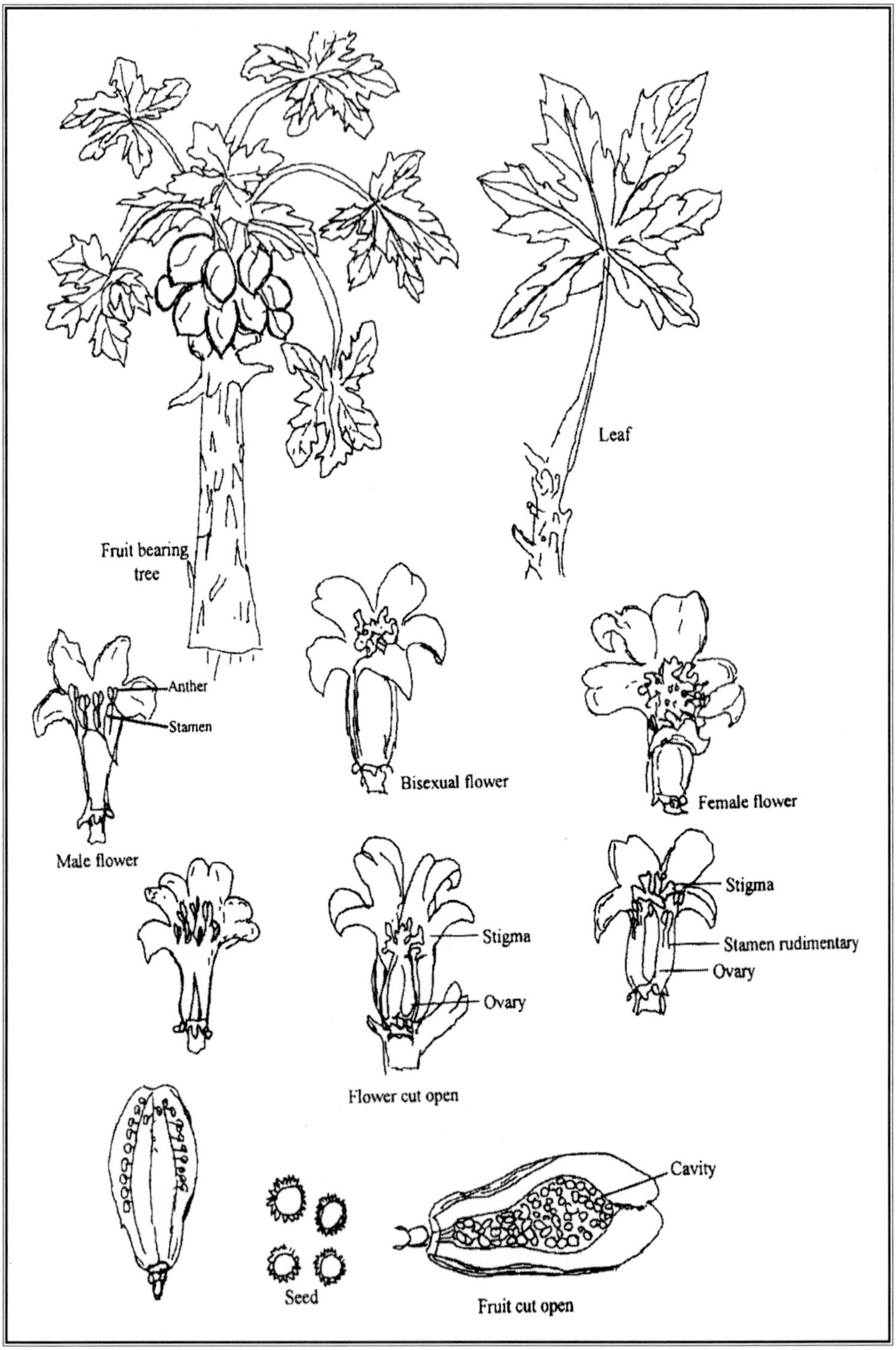
Leaf
Fruit bearing tree
Anther
Stamen
Male flower
Bisexual flower
Female flower
Stigma
Ovary
Flower cut open
Stigma
Stamen rudimentary
Ovary
Cavity
Seed
Fruit cut open

of beer and its digestive action too has been very well recognized. Another remarkable papaya character is the irregularity which it presents in the sex distribution. Being dioecious, staminate and pistillate flowers are produced on different plants, besides a number of intermediate form have also been observed.

The papaya is an important fruit of tropical and sub tropical regions of the world. Papaya is a very wholesome fruit and is placed next only to mango as a source of vitamin A precursor, also has vitamin B_2, C, riboflavin, carbohydrates and other minerals in substantial quantities. Papaya belongs to a small anomalous family Caricaceae which has four genera and about 31 species out of which edible fruits are present only in Genus *Carica*. Carica has some 22 specie but C. *papaya*, the common papaya is the only one specie which is extensively grown for its fruit.

Origin and Distribution

The papaya is considered to have originated in the tropical America. The Spanish explorers in the 16th century carried the fruit from its natural habitat to Caribbean (West Indies) and later to south east Asia. From here the crop further moved to India, Africa and Oceania and presently it is extensively cultivated all over the world especially in the tropical and warm sub tropical areas.

Uses

The entire plant is very useful to mankind in one or the other way. Papaya fruit can be employed to make jam, jelly, fruit salad, highly refreshing drinks, candies, marmalades or frozen fruits, green fruits as vegetable after cooking, sweet meat from flowers, papain a proteolytic enzymes collected from green fruits has wide spread uses, softening of meat, beverage, pharmaceuticals and food industries, preparation of drugs for wound treatments and for digestive ailments.

Description

Plants are herb like (2-10 m) contains latex vessels unbranched, or may be branched (injury to crown), cylindrical stem (1 foot in diameter) hollow inside with spongy fibrous tissues, stem marked with numerous sharp, prominent orbicular leaf scars which are smooth and grey in colour. Leaves are large upto 75 cm across, found in spiral manner and divided deeply into about 7 palmate lobes, each lobe is further pinnately lobed, greenish or purplish green, arranged spirally usually at the top/apex to form crown, palmately and deeply lobed, veins prominent, lobes deep and broadly toothed, petiole long (1 m) and hollow, which ultimately sheds to leave behind a large stem scar. Flowers are mainly of three types, male, female and hermaphrodite usually borne in the leaf axils on separate trees.

Male flowers are borne in panicles (25-100 cm long) which are erect / pendent, sessile calyx, small 5, toothed, generally cup shaped, corolla pale yellow, 2-2.5 cm long trumpet shaped, lobes 5 gamopetalous, yellow corolla often fragrant and spreading, stamens 10, generally in two whorls of five each, alternating with the petal lobes, filaments light yellow, anthers oblong, pistilloides also present.

Female flowers generally appears singly or in few flowered cymes, in the axils of leaves, larger than males measuring 3.5-5 cm across, rachis short and thick, more or less sessile on main rachis, calyx cup shaped with 5 narrow teeth, yellow green in colour, corolla 5 yellow, almost free, united at base, fleshy somewhat twisted, lanceolate in shape, stigma 5, fan shaped, deeply 5 lobed / cleft, sessile, ovary ovoid-oblong (2-3 cm long) has five united carpels, central cavity with numerous ovules on parietal placentation.

Hermaphrodite. These are of two types i.e. elongata and pentandria type

a) Elongata : Important features are, flowers in clusters with short peduncles, petals partially united, stamens 10 in two whorls / series, ovary elongate.

b) Pentandria : Flowers are similar to female flowers, but has only 5 stamens. Sometimes intermediate flowers appear, where stamens are carpelloid and result in irregular fruits. Type of flowers produced and their proportion vary according to environment and age of the tree.

Fruit is a large fleshy hallow berry variable in shape, 5 grooved, spherical, cylindrical, pyriform or ovoid-oblong measuring about 30 cm in length, may weigh upto 10 kg. Skin smooth (epicarp), green yellowish or orange in ripe fruit, flesh edible yellowish to deep orange-red, flavour pleasant, mild, seeds are small, round-globose, greyish black attached with numerous placentae in 5 rows to the central cavity, covered with gelatinous sarcotesta, which is attached to flesh by fleshy gelatinous stalk.

Brief description of some other papaya species which are of significance from horticulture point of view are detailed below.

Carica pubescens / candamarcensis : 2n = 18

Origin Commonly called as mountain papaya. Originated in the Andeans range at an altitude of 1500-3000 m extending from Panama to Bolivia.

Presently the crop is being cultivated in number of countries in the tropical region.

Uses has limited use, generally fruit is eaten fresh, but only when after it is sweetened, used to prepare jams and alcoholic drinks.

Description plants are shrub or small trees, but more branched than *C. papaya*, dioecious, but is smaller in all respect (leaves, fruits etc) to *C. papaya*. Characteristic feature is that the male flowers are borne on branched peduncles about 15 cm in length and the bigger female flowers appear on short , branched stalks. The ripe fruits are orange-yellow, acidic in taste with typical fragrance, flesh firm obovoid in shape measuring about 12 x 8 cm, seeds are many and are enclosed in juicy white covering (sarcotesta).

In growth habit and general appearance the mountain papaya resembles the tropical papaya but is inferior or smaller in all its plant and fruit characters. Plant is small (8-10 ft tall), leaves are smaller but deeply lobed, fruits 3-4 inches long, too acidic to be consumed fresh but can be made into jam and preserves. In short plant is handsome in form with straight trunk and without branches (but may be developed if the crown is damaged by frost). The specie presents hope to develop papaya suitable for cultivation at higher elevations (hardy to frost).

C. quercifolia Benth and Hook. Is another *Carica* specie indigenous to tropical America. Is said to be hardiest of all species, yields fruit which are inferior both in quality and size (2-3 inch) has potential to breed hardy papayas.

C. gracilis Solms. Plant short statured 4-6 feet tall, trunk simple high and slender, very shiny leaves five digitate, individual lobes sinuate lobed, the central one is 3 lobed. The entire leaf blade is suborbicular in form, petioled.

Cultivars

Papaya has number of cultivars which are cultivated widely world over. Some with salient features are described as below.

Washington : Grown widely all over the world. The stem, leaf petioles and basal part of the fruit close to the stem has purple pigmentation, fruit round ovate oblong in shape, large in size, weight may be around one kilogram. Quality and flavour good sweet in taste pulp orange in colour.

Honey Dew : This papaya cultivar is extensively grown in north India. Fruit has very few seeds, taste and flavour is good.

Coorg Honey Dew : Considered to be a chance seedling of Honey Dew, mostly the plants are dwarf in nature, prolific bearer, hermaphrodite but sometimes pistillate flowers are also found on the same plant, fruits are oblong, flesh thick therefore more edible portion, flavour pleasant, taste sweet quality good.

Pusa Majesty (Pusa 22-3) : Evolved at IARI, is gynodioecious line, also tolerant to viral diseases has very good keeping quality and withstands long transportation spoilage. Fruits are round with short apex, medium sized, skin smooth, flesh firm solid, yellowish in colour.

Pusa Dwarf (Pusa 1-45 D) : The characteristics feature is that plants are distinctly dwarf hence suitable for high density plantation, kitchen garden etc. Cultivar has been developed at IARI, is dioecious starts bearing earlier i.e. plant starts fruiting when about a foot tall. Fruit oval in shape medium in size each fruit weighs about 1-2 kg.

Pusa Delicious (Pusa 1-15) : Developed at IARI is gynodioecious cultivar with two sex types frequently found on the same tree i.e. pistillate and hermaphrodite. Fruit oval with a bulge near distil end and flesh orange coloured, eating quality very good with excellent flavour.

Co 2 : This selection was obtained from local type grown in Tamil Nadu (Coimbatore) has good table qualities and also produce high papain content. Plants are medium tall, fruit large obovate in shape, initially bright green later turns yellowish green in colour, flesh soft orange coloured, juice moderate.

❑❑❑

5
Pineapple

Genus : *Ananas*
Species : *comosus L. Merr*
Family : *Bromeliaceae*
Chromosome No. : 2n = 50, 75, 100

Pineapple (*Ananas comosus* L) 'Golden Queen' belongs to order Farinosae and family Bromeliaceae. The name is derived from genus *Bromelia* in honour of Olaus Bromel, a Swedish botanist. Herbs or sub-shrubs mostly epiphytic, has the distinction to be one of the major food plants of the world, which was selected, domesticated and developed by the early pre-historic people, who passed this valuable knowledge gathered for hundreds of years through earlier civilization. Pineapple developed to its present form in its natural habitat - New World, and was not know to the Old world before Columbus discovered America. Red Indians the local inhabitants of South America at that early time had already developed and named number of varieties which were partially selected by them or their ancestors based upon various desirable horticultural traits like fruit size and quality, earliness, long storage, and absence of seeds etc. In its natural habitat - tropics, at various places in tropical America wild pineapples are still found growing in abundance which in fact provided the foundation stock for future improvement. From this raw material a number of varieties were selected, and were characterized by small size, high seed number and with inferior eating qualities. These undesirable characters are no more found in the present commercially grown varieties. In every likelyhood pineapple developed through the process of evolution from more primitive ancestral forms which may have been recorded in the geological and evolutionary time units.

On the basis of variation in growth habit, the pineapple species have been divided into two groups - terrestrial and epiphytic. Pineapple belongs to terrestrial group, although it does possess some characters of epiphyte, its ability to store some water in their leaf axils, which helps the plant to resist periods of drought. The term pineapple is specifically limited and is used to refer the genera *'Ananas'* and *'Pseudananas'* and it is from these that the family has its common name. The pineapple have been separated from other genera of the family primarily on the basis of syncarpous type of fruit which is not found in any other fruit of the family.

Earlier the genus *Ananas* was said to be monotypic with only one specie and number of cultivars, but presently two genera *Ananas* and *Pseudananas* are well recognized. In *Pseudananas* syncarp bearing at maturity bears a minute inconspicuous coma of reduced squariform bracts, never producing slips at its base, plant producing elongate stolons at base, petals bearing appendages in the form of lateral folds. In *Ananas*, syncarp at maturity bears a conspicuous coma of filiaceous brats, frequently producing slips at its base, plant not producing stolons, petals each bearing two infundibuli form scales.

Origin and Distribution

The crop has its origin in South America. Later the crop was introduced in the East Indies (Malaysia, Indonesia etc) and Philippines by the visiting Spanish. At present the crop is grown throughout the tropics and subtropics. The crop is extensively grown in Brazil, Mexico, Puerto Rico, South Africa, Hawaii, Sumatra, Malaysia, Philippines and Thailand.

Uses

The crop is used in several ways, the best being eaten fresh, besides a number of delicious dishes are prepared from pineapple. The canning

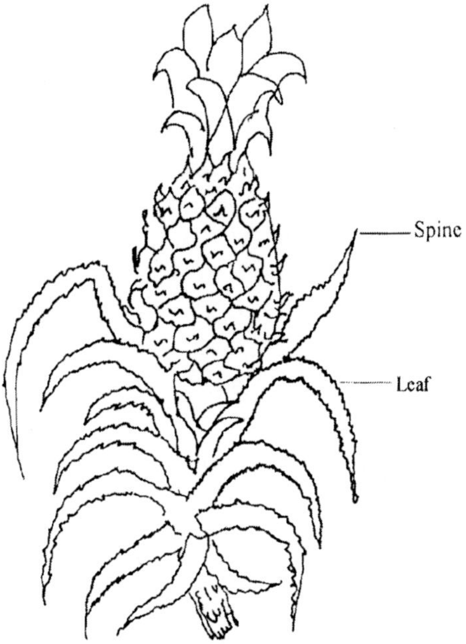

Plant and fruit

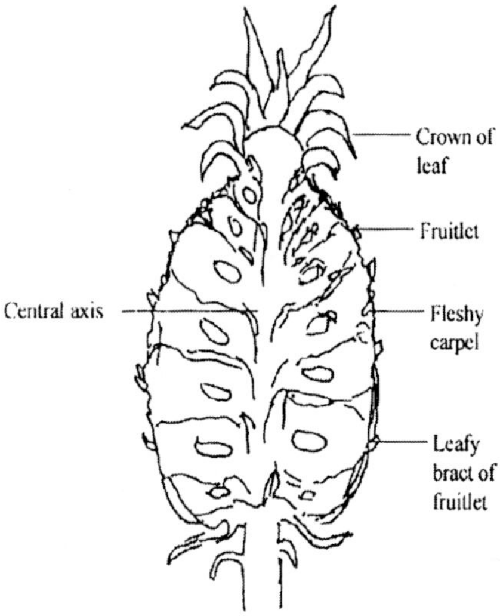

LS of fruit

Flower bud

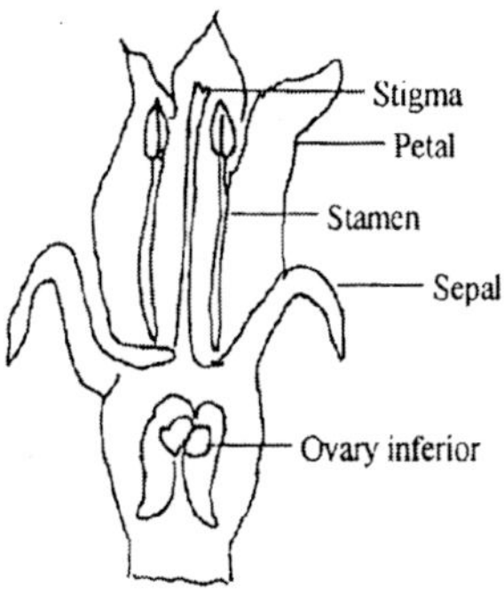

VS of flower

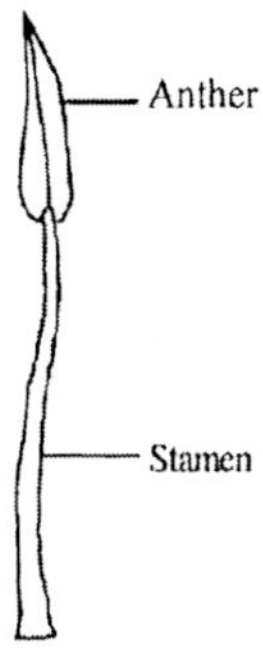

TS ovary

prospects of the crop were known at early times and in the present cultivation scenario crop is grown primarily for canning. The fruits are generally canned as slices, cube, titbits, spirals etc; jams, pineapple wine or vinegars. The left portion of the fruit after peeling is used as cattle feed.

Description

The plant of pineapple is perennial or biennial herb (50-150 cm tall). Leaves are typically sword shaped tapering at the apex and broader at the base. Leaf blade sides are upturned trough like shape to give more rigidity. Leaves may be about 1m. in length, sometimes even more, fleshy, upper surface grooved, margin entire, may be spiny too. Leaves arranged spirally, leaf bases clasp the main stem of the herb thickly throughout its length. Pineapple inflorescence is dense compact spike containing numerous, hermaphrodite (200 or more) sessile flowers, reddish purple in colour each ending with a pointed bract which secretes nectar. The pineapple is a monocot, therefore the flower parts are in multiple of three's, sepals three, short and fleshy, pale violet petals forms a tube, 3 in number and encloses 6 stamens, 1 style, stigma 3 lobed. The multiple fruit of pineapple is formed by the fusion of numerous berry like single fruits and the thickened axis of the main stem (inflorescence). Fruit generally seedless, flesh pale to golden yellow, white at core. The outer horny, thick hard rind results from the persistent sepals and bracts. No floral component abscises. The fruits are generally cylindrical, sometimes may be sharply or abruptly tapering at the apex. The fruit apex (top) is usually crowned by cluster of short pointed stiff spirally arranged leaves and are referred as 'crown'. Shoots emerging below the fruit on the main stem are called 'slips', whereas 'sucker' are formed in leaf axils at the base of the herb. All three types are employed as vegetative propagules for multiplication.

Detail of different species is as

A. bracteatus (Lindl.) Schultes. Red pineapple. This specie has uniformly green leaves with bright pinkish red fruitlet, bracts and upper peduncle and leaves with large elongated imbricate spiny bract that conceal the tops of the ovaries. Fruit seedy but edible.

A. ananassoides (Bak) L.B. Sm. Leaves are long narrow recurving upto 1 inch wide spiny scape, slender, elongated terminating in a many flowered inflorescence. Flower bract serrulate. Fruit is a many seeded syncarp of 6 inch long. Has one botanical variety Ananus which is almost seedless and is half the size of the typical form and other parts are similarly reduced.

A. fritzmuelleri. Syncarp well over 15 cm long at maturity with copious palatable flesh, scape stout and usually short. Leaf spiny recurved towards the base, floral bracts pale green at maturity, petals have vertical folds.

A. erectifolius. Syncarp 15 cm long or usually much shorter with scant unpalatable flesh at maturity, scape elongated slender. Leaves stiff erect, entire except for a long terminal spine 3-5 cm wide.

In pseudananas the petal appendages consists of a pair of long cigar shaped fleshy ridge about ¼ the length of the petal. They are produced along the inner edges and overlap and partially enfold the filaments of the epipetalous stamens. They are thin quite different from the flatter ridge of Ananas which bears delicate fabricated funnel shaped appendages scape.

Pseudananas sagenarius [(*P macrodontes* (E. Morr.) Harms)]. Plant of this specie is characterized as leaves are about three feet long some even more in length and three inch wide with spines which are one inch long or more. Inflorescence about seven inch long and three to four inch in thickness, individual flower rose coloured about 2 inch long.

In pineapple there is no definite or standard botanical or horticultural basis of cultivar classification. However fruit colour, shape, flat or sharpness of 'eyes' and spine, character of leaves often have been taken for classification. Presently the pineapple cultivars have been classified into Spanish, Queen, Cayenne, Abacaxi and Maipure group. Important horticultural traits of the individual group are summarized here under as.

Spanish. Leaves are spiny, fruit globose long weighs about 1-2 kg, eyes are deep, rind deep orange to red tinged, flesh pale yellow to whitish, core large and fibrous with spicy acidic taste, considered very good for export and local consumption with fair canning properties.

Cayenne. Leaves generally smooth with few spines only at the tip, shape cylindrical, slightly tapering at tip, eyes are usually flat weighs 2-2.5 kg, rind colour deep orange flesh pale to yellow, core medium, flesh tender with low fibre, juicy sweet, with mild acid. Rated very good for canning, export potential is fair.

Queen. Leaves are usually spiny. Fruit shape conical, tapers at tip, eyes are deep set, weighs half to one kilogram. Fruit colour yellow, flesh also yellow core small with low fibre content, sweet in taste and less acidic compared to Cayenne. Rated to be good for export and local consumption, fair for canning.

Abacaxi. Plant leaves are spiny, fruit conical in shape and weighs about 1.5 kilogram on an average. Fruit colour yellow, core very small flesh pale yellow to whitish, fibre low tender, sweet juicy in taste. Rated as fair to poor for canning and export, but relatively good for local market.

Maipure : Leaves are smooth fruit cylindrical ovoid to cylindrical in shape, fruit yellow to dark orange with reddish tinge flesh whitish to deep yellow, core small to medium in size, fibrous, tender and juicy, but more sweeter than Cayenne. Rated fair to poor for canning and export but good for local market.

Cabezona : A tropical type, fruits mainly grown for fresh fruit industry, extensively grown in Puerto Rico.

Horticultural traits of some commercially important cultivar are described below.

Kew or Giant Kew : 'Giant Kew' is a mutant of 'Kew' most of the characters are same except that fruit size is bigger than the Kew. Plants are vigorous, leaves are trough shaped margins almost entire. Fruits are large sized ranging 1.5-3.0 kg and sometimes even more cylindrical slightly tapering at apex eyes are broad and shallow good for processing. Fruit dark blackish green when unripe, turns orange yellow when ripe, flesh light yellow, pale at core, almost fibreless very juicy sweet with pleasant aroma. Commercially very important in India.

Queen: Leaves are spiny margin serrated with closely arranged spines. Fruit small in size average weight ranging from 1-2 kg, cylindrical in form, fruitlets or eyes are small centre of eye is protruded, deep set and irregular, ripens uniformly and keeps well, juicy sweet, flesh fibreless, yellow. Rated as an outstanding table cultivar, but unsuitable for canning.

Mauritius : Fruits are medium sized, mid season cultivar, has two strains yellow and red skinned. *Yellow type* fruits are oblong, slopes slightly towards apex, green in colour when unripe and deep yellow when ripe. *Red type* Fruits are broader at base and tapers abruptly at apex, red when ripe, flesh reddish, fibrous and more sweet in taste than yellow.

Abacaxi : Cultivar is grown for local market in Brazil. Plants are erect, leaves with spiny margin and tip. Fruit broader at base and narrow at apex or crown, fruitlets small average fruit weighs 1.5 kg. Flesh pale yellow when ripe paler at core, juicy, sweet, flavour good.

Singapore Spanish: Leaf margins are almost smooth except few spines at leaf tip, fruits cylindrical hence suitable for canning, weighs about 2 kg, reddish-orange when ripe flesh bright yellow, core light yellow, small and fibrous, flavour aromatic.

Red Spanish : Fruit mostly grown for fresh consumption, leaves spiny and long, average fruit weighs 1-2 kg, fruitlets large and few, firm orange-red in colour, flesh pale yellow when ripe, flavour pleasant.

□□□

6
Cashew

Genus : *Anacardium*
Species : *occidentale* L.
Family : *Anacardiaceae*
Chromosome No. : 2n =42

This delicious nut fruit of family Anacardiaceae is unique in number of ways firstly for fruits brilliant shades of colour (yellow to dark red-scarlet) characteristic aroma makes it one of the most enticing of tropical fruits. Secondly it multiplies and grows very fast and produces in abundance its attractive handsome fruits. Thirdly the cashew fruit is very peculiar for at the first look the swollen peduncle and disc (cashew apple) is taken to be the real fruit, a fact which is not true. The real fruit of commerce is the kidney shaped nut attached to the lower portion of the cashew apple. The English name cashew is adapted

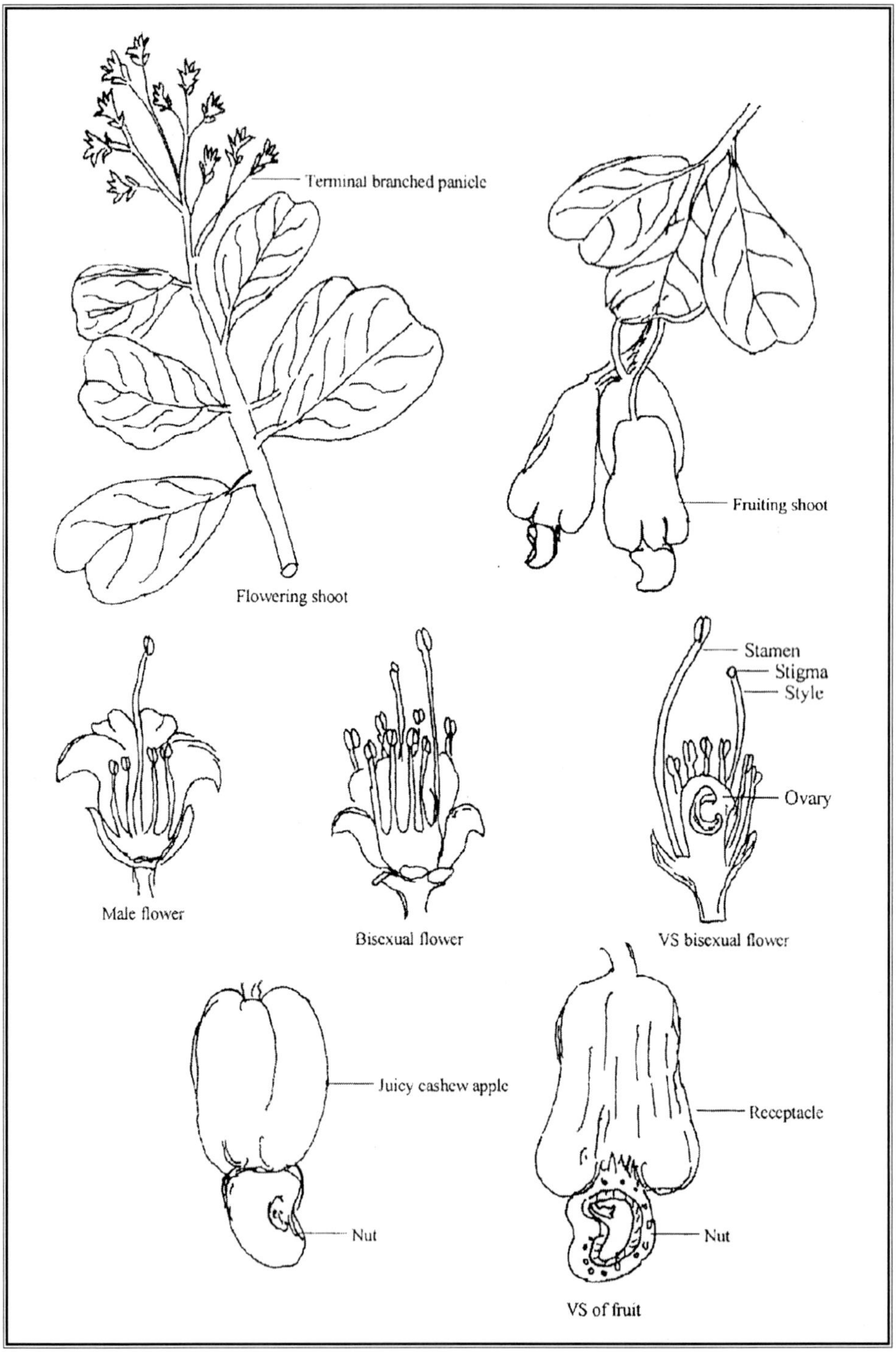
Terminal branched panicle
Flowering shoot
Fruiting shoot
Male flower
Bisexual flower
Stamen
Stigma
Style
Ovary
VS bisexual flower
Juicy cashew apple
Nut
Receptacle
Nut
VS of fruit

from Portuguese word 'Caju'. India is one of the major cashew producer and exporter.

Cashew is a member of Anacardiaceae and is placed with the natural order Sapindales, is an evergreen tree comprises of 60 genera and 400 species of tropical trees and shrubs. In India after its introduction by the Portuguese in the 16th century was first planted in Goa to Malabar Coast in Kerala by the Spaniards and to the other centers in India. Presently it has naturalized throughout the west coast. Cashew has 20 species in central and south America, but only *occidentale* 'Kaju' was the one specie which was introduced to outside world and is described in detail. Other species are *A. brasilience A. humile A. parviflora, A. rhinocorpus, A. gigantum A. nanum, A. excellsum, A. pemilum, A. mediterraneum, A. marcocarpum* etc.

Origin

Cashew is considered to have its origin in the north eastern Brazil. From here the crop was later introduced into south and central America. The Portuguese with the desire to discover and conquer new areas brought cashew to India and east Africa in the 17th century. During the same period the Spanish took the crop further to Philippines. Today cashew is grown in tropic and subtropic areas of the world and major cashew producing countries are Brazil, India, Tanzania, and Mozambique.

Uses

Cashew fruit comprises of two parts i.e. cashew apple and cashew nut. The cashew apple is swollen fleshy fruit stalk, astringent used to make squashes, wines from fermented juice (at harvest) later distillation yield strong alcoholic drinks, pulp as preserves. The seeds/kernel is of economic importance and is rich in oil. Cashew shell oil is a vesicant, used as waterproofing agent and as preservative Seed is separated after roasting of cashew apple which also reduces astringency, cakes after oil extraction makes good cattle feed, seed coat as poultry feed. Kernel is delicious, used in bakery and confectionery products. All plant parts are used in traditional medicine, for skin ailments, mouth washes, gum exudation from cut parts yields an adhesive, sap from bark yields indelible ink.

Discription

The cashew plants are tall trees (12m or more), evergreen, crown spreading, dome shaped. Has deep root system (taproot - 3-4m and sunken roots 6m. deep). Stem producing number of branches. Leaves are simple, alternate, leathery (coriaceous), obovate to obovate-oblong measuring 20x15cm., usually with reddish-brown tinge when young, later turning dark green, glabrous, shining, mid rib and veins more conspicuous on the lower side, petiole medium long, flat on the ventral side, base swollen. Inflorescence appears terminally drooping on branched panicle, measuring about afoot in length, hermaphrodite and male flowers are fragrant. Individual flower has 5 calyx, pubescent, lanceolate to oblong ovate in shape, corolla 5, reflexed linear lanceolate, usually whitish at anthesis, turns pinkish- red later, stamens usually 10. In the hermaphrodite flowers there are 9 short and 1 long stamen, only long stamens produce viable pollen, others do not. The number of stamens vary from 7-9, of which 1-3 are about 1cm long and the remaining are less than half centimeter in length. Style single, simple, exerted from the corolla and reaches the same length as that of the long/ fertile stamen. Fruit is a typical kidney-shaped nut measuring about 3x 1.5 cm., pericarp gummy/ resinous grey -brown in colour. The enlarged and swollen pedicel form the cashew apple, shiny, red to yellow, juicy, soft, pear-shaped (10-20x5-8cm.). Seed / kernel has reddish brown seed coat (testa), with two large white cotyledons and a small embryo, usually kidney-shaped. The kernel without testa is cashew nut.

The cultivars of cashew are divided into two botanical varieties Americanum and Indicum on the basis of cashew nut and apple ratio and further two types on the basis of apple colour red or yellow, besides *A. giganteum* (Suriean) a type with very large apple is known.

Cashew cultivar with important horticultural feature grown in India are described as below.

BPP-1: Is of hybrid origin (Tree No 1 x Tree No 273). Cashew apple medium in size with juice recovery of 68 per cent. Nut medium sized, 5g. in weight shelling of about 28%, nearly 6-8 fruit set per panicle, perfect flowers varies about 13%. Plant at an age of about 25 years produces 15-17 kg nuts.

BPP-2: Has same parentage as BPP-1. At about 25 years of age results in 18-20 kg/tree produce, nuts are small 4 g in weight, shelling percentage about 24-26, fruit set/ panicle 7-10%, perfect flowers nearly 7% with protein content nearly 20%.

Ullal-2: A late flowering cultivar, produces 18 kg/tree nuts at an age of 25 years, harvest duration small (80-85 days). Nuts somewhat bigger, 6 g in weight, shelling nearly 30%, perfect flowers about 7%.

Vengurla-4: Is of hybrid origin (Midnapore Red x Vetore 56). Nuts are bold, about 8 g in weight shelling percentage high nearly 30-32%, cashew apple is bright red in colour, fruit set 5-6%, and bisexual flower/panicle nearly 35%.

Vengurla-1: Perfect flower percentage/panicle is about 8%. Nut weight 6 g. and shelling percentage nearly 31%, cashew apple yellow coloured. Plants are compact with profuse branches.

❑❑❑

7
Coconut

Genus : *Cocos*
Species : *nucifera* L.
Family : *Palmae*
Chromosome No. : 2n =32

"Tree of heaven, Tree of life, Mankind's greatest provider in the tropics"

The generic name Cocos is after Portuguese for monkey which refers to the end of the nut being like a monkey's head. The genus comprises of beautiful palms and grows to majestic form in its natural habitat usually unarmed trees with pinnatisect leaves. Spadices appear in leaf axis grown widely in almost all the tropical countries. Coconut is often considered as Kalpa Vriksha (Tree of Heaven) for it not only supplies drink, food and shelter but also provides raw material for number of industries.

The coconut is considered as the most important of all the cultivated palms. The plant is a typical tropical in its adaptation and performs best where 72^{0}F temperature prevails with not much variation in temperature, and with an annual rainfall of over 40 inches. Under these suitable climatic conditions coconut in cultivated commercially in large grooves. The water, milk of juice, called as toddy/todi from the flower stalks has nearly 16 per cent sucrose which is employed in production of sugar and alcohol.

Coconut's principal and commercial product is the copra which is the hardened endosperm or dried meat of the fruit and is commercially used in the production of different oils. Toddy or beverage is made from the sap which is had after making a cut in flower clusters may also be source of sugar, coir fibre from the husk of the nut used to make brushes, mats etc. The coconut is so intensively cultivated in the tropical and subtropical areas of the world that its native origin is obscure. The coconut fruit is so buoyant and skin very water proof and it is said that it can float across the world's ocean without an adverse impact on its power to germinate.

Original home of coconut is considered to be the islands near the Cocos Island (Panama), the place which was covered by coconut trees. In every likelihood through regular current the nut may have reached Indian Archipelago.

Origin and Distribution

The genus is considered to be monospecific as no true wild coconuts are known. Coconut occurs throughout the inter-tropical zone. Exact place of coconut origin is disputed. On the basis of spread of coconut both east and west ward (Central & South America), south east Asia is considered to be the main centre of coconut diversity. Palms near Atlantic are of recent introduction. The Spaniards found coconuts on the western coast of Panama and none on the either side of Atlantic at the time of Columbus. It is therefore believed that American palms have recently arrived from south east Asia via Africa, further Portuguese introduced it to Brazil and Spain to Puerto Rico in the fifteenth century.

Uses

Coconut is very useful to mankind, almost every part of Cocos is used in one or the other form. Such is the significance of coconut that Burkill (1966) described it as "One of the nature's gift to man", and in its absence human life would be intolerable and sometimes impossible. It yields nutritious health, drink, food, medicine, fibre, timber, mats, thatch, fuel etc. Coconut oil is used

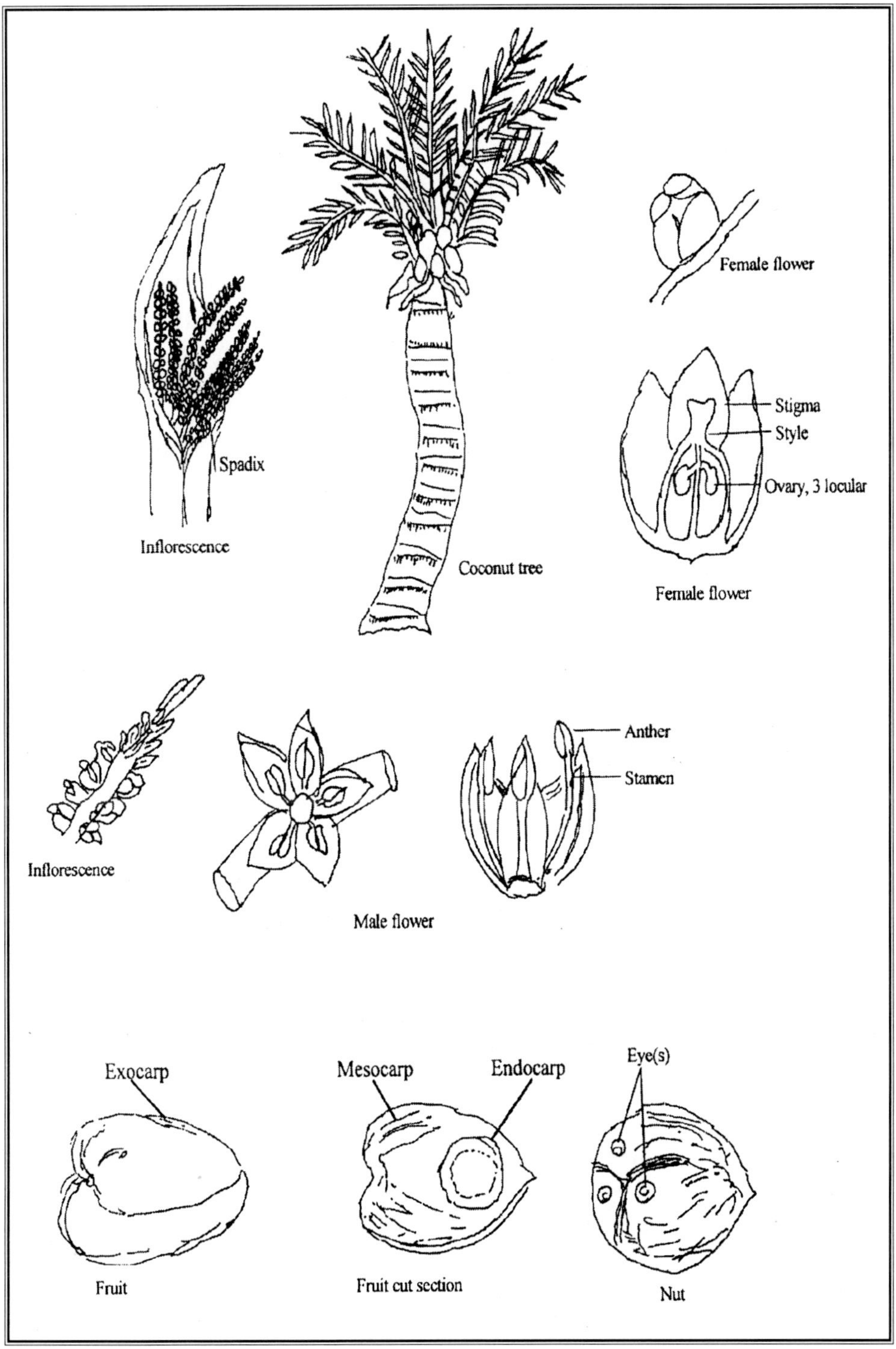
Spadix
Inflorescence
Coconut tree
Female flower
Stigma
Style
Ovary, 3 locular
Female flower
Inflorescence
Anther
Stamen
Male flower
Exocarp
Mesocarp
Endocarp
Eye(s)
Fruit
Fruit cut section
Nut

in cooking, anointing the body, soap making, lubrication and illumination. It yields vegetable oil, copra yields 60-68% oil, oil resistant to oxidative rancidity because of low levels of unsaturated fatty acids, used in resins detergents and cosmetics, copra residue serve as cattle feed. Coir is had from husk. The husk after prolonged soaking (6-12 months) in saline waters fibre is extracted by beating with wooden mallet. Quality fibre is employed to make mats, mattings, rugs, carpets and inferior for twines and ropes. Endocarp/shell used as fuel, half shells as bowls, measures, scoops, cups can be carved, polished and decorated, used to make bangles, combs, buttons, musical instruments, fillers in plastics, wood tar on distillation. Toddy from unopened flower/spathe is sweet, 15-20°B TSS, 17.5 % sucrose is drunk or boiled to make palm sugar or jaggery, used to make vinegar, fermented and distilled to make arrack. Palm cabbage from freshly cut terminal bud is relished as raw or cooked, leaves used as mats, hats, baskets, thatching, screens, walls of buildings.

Description

Plants are tall growing, unbranched with one growing point, cambium absent, trunk formed after several years when apical meristem is developed fully, trunk medium thick 20-40 cm in diameter, columnar erect, sometime leaving light grey- brown scars of leaf, peduncles persistent and very prominent on trunk, plant age determined by counting leaf scars (12-14 leaf fall/year). Leaves borne in terminal radiating crown, 25-35 opened leaves with a central bud, young leaves are entire, divided when adult, becomes compound due to splitting between the veins, the leaflets being carried apart by subsequent growth of midrib. A healthy plant has about 8-10 leaves from which fruit bunches are harvested, 10-14 which support fruit bunches at different growth stages, 10-12 opened leaves with axillary spadices at different stages of growth, arrangement is such that each gets sufficient quantum of light. Leaves are large , peripinnate, 4.5-6m long, weighs about 10-15 kg, petiole large stout, sheath clasping at above, grooved above, help to bring water to the crown, pinnae 200-250 in number 50-120 x 1.5-5cm, sessile, linear lanceolate, apex acute, cuticle thick, stomata 200 per mm square, mainly on lower surface. Inflorescence monoecious, each spadix has few female and numerous male flowers, borne singly in leaf axil. Flowering takes place when palms are 6-12 years old, flowers protected by 2 sheaths which form spathe, when fully grown measures 1-1.5m long and 15 cm in diameter. Inflorescence pressure inside tears the spathe along ventral side, inflorescence emerges and spathe falls off. Each inflorescence is 1-2 m. long, central axis with 30-40 laterals, each having 200-300 male flowers, opening from tip to downwards, has one or more female flowers at the base. Male flowers borne singly or in cluster of 2s or 3s, sessile

0.7-1.3x0.5-0.7 cm, light yellow, perianth 6 segmented in 2 whorls, outer 3 small, inner 3 splits longitudinally, pistil rudimentary. Female flower globose, large 2-3 cm across, leathery, concave, persistent, ovary large, 3 locular, globosely 3 sided, stigma 3, triangular, terminal white, each with one nectary at base. Each inflorescence may have 20-40 female flowers. Fruit is a large fibrous drupe, 20-30 cm long. The nut, endocarp and seed measures 10-15 cm in diameter, weighing 0.5-1.0kg. Fruit round or ovoid, colour and size variable. Exocarp smooth, tough, hard when young, green, yellow orange, red or brown, greyish-brown on drying. Mesocarp fibrous, light brown, 4-8cm thick. Endocarp/shell stony hard, dark brown three distinct line on the surface, 3 eyes at basal end, 1 large and soft (embryo is embedded below it) other two with hard compact fibres (aborted carpels). Seed one, appressed to endocarp, white, firm, oily, 1-2 cm thick, forms the edible parts, provide copra and oil. Embryo small 0.5-1 cm, embedded below the larger eye, large cavity inside the seed, partially filled with coconut water.

The coconut fruit has also the essential ingredients i.e. carbohydrates fat, protein, minerals like calcium, phosphorus, iron and non is in very high quantity. The fresh kernel however is rich source of fat (37%) carbohydrate (10%) protein and minerals (4% each).

In the genus *Cocos* species *nucifera* is the most important horticulturally although the number of species is very large and many of them are valued as ornamental palms. Pomological description of few species from their ornamental as well as edible fruits point of view are given below.

C. plumose. Native to Brazil. Has attractive foliage, leaves are drooping pinnate 3-15 feet long, pinnae together in bunches 1-2 feet long, round pointed dark green above shiny below, stem column-like stout and straight. Flower waxy, nut orange coloured and enclose edible pulp.

C. australis. Native to Paraguay and Argentina. A very highly ornamental palm, leaves pinnate with large number of linear spiny pinnae. Stem straight erect, columnar.

C. Ramanofflana. Native of Brazil. Highly attractive, decorative plant, leaves are long drooping and dark green in colour.

C. schizophylla. Indigenous to Brazil. Leaves pinnate dark green spreading petioles bordered with red stout spines at the edges.

Cultivars

The coconut cultivars have been grouped into two distinct types i.e. tall and dwarf. As there is no crossing barrier between the two, natural crossing has resulted in genetically heterozygous forms that differ from one another widely. Description of some established cultivars both in tall and dwarf types is given below.

Tall cultivars

West Coast Tall. Very popular Indian cultivar grown extensively along west coast, starts bearing early 6-8 years after plantation. Nuts are nearly round to oblong, initially green at maturity turns to yellowish orange to brown, produces nearly 70-80 nuts/tree/year. Meat or copra is 150-165 gm in weight and oil content nearly 70-73% of weight.

Lakshadweep Micro. Cultivar is native of Lakshdweep Island. Bears nearly 175-180 nuts/tree/year, which are small in size, has high percentage of flowers, setting is also high button sheeding low, is a distinct and desirable feature. Copra content is nearly 70-75%, oil by weight is around 75%.

Kappadam : Take nearly 10 years to start bearing. Nuts are large ellipsoid in shape though shy bearer, 60 nuts/tree/year but has very high copra content of 285 g per nut.

Macapuno : Grown extensively in Philippines tall growing, special trait of this cultivar is that in some nuts the whole endocarp is full of jelly like endosperm which is soft and is a delicacy.

Dwarf cultivars

Dwarf Green : Name suggests palm to be dwarf and green, also petioles and nuts are markedly dark green. Trunk thin, crown with nearly 20-25 small leaves, flowers early (3rd year) and each tree produces 110-120 nuts per palm annually. Nuts are small oblong beak like at distal end, copra 90 g per nut, leather like, oil 74%.

Dwarf Orange. Much like Dwarf Green except that petioles, spathes and nuts are orange in colour. Trunk thin crown compact, leaf blades are narrow. Starts bearing in third year, nuts small to medium spherical low in oil, mean copra nearly 100 g/nut better in yield to Dwarf Green.

Ganga Bondam. Palms start flowering in fourth year, semi tall type, copra of high quality though nuts are medium sized, produces nearly 65-70 nuts/plants/year, oil content is around 70%.

Coconino : Philippines cultivar, dwarf type, commences flowering in 3-4 years time, produces nearly 100nuts/plant/year. Copra content per nut is high about 260 g.

❑❑❑

8
Date Palm

Genus : *Phoenix*
Species : *dactylifera* L.
Family : *Palmae (Sub family-Phoenicoideae)*
Chromosome No. : 2n=36

Date (*Phoenix dactylifera*) Finger bearing is derived from Greek word Phoenix. This ancient plant has been named phoenix after the mythological bird which was native of Egypt and also very likely after 'Phoenicia', an old country on the coastal regions of Syria where palms were found growing in plenty. The date palms are usually planted as ornamental for their luxurious crown which is spreading at the top. The trunk and crown is covered with stiff pinnate leaves, suckers or offshoots at the base, markedly transforms the site or scene and provide an exotic charm.

To the dwellers of desert, date palm from prehistoric times has been esteemed for its economic importance, both as a source of food and shelter and is a high source of nourishment. In the Persian Gulf region the Arabs have accorded religious honours for its significant place in food supply and the intoxicating drink prepared from its sap which very often is referred as "drink of life". The best quality 'soft dates' are said to have sugar contents of about 60 per cent. A diet of date is rich in carbohydrates and lack in fats and proteins. Dates taken with milk forms an ideal diet and some Arabian tribes totally live on this combination for months together during their voyages. The dates of Al-Madinah in Arabia are supposed to be the best in the world as it was the home of the prophet Muhammad who himself was a great admirer of the fruit.

Date palm at very early times became naturalized in the northern India, northern Africa and southern Spain. In the 4th century BC when Alexander the Great invaded India, date palm was already introduced in the Indus valley. Later Muslims did the same in the 8th century AD.

The monocotyledonous date palm belongs to family Palmae (2n = 36) with about 12 species of genus *Phoenix* which are native to tropical and subtropical parts of southern Asia and Africa. Fruits are rich in sugars, minerals like iron, calcium, potassium, nicotinic acid, with relatively small contents of copper magnesium, sulphur, chlorine, vitamin A, B and B_2. Fibre, fat and pectic substances are less than two per cent.

Origin and Distribution

The progenitor and centre of origin of date palm is not very well established. Most likely place of its origin appears to be Mesopotamia where records indicate its early cultivation (5000 years). In middle East, and northern Africa the crop is extensively cultivated, high production being in Iran, Iraq and Saudi Arabia. For centuries, dates have been staple food of local people. Later invaders and missionary people introduced it to other tropical regions, due to lack of pollination, cultivars were rendered completely sterile. Presently crop is successfully grown in the tropics of southern Arabia, Sahara, and arid areas near by like Sudan etc. Absolute dry climate is very essential during fruit ripening, accompanied with high temperatures of 30°C (average) with low humidity.

Uses

Date palm has a wide range of uses, ancient records show that it has some 800 uses. All plant parts are effectively used in one or the other way.

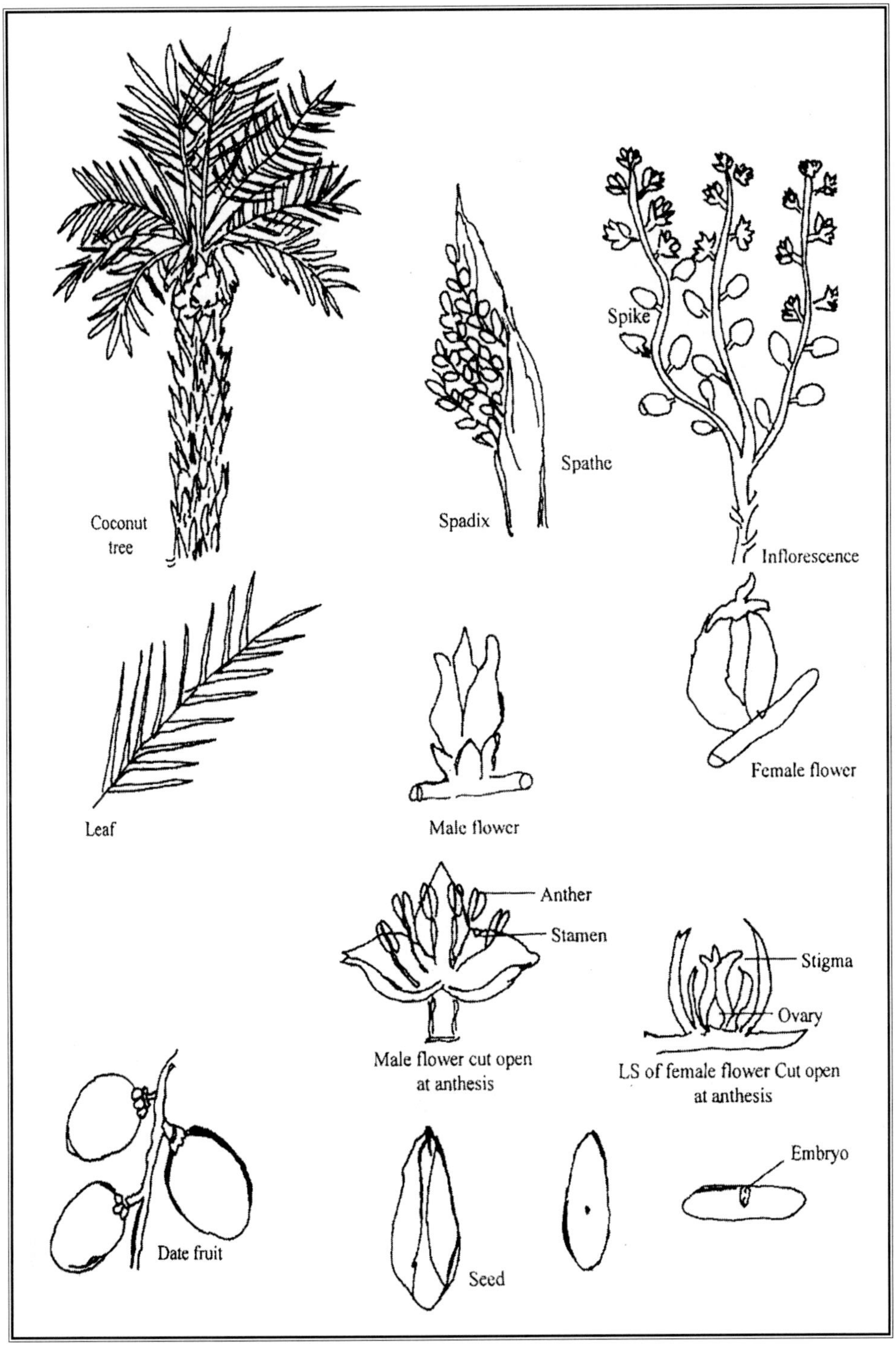
Coconut tree
Spadix
Spathe
Spike
Inflorescence
Leaf
Male flower
Female flower
Anther
Stamen
Stigma
Ovary
Male flower cut open at anthesis
LS of female flower Cut open at anthesis
Date fruit
Seed
Embryo

The sweet ripe dates are usually eaten fresh, tree trunk used as poles in building, leaves for thatching, to make mats, hats, baskets, bags, decoration articles etc., fibre is extracted, young spathes are tapped for palm wine.

Description

Plants are dioecious, tall trees (20-25m tall), unbranched, suckering at base, trunk thick, 40-60 cm across, sometime more in diameter, protected by persistent leaf bases, usually straight, terminal spreading crown with 100-120 leaves. Leaves long (4-7m), arranged spirally on the trunk, pinnate, bluish or grey green shiny leaves, petiole long, thick and rigid, hollow, base broadened. Nearly 12 new leaves produced every year. Younger leaves/pinnae shiny, long (20-40 cm), lower pinnae modified into stout spines on petiole. Inflorescence dioecious, appear in leaf axils, present in deciduous spathe, excessively branched, male inflorescence generally erect 12-15 cm long with 100-150 branches, flowers many, creamish, about 8 mm long, each with 6 stamens. Female inflorescence less branched (10-30), elongated becomes pendulous as it develops, long or globular. Both male and female spikes are protected by spathes which splits longitudinally and release the inflorescence as it matures. Perianth 6 in 2 whorls of 3 each. The outer 3 cup shaped lobes are smaller in male flower than in female flower. Stamens 3 inner sepal segments open and six stamens are exposed, filaments short with long flat anthers, centre is with rudimentary ovary. Ovary superior with 3 free carpels each has short hooked stigma. Fruit ovate-cylindrical (2.5-6x2.5 cm), pericarp fleshy yellow to reddish brown, sugar very high (60-70%) comprising mostly of sucrose, ripens in 6 months time, seed single oblong/cylindrical deeply grooved on ventral side, embryo embedded on dorsal side, endosperm very hard. In dates metaxenia, tissue outside the embryo sac are affected by pollen source whereby fruit size, quality and ripening time are significantly influenced, fruits generally harvested before fully ripe, or dried to produce dry dates.

Fruiting may take place place after 3-4 years of planting, full cropping occurs when plants are 6-8 years old. On the basis of water content in fruits, cultivars are classified as soft, semi dry or dry etc.

Description of some important Phoenix species other than *dactylifera* which are important horticulturally are briefly described.

P. sylvestris Roxb. Has single stem, the fruits are astringent and dry, usually used as a source of sugar and toddy. The crown is tapped from the side to collect the sap, without adverse effect on plant for number of years.

P. reclinata Jacq. (wild date palm). Is native of marshy Savanna areas of tropical Africa, suckers profusely. Young leaves/pinnae are used to make hats, mats, bags etc., also tapped for palm wine. Fruit resembles small dates and are eaten.

P. canarienses. Native of Canary Island, trunk stout, leaf bases adhere for many years, petioles short strong, leaf pinnate form dense crown of about 200 leaves 17-20 ft long, flower stalk about 6 ft long, much branched, fruits in clusters, ovoid globose 1 inch in diameter seeds wrinkled.

P. sylvestris (Latin of Wood of forests). Is handsome attractive large palm, fast growing and in short time attains an impressive height, leaves usually grey-green, sometimes almost bluish. Trunk straight covered with leaf bases, petiole short spiny, leaf pinnate 10-15 ft long 2.5-3 ft wide, crown heavy, leaflets 6-18 inch long, dark grey-green with whitish bloom. Flower stalk 2-3 ft long erect covered with two boat shaped spathes. Flower white, scented. Fruit small 1 – 1.25 inch long, oblong, orange yellow, astrigent, seed 2/3 inch long rounded at ends.

P. Roebelenii O' Brien. Native to Assam and Cochin China. Leaves are about 1 feet or more long, leaflets 5-7 inch long, dark green glabrous, mostly falcate and not sharp or spinous at tips. Is a dwarf palm.

P. paludosa Roxb. Native of Tropical Asia. Plants are tree like, trunk height and girth more than usual palm, leaves long 8-10 feet long, leaflets 1-2 feet long alternate and opposite with filiform tips, petioles 3-5 feet long and thin armed with many spines. Sheath fibrous, fruit purplish black.

P. pusilla Gaertn. (*P. farinifera* Roxb). Wide spread in different regions of Sri Lanka and S. India. Plant is shrub like with short trunk, 4 feet tall and 6-8 inch across, very fast growing forming very thick masses. Leaves dark green, glaucous, 3-5 feet long, leaflets quadrifarious pungent and stout, yellow cushion at base. Inflorescence 1-2 feet long, fruit about three quarter inch long, edible, black when ripe. Central portion of stem is full of starch as in other palms.

P. humilis Royle (*P. ouseleyana* and *P. pedunculata*). Native to India and central China. Plant small 6-12 feet, tufted, leaves somewhat shiny, leaflets/pinnae scattered and interruptedly fascicled, margin veins faint, peduncles elongated like that of *P. acaulis.* Fruit small ovoid ½ inch long, pulp edible. Has botanical var. Hanceana, 3-4 feet high trunk with greyish green foliage.

P. acaulis Buch. Native to India more abundant in north and central Bengal. Stemless plant, trunk is formed by bulbiform caudex, 1 feet or less in diameter that arise above ground, leaves 6 feet shiny, pinnae about 1.5 feet long with strong marginal veins, petiole stout and spiny, spadix 1 feet long, fruiting peduncles short. Fruit edible elliptic-oblong red to blue black in colour.

P. zeylanica Hort. Ceylon Date. Straight trunk attaining maximum height of 20 feet with reminents of leaf bases, leaves somewhat short with many stout pinnae 1 feet extending at right angles, bright green in colour with sharp edges. Fruit edible obovoid-oblong, ½ inch long red, later turns to blue or dark purple, sweet in taste.

P. reclinate (syn P. *pumila Hort. P. natalensis, P. spinosa, P. spinifera P. senegalensis*). Indigenous to tropical Africa and south to Natal. Produces number of stem in a clump, trunk 20 feet, but if suckers are removed plants become more tall, has prominent leaf scars, soft pubescence below, spine single or in pairs. pinnae sub opposite or alternate in 2-3, 1 feet long firm with sharp stiff tips. Staminate flowers slender and sharp. Fruit ovoid-ellipsoid brown or reddish in colour.

Cultivars

Although the number of date cultivars is around 1000, but only a few are cultivated commercially all over the world. Based upon consistency of flesh, dates are grouped as soft, semi dry and dry.

Soft Cultivars

The cane sugar is converted to invert reducing sugar (also called as Invert sugar dates) at ripening time e.g. Halawy, Khadrawy Shamran, Saidy.

Semi dry and dry cultivars. These types retain much of sucrose on full ripeness of fruits, hence called as 'cane sugar' dates. Nearly one third of the total sugar could be sucrose in cv. Deglet Noor (semi dry) and 'Thoory' (dry type).

Halaway : Early cultivar, tolerates rain more effectively than other cultivars, fruit medium-small, shape oblong apex obtuse, total soluble solids range from 27-43 per cent, low in astringency at doka stage and yellow in colour.

Khadrawy : Plants are small or dwarf palm, fruits small to medium in size, ovate or oblong in form, light yellow at doka stage and fruits mature after Halawy.

Shamran : Cultivar somewhat more tolerant to high humidity than other important cultivars, fruits relatively bigger in size oblong, oval oblong yellow at doka stage has red line at fruit base. Mid season variety.

Barhee : Fruits are small to medium in size, round ovate in form. At doka stage fruits develop yellow colour, almost free of astringency, mid to late in ripening.

Zahidi : Cultivar is known to resist, rain and high humidity, but fruits are somewhat small to medium in size, obovate in shape yellow at doka stage.

9

Annonaceous Fruits

Genus : *Annona*
Species : *cherimola Mill.*
Chromosome No. : 2n = 14, 16

Other species

A. squamosa 2n =14, 16 L.
A. reticulata 2n =14, 16 L.
A. muricata 2n =14, 16 L.

Collectively all the Annona fruits technically are syncarpous is composed of more or less numerous carpels coherent with the fleshy receptacle. Fruit has been described as 'Deliciousness itself' compared to pineapple, mangostean and cherimoya the finest fruits of the world. Its native place is the mountains of Ecuador and Peru as terra-cotta models of cherimoya fruits are frequently obtained from

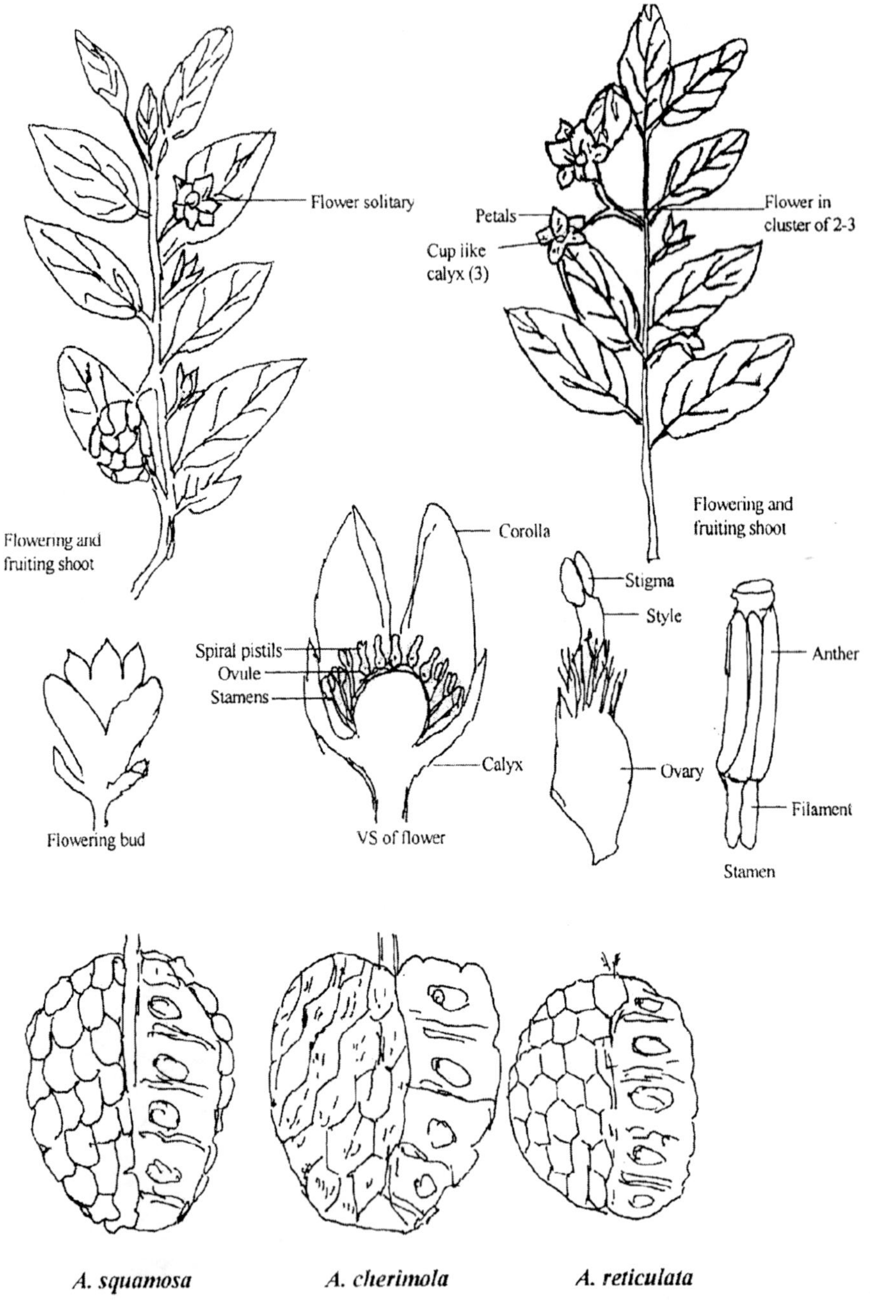
Flower solitary
Petals
Cup like calyx (3)
Flower in cluster of 2-3
Flowering and fruiting shoot
Corolla
Flowering and fruiting shoot
Stigma
Style
Spiral pistils
Ovule
Stamens
Anther
Calyx
Ovary
Filament
Flowering bud
VS of flower
Stamen
A. squamosa
A. cherimola
A. reticulata

FRUITS

prehistoric graves in Peru. The word cherimoya is from the Peruvian name 'cherimuya' meaning cold seed. Although originated close to the equator the fruit attains best quality or performs better where climate is cold and relatively dry indicates it to be a sub tropical fruits. The family which has these fine tropical fruits also has papaw or pawpaw (*Asimina triloba* (L) Dum.) which is the only temperate climate fruit of Annonaceae. The genus has some eight to ten species all produce edible fruits. The papaw fruit taste resembles that of sweet avocado. On account of its bearing in cluster and fruit shape like that of papaya it is also called as Indian banana.

Origin and Distribution

The cherimoya is native of tropical highlands (Andeans) of Peru and Ecuador. They are of hybrid origin. From its habitat in South America the cherimoya spread northward into Mexico, later into West Indies southern parts of south America. From here to some islands near the African coast to Mediterranean region, later to India and Polynesia. The crop is naturalized in many parts of Mexico and Central America and occurs in abundance at an elevation ranging between 3000-6000 feet and in Guatemala upto 8000 feet.

Uses

Fruits usually are eaten fresh, the sweet fragrant pulp is used to flavour milk or to make delicious drink. Puree is consumed after squeezing the pulp through a sieve. It can be made into fruit jelly, juice nectar or syrup. In Indonesia sweet cake is made by boiling the pulp and adding sugar till the mixture hardens. Young fruits with soft seeds are used as vegetables. Mature firm fruits are made into candies. The seeds and mature fruits also have insecticidal and vermicidal properties.

Description

The plants of cherimoya are tree, medium in height (3-10 m). Leaves simple, dark green, shinning above, ovate to ovate lanceolate , or may be obovate or elliptical, large(20 cm x 8 cm) in size, dense/tomentose, brownish velvety pubescence underneath, petiole small. Flowers appear in leaf axils at the base of branchlets, generally single or solitary, fragrant, about an inch across but sometimes in small clusters of 2-3 on short dense pubescent peduncles on current season growth, hermaphrodite, yellowish, protogynous, regular, pedicelled with small single bract triangular and pubscent. Petals 6, the 3 outer petals are pale yellow with purple spots inside at the base, oblong linear in shape, inner 3 petals are reddish to purplish, very small, stamens numerous, fleshy and free, arranged, spirally at the base of the conical receptacle. Gynoecium has many free pistils on the upper part of the receptacle each having single ovule, stigmas are sessile and glued together. Fruit is a syncarpous (pseudocarp) formed by carpel fusion and receptacle forms the fleshy mass. The fruit shape and appearance variable, round to ovoid, or typical depression at stalk end forming heart shaped fruit with projection (tubercules) or rather smooth surface with U shaped areoles, measuring about 10-15 cm in diameter. Edible part aril / pulp is white melting free from seeds. Seeds obovate, obliquely truncate, slightly compressed, testa thin, wrinkled and membranous. Flavour resembles that of banana and pineapple sub acidic in taste.

A. muricata L. Sour sop. : Native of tropical America and is one of the first fruit trees introduced to the old world. The Spanish further introduced it to the Philippines. Presently it is being cultivated successfully in most of the tropical regions of the world. Plants are generally shrub or small tree, may attain a height of about 10 meters, low branching. Leaves large, simple, dark green, shiny, oblong obovate, with short pointed apex (acuminate). Inflorescence short in 1-2 flowered clusters, flower regular, greenish yellow in colour, pedicel long (2.5 cm), calyx 3 in number, persistent, small(4 mm) triangular, corolla 6, the three outer are broad ovate(4-5 x 2-5 cm) inner 3 petals slightly smaller (4 x 3.5 cm), short clawed at base, stamens many and are placed on a raised torus filaments tomentose, ovaries many, heavily pubescent. Fruit is largest among Annonas, pseudocarp (30-35 x 15-20 cm), dark green broad ovoid or ellipsoid covered with spines (5-6 mm long), pulp white, juicy and fleshy, seeds numerous, shiny blackish brown, obovoid measuring about 2 x 1 cm in size.

In India the fruit is called as sitaphal (fruit of Sita) or sharifa meaning noble and predominantly is a dessert fruit. The fruit is believed to have an Asiatic origin as trees are found growing wild in many part of India, its name occurs in Sanskirt writings, further carvings and wall paintings from the ancient ruins of Ajanta all suggest Indian origin.

Sugar apple (*A. squamosa* L.) plants are small may attain height of 15-20 ft, heavy bearer. Leaves are pale green on both surfaces, shiny, petiole slightly pubescent, lanceolate or oblong lanceolate, base narrow (acute), apex also narrow or with short pointed tip, flowers occur in leaf axils on current growth singly or in clusters of 2 or 4. Individual flower small measuring about one inch across, greenish yellow, the outer 3 petals are thick, oblong tips rounded, the inner 3 ovate and small. Fruit 2-3 inch in diameter, round ovate or conical, surface tuberculate with significant white bloom. The flesh or pulp custard like white, sweet slightly acidic. The carpels are loosely or not at all fused with each other contain black seeds.

Sugar Apple (*A. squamosa* L.) has tropical climatic requirements and resembles that of soursop and bullock's heart but differs from cherimoya which prefer subtropical conditions. As regards production precocity and fruit quality, this tropical annona is rated to be the best. In India and else where is considered as an important fruit and is widely distributed throughout the tropics. It loves hot and relatively dry climates that prevail mostly in the low lying interior plains of tropical regions hence withstands drought better than compared to many other fruit crops.

Bullock's Heart (*Annona reticulata* L.) the Bullock's Heart in India *A. reticulata* is called ramphal (fruit of Lord Rama) widely grown fruit but of low value as it lacks flavour and quality compared to that of cherimoya and sugar apple. Tree semi deciduous, 20-25 feet tall. The leaves are shiny oblong lanceolate to lanceolate. Flowers borne in small cluster on new branchlets, floral morphology similar to other annonas. Fruits are typically heart shaped, large sized surface smooth usually reddish brown or reddish yellow when ripe with rhomboidal or hexagonal areoles, flesh white with numerous brown seeds, granular, skin thin sweet in taste.

Annona diversifolia (Saffod). The Ilama was identified by the end of 16th century is considered to be one of the finest Annona fruit. The climatic conditions are similar to those of sugar and custard apple and prefer hot climate. Trees are slender starts branching close of the ground, trunk not thick, foliage resembles that of *A. squomosa*. Leaves broadly elliptic to oblanceolate (narrow at base rounded at apex) 4-6 inches long. A distinct rounded/orbicular leaf like bract present at the base of smaller branchlets is a distinguishing character of this specie. Flower small 1 inch across maroon coloured. Fruit round conical or oval with thick bloom, surface may be smooth or rought depending on the carpellary wall wheather raised or suppressed. Fruit colour may be green or pink, in green varieties the flesh is white and pink tinged in pink types. Seeds are large and numerous, flavour is sweet.

A. glabra L. Pond apple, mainly used as rootstock for other annonas and withstand some frost, loves swampy condition, grows vigorously 45 feet or more tall. Flowers are large, outer petals cream coloured inner white outside and blood red from inside. Fruit surface smooth yellow skin leathery, initially green later turns yellowish, surface with distinct areoles, flesh also yellow or creamish when ripe, not edible.

A. longiflora Wats. Wild cherimoya of Jalisco. Plant is a shrub or a small tree varying in height from 3-10 feet. Branches like that of cherimola i.e. young growth pubescent, leaf scar prominent. Leaf broad elliptical to obovate-elliptical, generally rounded but may be acute at base and rounded at apex, velvety on the undersurface. Flowers are with short peduncles pubescent on outer side with fine soft hairs, outer petals leathery linear-oblong or oblong lanceolate, creamish or white with dark purple or blackish spots at the base, inner petals very small or missing. Ovaries with rufous hairs and styles finely puberulent. Fruit globose-ovate surface with reticulate areoles or protuberances, pulp white, flavour similar to that of cherimola.

A. montana Macfadyan. Mountain Soursop. This specie is more hardy than *A. muricata.* Tall tree 45 feet or more. Leaves like that of soursop, shiny above and pubescent below. Flowers occur either singly or in pairs. Fruit broad ovoid or spheroid, size variable, from a size of an ordinary orange may be upto 6 inch in diameter. Initially green later turns to yellowish. Skin shiny with many erect short spines, pulp not edible.

A. purpurea Mocino and Sesse, Negro-Head. Tree small to medium in size. The young twigs/branches are with dense reddish pubescence, but becomes shiny later. Leaves usually large, somewhat leathery, surface uneven oblong-elliptical to oblong-obovate in form, apex acuminate, round or connate at base. Petioles thick and short. Flowers sessile, single enclosed in an involucre, calyx 3 lobed, velvety on outside, 3 other petals very thick, velvety on outside and purple within. Inner petals thin and overlapping and forms a dome-like cover over essential parts, white outside purple within. Stamens velvety, carpels distinct earlier, ovaries, crowded by fine hairs. Fruit globose-ovate with sharp pointed protuberances.

Cultivars

No standard recommended cultivar of Annona is grown in India, however in cherimoya some botanical forms have been obtained through careful selection varying in fruit size, seed number and flavour which are described briefly as

1 *Smooth cherimoya :* mostly cultivated in Mexico, fruit skin smooth and is considered as one of the finest cherimoya cultivar

2 *Finger printed cherimoya :* has very less number of seeds, quality excellent sweet in taste, pulp juicy with aromatic flavour.

3 *Tuberculate cherimoya :* the projections on the fruit surface are round pointed, a very common Annona fruit of Peru.

4 *Mamomillate cherimoya :* Such type of cherimoya are common in the Nilgiri hills of India, the projection of mammillae is truncate type.

Cultivars

Cherimoya

Ryerson : Originated in California parentage unknown, fruit regular in shape, skin smooth thick and tough, transports well.

African Pride : Originated in Deepdale Natal, S. Africa. Appears to be a hybrid between *Annona squamosa x A. cherimola,* fruit is large, dull green in colour, smooth to medium rough, quality good, seeds are few. Tree usually small in size, bears two crops in Florida, first crop in Sept. - Oct. and second in Mar. - April.

A. Squamosa Cultivars

In India red strain of A. squamosa var. Sangareddy is grown. Tree small, leaf midrib, flowers and leaves, petals, fruit furrows and pulp all are pink tinged. Poor yielder quality not very good.

Mammoth : Tree dwarf type, branches are drooping and profuse, breeds true to type from seed. Leaves are narrow, light green, produces heavily. Fruit light russetted green in color, pulp white, very sweet, juicy, seeds 20-40, keeping quality good.

Balanagar : Tree vigorous and large, leaves dark green, thick and large medium cropper, seedy (40-80/ fruit), fruit medium-large, areoles large and rough, tuberculate, furrows are deep with cream-yellow margins, skin greenish, pulp white, somewhat leathery adhering to seed, juicy, flavour excellent.

Booth : Discovered in California, parentage unknown. Tree bears quite heavily in some places, skin somewhat smooth, ripens late in season, quality good.

Whaley : Originated in California, parentage unknown, Fruits are large, light green, quality good has tendency to develop a membrane around the seed.

Carter : Originated in California, parentage unknown. Tree productive bears heavily sets well even without pollination, resembles 'Ryerson'. Fruit long conical, smooth ripens in June. Keeps well, transports well.

❑❑❑

10
Guava

Genus : *Psidium*
Species : *guajava* L.
Family : *Myrtaceae*
Chromosome No. : 2n = 22, 33

Guava (*Psidium guajava*) for its high adaptability to varied soil and climatic conditions along with plants hardy nature, in short period of time has naturalized so much to the Indian condition effectively that today it is a common backyard fruit in almost every house. The round topped plant with wide crotched branch and light green foliage has an added advantage as landscape plant. From horticulture perspective it is one of the most common fruits grown commercially in India and is ranked next to mango, banana and citrus in respect of area and production. Guava

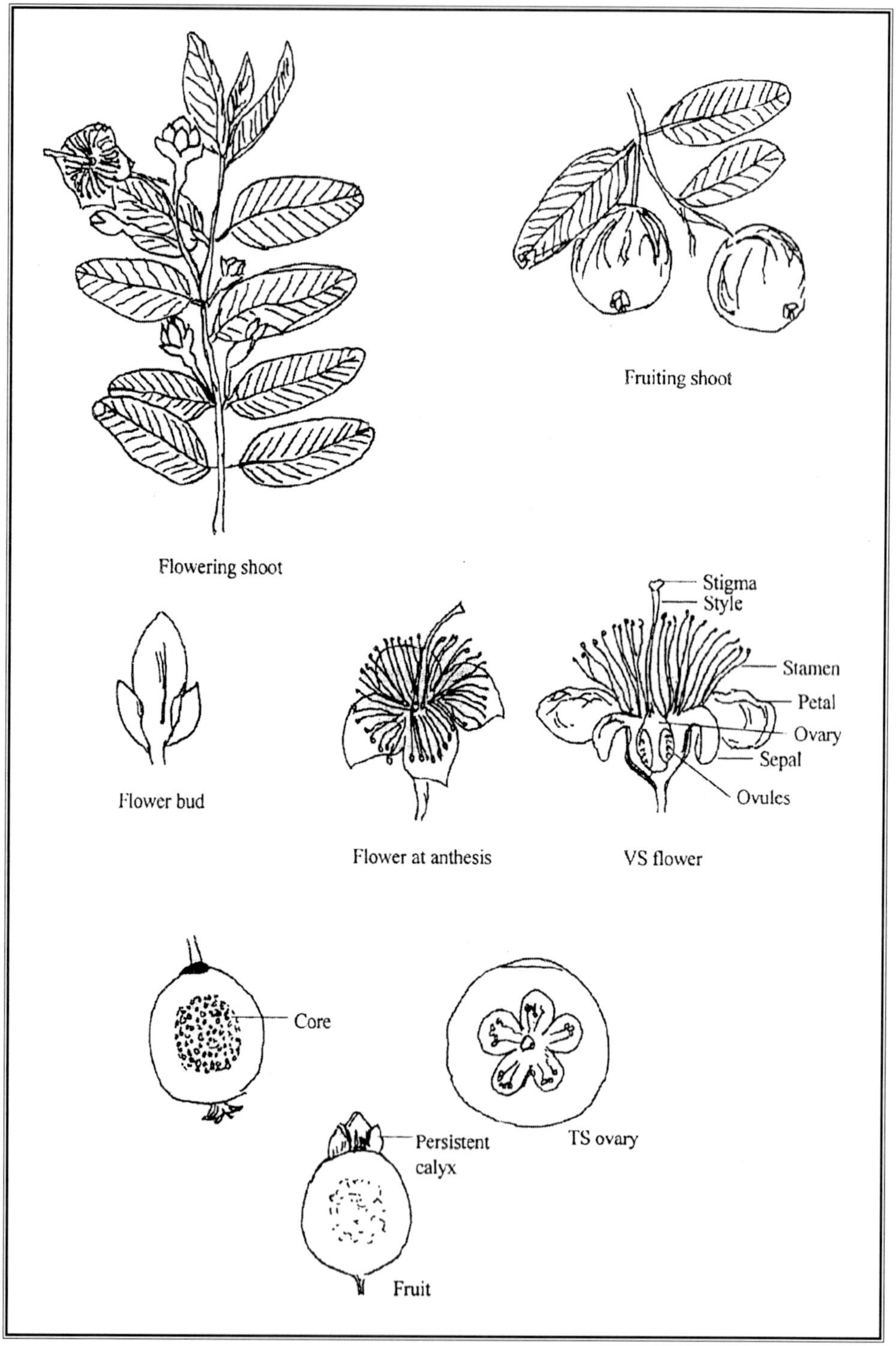
Flowering shoot
Fruiting shoot
Flower bud
Flower at anthesis
Stigma
Style
Stamen
Petal
Ovary
Sepal
Ovules
VS flower
Core
Persistent calyx
Fruit
TS ovary

fruit is often referred to as apple of tropics probably as it is the only tropical fruit which matches the high nutritive values of more commercially important temperate fruit apple. In the present tropical fruit culture scenario guava is considered to be the precocious, most prolific and highly paying fruit crop and the horticulturist admire this specie as it is least demanding in its cultural requirements.

Guava besides being a wholesome fruit is a store house of pectin, minerals (Ca P and Fe) carbohydrate, fibre, riboflavin, vitamin B and vitamin C in particular which is ranked third in fruit after Barbadose cherry and aonla. Further powder of dehydrated guava is a good source of vit. C and its content are almost 2 to 5 times more than that present in the juice of fresh oranges.

The family Myrtaceae comprises of a number of unique very diversified and interesting fruits for their aromatic fragrance present in all plant parts, i.e. leaf, bark, flower and fruit and also on account of numerous long stamens which the individual flower bears. Four fruits namely guava, jamun, pineapple guava and clove are important horticulturally.

Origin and Distribution

Guava has its origin in the tropical America more so in the area extending between Mexico and Peru. The Spanish invaders took it to Philippines and Portuguese in their various voyages introduced it in India from the West. On account of its wide adaptability this fragrant crop has naturalized all over the tropics and subtropics.

Uses

The flavoured fruit best are eaten as raw both when ripe and unripe, potential source of vitamin C and pectin, pulp (without seed) is made into jam, jellies, nectar, juice, pies, cakes, stewed and preserves, eaten as sweet meat, halves can be canned. Guava jelly of firm consistency and musky flavour is a highly priced commodity, leaves are cooked to impart flavour, used for tanning and dyeing purposes, wood strong and is employed for carpentry purposes.

Description

Guava plant is a shrub or a small spreading tree, may attain height of 10 m, branches more or less erect generally appear from the base of the trunk. Trunk quite thin rarely and often multicoloured bark, Bark green to reddish brown (when young), bright and smooth, later dry dead flakes (bark) fall off its own. Young shoots pubescent, ridged and 4 angled. Leaves simple, almost sessile, opposite, light green, elliptical to oblong, glabrous above, soft pubescence below, veins compressed above, raised and prominent below shortly stalked, nearly sessile, petioles short thick (3-10 mm) slightly grooved. Flowers appear in leaf axil of current growth, solitary or in clusters of 2-3 flowered cymes, measures 3 cm in diameter attractive and hermaphrodite. Calyx lobes persistent irregular usually 4-6, green pubescent (1-1.5 cm long) calyx tube completely encloses the flower end, petals 4-5, obovate, white (2cm long), stamens numerous with long filament (1-2 cm long) and short rounded anthers, usually present in groups around the top of the ovary. Pistil, ovary inferior with 4 united carples, each containing numerous ovules in axile palcentation. The style extends beyond the stamens, stigma capitate or broom like. Fruit is a berry, round / ovoid, globose or pyriform with persistent calyx at the stylar end, skin green yellow or red, smooth densely dotted and glabrous, surface bumpy, flesh white yellow, pink or red, core with numerous stone cells, soft to very hard, yellowish, taste with typical flavour sour sweet and aromatic. Seed small light yellow or yellowish brown which makes the central core.

Other important related species are.

P. cattelianum, strawberry guava, fruit is small, flavour aromatic resembles that of strawberry, fruits are smaller.

P. guineense, Brazilian guava.

P. fredrichsthalianum, Costa Rican guava

The genus *Psidium* has some 150 species, and those important horticulturally are described:

P. guineense Swartz. (Brazilian/Guinea guava). The specie is characterized by small fruits usually greenish yellow, hard when ripe, flesh white and contain numerous small seeds, flavour sub acidic, quality poor. Young branchlets finely pubescent and compressed, cylindrical.

P. montanum Swartz. (Mountain guava). Plants are shrub-like, tree top rounded, branchlets 4 angled, leaves oblong, oval, apex acuminate, spiny, peduncles many flowered fruit round.

P. friedrichsthalianum Niedenzu. Costa Rican Guava. Medium tall plant 25-35 feet, branches thin and slender smooth bark brown to dark brown. Young branchlets of same colour and pubescence. Leaves smooth deep green shiny above pubescent below, mid rib prominent below oval or oblong oval, apex pointed, base narrow. Flowers appear axillary on young branchlets, solitary, petals 5, ovary 5 loculed. Fruit globose small and sour in taste.

P. molle Bertol. Sour Guisario, Guayaba Acida. The specie is wide spread in Mexico and Central America. Small tree or a shrub, fruits heavily, growth slender, under surface of leaves, young branchlets and fruit stalk with reddish velvety pubescence. Leaves oval oblong, apex rounded base narrow pointed or rounded, light green with slight pubescence above. Flowers occur in clusters of threes in leaf axils, ovary 4 locular, fruit pale yellow when ripe, globose, pulp white with numerous seeds, acidic in taste.

P. araca Raddi. Brazilian Guava Araca Do Campo. Found in plenty in uplands of Brazil. Usually a large shrub, young growth highly marked with hairy growth, leaves large oblong-oval with soft velvety pubescence above and dense pubescence below. Veins conspicuous and reticulate. Flower axillary and in clusters of 1-3. Fruit ovoid or oblong yellow in colour, sweet when ripe.

Formerly the common guava on the base of fruit shape was classified as

P. pyriferum - Guava with pear shape and are called as pear guava

P. pomiferum - fruits are round shaped-apple guava.

Other species which require horticulture attention are *P. molle, P. pumilium P. chinensis, P. araca P. polycarpum P. cujavillais P. corecium* etc.

P. cattleianum Sabine Strawberry guava, also called as cattleye guava. Has flavour that of strawberry hence the common name. Usually a shrub or a small tree about 20 feet tall, bark not squarish smooth greenish brown, short shoots shiny, leaves obovate-elliptic 2-3 inch long dark green thick, shiny, leathery. Flowers occur axillary usually solitary, calyx tube turbinate with 4-5 broad oblong calyx lobes, corolla thin and obovate, style slender Ovary 4 chambered. Fruit round or obovate purple red in colour, 1-2 inch across, skin thin, flesh soft, core white with numerous hard seeds. Has one important botanical variety *P. cattleianum* var. lucidum, Hort., fruit are sulphur yellow in colour has delicate flavour, tree large in size.

Cultivars

Guava has large number of well defined cultivars which show variation in tree form, fruit shape, size, colour, flesh colour and seedless and seeded forms etc. Salient features of some important guava cultivars are detailed below.

Lucknow-49: Also popular as Sardar Guava, is a selection which was made at Pune. Plants are vigorous in growth, tree top rounded produce branches profusely, leaves large elliptic-ovate to oblong, light green, fruit round or short globose, skin primose yellowish with few red specks, flesh white, seed core small, sweet in taste with excellent storage quality.

Allahabad Safeda : The cultivar resembles L-49 and commercially grown in north India. Trees are vigorous, medium in height, branches profusely, shoots are long, tree top/crown compact and rounded. Foliage dense, leaves light green long elliptic oblong, fruit round medium in size, skin yellowish, flesh white, keeps well for long time.

Banarsi : Keeping quality though not very good but is one amongst few very sweet guavas that are grown commercially in India. Tree somewhat tall growing, branches spreading, crown broad, leaves light green elliptic-oblong with short pointed apex, base rounded. Fruit roundish yellow in colour, flesh white.

Chittidar : Spotted guava. Fruit subglobose light yellow/straw yellow, surface smooth with conspicuous scattered red dots on the surface. Sweet in taste with good keeping quality. Tall growing tree with spreading branches, crown rounded. Leaves light green pubescent underneath, elliptic-ovate to oblong in form, leaf base obtuse, apex shortly pointed.

Seedless : This cultivar is known for its excellent keeping quality, withstands transportation. Tree tall growing with long trunk, branches upright, leaves light green, slightly pubescent below, elliptic oblong. Fruit globose or short oblong small sized, skin smooth, shiny straw yellow in colour, flesh white and creamish on ripening, core small flesh thick, seeds few but bearing is poor

Red Fleshed : The flesh colour is pink sometimes red in colour-hence the name. Tree tall vigorous with spreading branches, crown open, leaves elliptic-oblong light green pubescent below. Fruit round-ovate yellow, smooth glossy, surface with conspicuous red scattered dots. Quality wise inferior to yellow green guavas. Other important red flesh cultivars are Super Acid, Hybrid Red Supreme, Africa Florida, Kothrud, Patiollo but all are low in vit C content and poor in quality to white fleshed types.

❑❑❑

11 Jamun

Genus : *Syzygium*
Species : *cumini* (L.) Skeels
Family : *Myrtaceae*
Chromosome No. : 2n = 40

Syzygium cumini Jamun, Genus name *Syzygium* in from Greek word united, referring to the calyptra petals and is an important minor fruit of considerable commercial significance besides being tall growing with evergreen handsome foliage is often used as road side avenue plantation to provide shade. As with other genera this fruit crop is indigenous to the East Indies, is cultivated for its fruit which is refreshing when taken with salt in scorching summer heat in India. The plants are small evergreen with attractive glossy light green foliage found growing along river sides and also planted in avenues.

Flowering shoot

Fruiting shoot

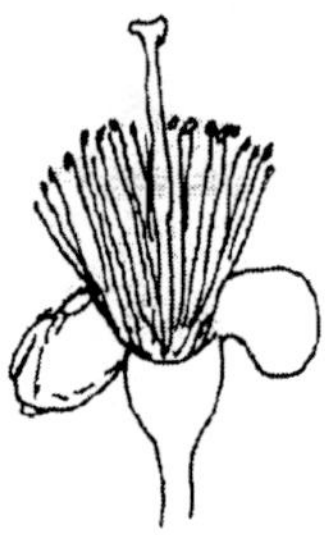

Flower

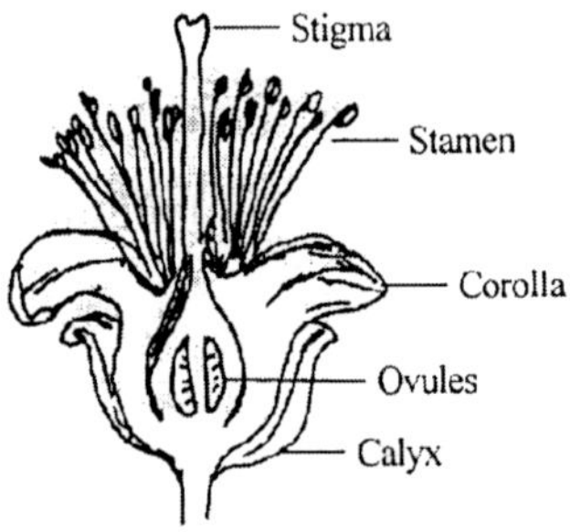

VS of flower

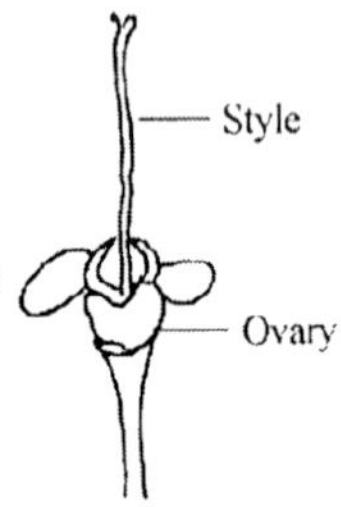

VS flower stamens and petals removal

Fruit

VS of fruit

Seed

Jamun is among the few fruit crops which is native to India or the East Indies and with the passage of time it has been introduced successfully in number of tropical and subtropical countries of the world. Being hardy in nature it grows successfully without much care in the marginal marshy and neglected areas. The maximum scattered Jamun trees in India are concentrated mostly in the tropical and subtropical region extending from Indo Gangetic plains in north to Tamil Nadu in south, Gujarat in west to Bihar in east. Also occurs in the foothills of Himalayas.

Jamun is rich in its nutritive value and is known particularly for its high iron source apart from normal contents of sugars, protein, carbohydrate and mineral constituents The jamun fruit with sub acid taste and spicy flavour is mostly eaten fresh.

The genus *Syzygium* has been separated from genus *Eugenia* on the basis of that inflorescence is generally paniculate-cymose, cotyledons of the embryo are free, the seed coat adhere loosely or tightly to the pericarp.

Origin and Distribution

Jamun or Jambolan is among the few crops which are credited to have their origin in India. Jambolan is native to the subtropical Himalayan region including India, Malaysia, Sri Lanka and Australia. It is a common tree in the south east Asia and now is grown all over the tropic and sub tropical parts of the world.

Uses

The ripe jambolan fruit which are astringent with sub acidic taste are eaten fresh. In India and other south east Asian countries the fruits are salted to lessenthe astringency before eating. Juice ,jam and wine can also be made. Plant is hardy and spreading, therefore good for wind break purposes. Seed has medicinal properties against dysentery, diarrhoea and other ailments, quality honey from flowers, gargle are made from tree bark.

Description

Plant are vigorous tall trees (30 m), branches low on trunk, spreading, tree top globose or irregular. Bark usually dark grey, rough on basal part of tree, light grey and smooth upwards. Leaves simple dark green, thick, coriaceous, opposite broadly obovate elliptic to elliptic –oblong, round or cuneate at base, apex tapering or blunt, petiole long (1-4 cm) , margin entire, pinkish when young. Leaves when crushed produce turpentine flavour. Flowers in panicles, borne singly or in dense clusters (12 cm long) usually on leafless branches. Individual flower is with a yellow disc, fragrant, small, sepal broadly companulate with irregular dentate above, petals 4 greyish white to pink free, round (orbicular), stamens many, white about 5 mm long, style longer than stamens (6-7 mm) white, ovary 2-3 celled. Fruit is a berry, dark violet, round /ovoid to oblong, very often with distinct curvature, calyx persistent, in clusters of about 35-40, pulp grey yellow to violet, juicy, astringent, seeds 0-5, oblong, green to brown in colour.

Syzygium comprises of about 1000 species most of the species produce edible fruits some are known for their medicinal and ornamental value.

Apart from Jambolan, Java Plum, Jambu (S. cumini(L). skeels). others important syzygium species along with their important characters are summarized below.

S. fruiticosum. Is a wild specie, fruits are edible small dark purple stringent even when ripe, seed prominently elongated. Tree is large evergreen quick growing sturdy, hence very useful as wind break.

S. jambos, Alston (Rose apple or gulab jamun or Malabar Plum). Is native of East Indies and from here was introduced to all parts of tropics. Plants are of high ornamental value and fruit being rich in pectin content is used in making of preserves, jellies and confectionery products. Trees are tall (25-30 ft), leaves are thick shiny/glossy 6-8 inches long, oblong- lanceolate coriaceous apex acuminate, flowers occurs in short terminal racemes on young branchlets which are greenish white. Flowers showy in few flowered clusters 2-3 inch across, stamens very long and numerous and hide other floral parts. Calyx tube turbinate, corolla 4, obovate in form. Fruit small ovoid 1-2 inch long with persistent calyx at apex, whitish green to light yellow, flavour like that of rose, appealing, flesh juicy crisp, sweet, seed solitary present loosely in the cavity.

S. aqueum (Burm. F.). Alston. Indigenous to Malay peninsula and Borneo. (Water Rose Apple). Trees are medium tall about 30 feet, leaves ovate 6 inch long or more, coriaceous and cordate, flowers appear in terminal or axillary 3-7 flowered clusters, flower variable in colour white, red or light purple. Fruit with number of seeds, berry white or red, 1 inch across.

S. oblatum (Roxb). Found in North India, Burma and China. Flowers in terminal and axillary panicles, somewhat large in size, petals united to form a cap like structure. Fruit small, of the size of cherry, berry globose. Plant is a tree, leaves broadly lanceolate or oblong, veins finely divided.

S. malaccense (L). Merrill L.M. Perry. Native of Malaya Archipelago. Malay Apple, Rose apple. Large fruited rose apple. Regarded to be most beautiful tropical tree, grown as an ornamental as well as for fruits which are eaten raw, cooked, as wine or in preserves. Flowers occur on previous season's growth, showy in few flowered cymes, purple red in colour. Fruit nearly 2 inch long, pyriform, seed brown and large. Tree tall, about 40 feet, leaves dark green very shiny coriaceous ovate oblong 6-12 inch long.

Cultivars

Despite variability jamun has no standard or named cultivars. Number of seedling strains showing variation in fruit shape (oval oblong) colour (purple deep purple, bluish black) size(medium, big) and other characters like earliness, sweetness, pulp, pulp to stone ratio, juiciness may provide a strong base for selection of better types. The most popular and commercial cultivar in north India is a selection named 'Ra Jamun'. The fruits are dark purple at maturity, attains bluish black colour, big sized, pulp purple pink, sweet and juicy with sub acid spicy flavour, stone small. Fruits ripen in summer (June-July), gives refreshing feel when taken with salt.

□□□

12
Pineapple Guava

Genus : *Feijoa*
Species : *sellowiana Berg.*
Family : *Myrtaceae*
Chromosome No. : 2n = unknown

Origin

Feijoa is native of subtropical South America, widely distributed in western Paraguay, parts of Argentina southern Brazil and Uruguay.

Uses

All parts of the plants are edible, the fruit has great resemblance to guava in respect of its appearance and size and other important pomological traits. Usually consumed as dessert or cooked, petals being thick are used in salad, on accounts of its high fruit flavour is also made into jams, jellies etc., besides these fruit juice is used to add aroma to non-alcoholic beverages.

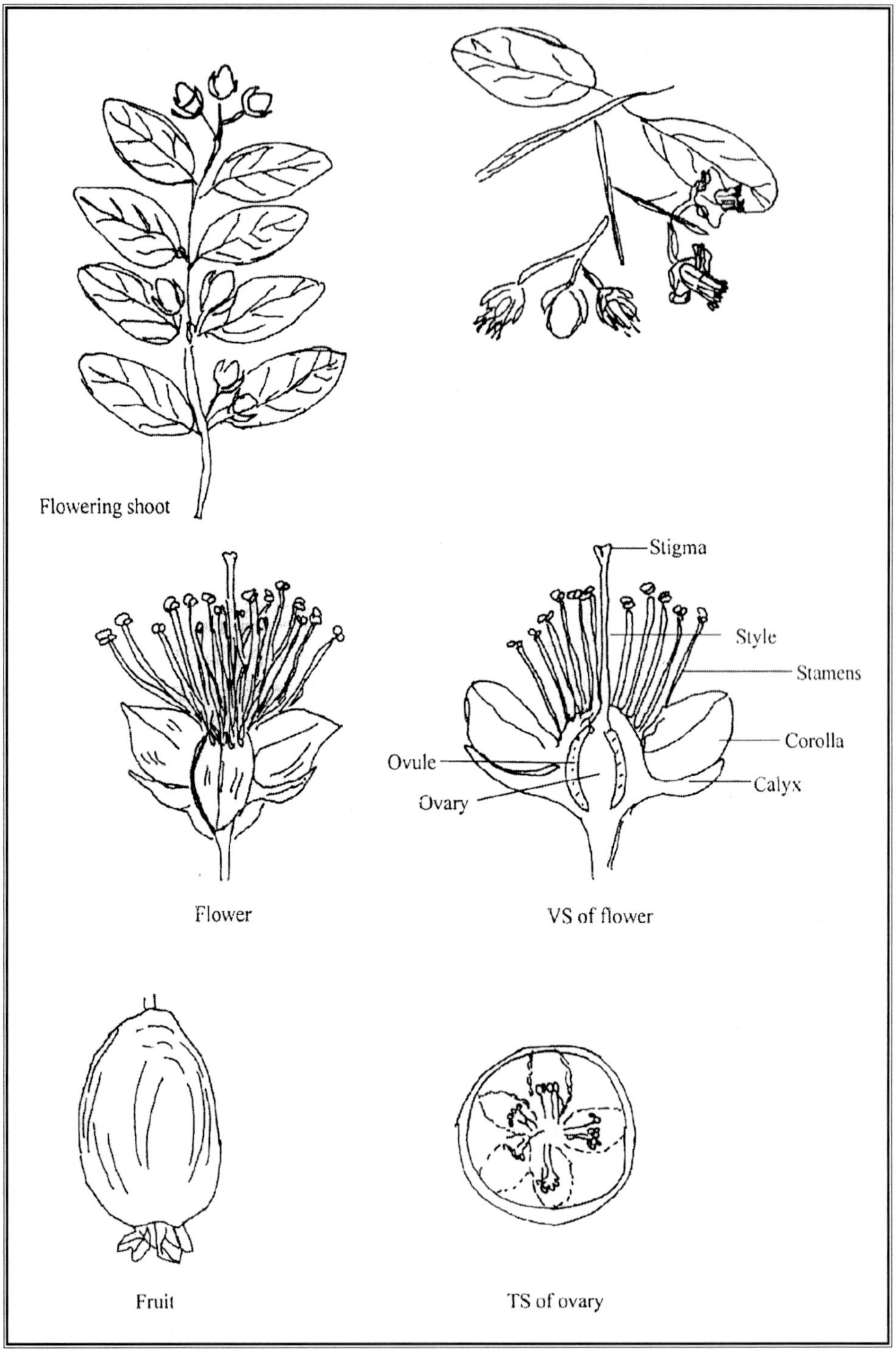
Flowering shoot
Stigma
Style
Stamens
Corolla
Ovule
Calyx
Ovary
Flower
VS of flower
Fruit
TS of ovary

Description

Feijoa O. Breq. Has two species *F. coolidgei,* which has no botanical standing and the other species *F. sellowiana* is planted for its edible fruit and also has ornamental value is regarded to be hardier than many of subtropical fruits. At temperature of 15^0C and even less it has withstood injury caused from low temperature. The crop prefers dry climate which is free from very high temperature. The *feijoa* is known for their small sized seeds, plenty of flesh and its delicious flavour suggestive of strawberry and pineapple but in a proper ripened fruit the aroma is very delightful.

Plants are usually shrub or small tree (15-20ft), leaves shiny ,oval oblong in shape, green above and silvery grey beneath (highly pubescent tomentose). Early growth scaly. Flowering occurs in leaf axil and terminally on current growth. Flower large 1-1 ½ inches across, calyx thick 4 green pubescent, petals 4, white outside and purplish crimson within. Fruit, shape variable oval oblong or round measuring about 2-3 inches in length with persistent calyx lobes (as in Guava) usually dull green is color, occasionally with red tinge whitish pubescent / bloom persistent. Fruit flesh whitish distinctly granular with very small 20-30 seeds surrounded by some what jelly like sub acidic pulp, taste pleasant with distinct delightful flavour resembling that of strawberry and pine apple.

❏❏❏

13
Clove

Genus : *Syzygium*
Species : *aromaticum*
Family : *Myrtaceae*
Chromosome No. : 2n = unknown

Origin and Distribution

The wild species are found growing wild in New Guinea and its cultivation was found on some islands of Moluccas. By the 17th century its cultivation has spread to number of other islands and from Moluccas crop reached Asia. The Britishers further introduced it to in India, Sumatra, Sri Lanka and other countries of south east Asia. Through French, the crop reached Mauritius which further spread to Madagascar, Zanzibar (Africa) and finally reached tropical America (Brazil).

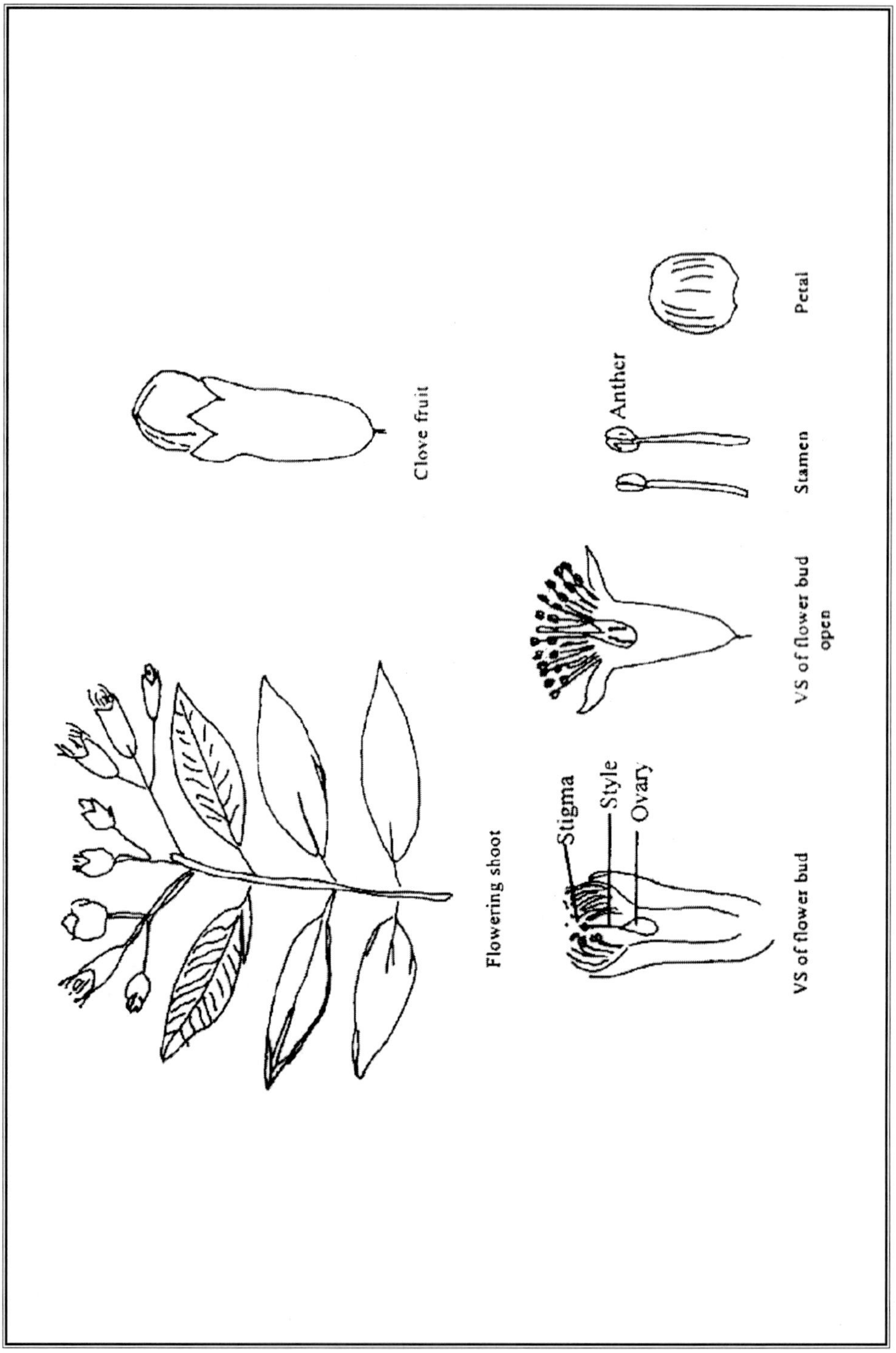
Clove fruit
Anther
Stamen
Petal
VS of flower bud open
Flowering shoot
Stigma
Style
Ovary
VS of flower bud

Uses

Chinese valued clove as a spice. In India, regarded as an important spice, dried clove buds or after grinding are used to flavour food, valued for medical purposes (flower and fruit) in south east Asia. Yields an oil oleoresin extensively used to flavour food and in perfume industry. Phenol eugenol is the major ingredient of oil, employed extensively in the manufacture of detergents, soap, toothpaste and pharmaceutical products.

Description

Clove plant is a tall (20 m), evergreen, slender, low branching spreading when young and cylindrical in shape as growth ages. Growth appears in flushes, determinate in nature, canopy dense with many fine short shoots or twigs. Leaves are simple, opposite, shining, glabrous, coriaceous (leathery) with gland like dots, petiole small, red tinged , shape elliptic to oblong-obovate. Flowers appear on terminal panicle/cymes, flower number variable, 5-20 may be as high as 35-40. Flower bud that make the commercial clove is about 1-2 cm long, flowers usually bisexual with reddish fleshy hypanthium , calyx 4 in number, fleshy and triangular in shape, corolla 4, stamens many with short style, stigma bilobed. Fruit is a berry, dark red in colour, ellipsoid-obovoid, with one oblong and long seed.

❑❑❑

14
Surinam Cherry

Genus : *Eugenia*
Species : *uniflora* L.
Family : *Myrtaceae*
Chromosome No. : 2n = 22

Eugenia uniflora, L. Is another important genera in family Myrtaceae which produces edible fruit. The fruit is known by several names like Pitanga, Surinam cherry, Cayenne cherry and Florida cherry. The word pitanga is derived from Tupi meaning 'piter' and 'anga', scent or odour. From its native home in Brazil the fruit was introduced in India at an early time probably by the Portuguese but is not cultivated commercially. The shiny deep green foliage is very attractive and the crushed leaves results in a pungent but agreeable odour which drives away the flies.

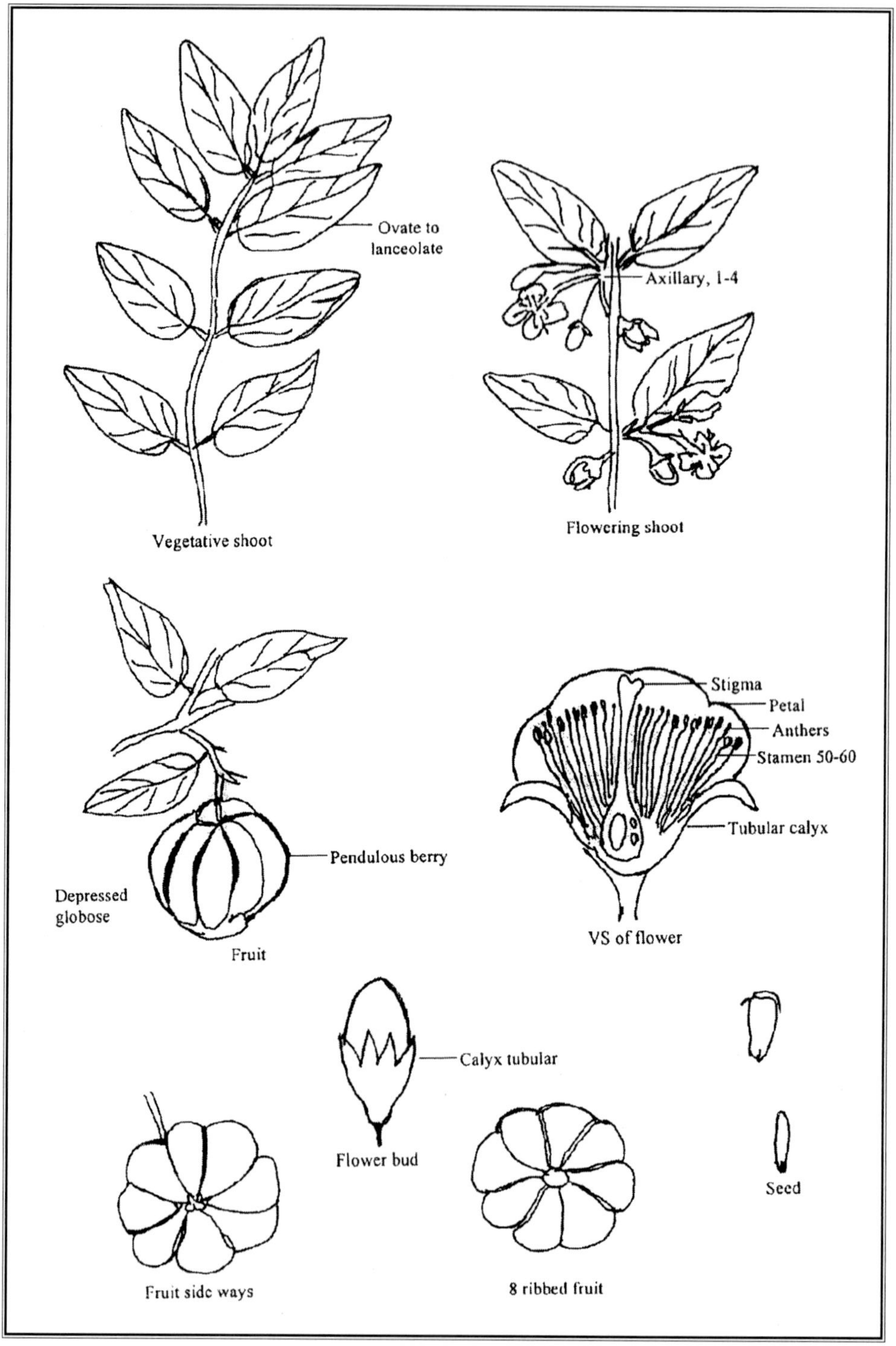
Ovate to lanceolate
Vegetative shoot
Axillary, 1-4
Flowering shoot
Pendulous berry
Depressed globose
Fruit
Stigma
Petal
Anthers
Stamen 50-60
Tubular calyx
VS of flower
Calyx tubular
Flower bud
Seed
Fruit side ways
8 ribbed fruit

Origin and Distribution

Surinam cherry or Brazil cherry is regarded to have been originated in south America extending from Surinam, Guyana, French Guyana way down to southern Brazil, Paraguay, Uruguay. The crop later moved to south east Asia and is cultivated to limited extent in Philippines, Malaysia and Java. The crop is grown all over the tropic and subtropic parts of the world.

Uses

Occasionally grown as ornamental plant (fruit bright coloured) but mainly for its edible fruits, eaten fresh or processed into pickles, jams, jellies etc; distilled liquor, juice fermented to make wine or vinegar. Pungent oil from leaves used as insect repellent, leather treatment (tannin from bark). Essential oil extracted from leaves comprises of terpenes, polyterpenes, sesquiterpenes, geraniols, citronella etc.

Description

Plant is an evergreen shrub or a small tree (7 m tall), branches usually slender spreading, may sometimes be crooked. Leaves opposite, ovate lanceolate slightly heart shaped (cordate), apex short pointed (acuminate) or obtuse, glossy glabrous, distinctly dotted, bronze coloured when young, on maturity turns dark green in cold conditions, develops red pigmentation. Flowers creamish white (1 cm diameter) and typically fragrant, appear in clusters of 1-4 in leaf axils, calyx tube-like/tubular 4 lobed, 8 ribbed, petals white, small (about 1 cm long), stamens numerous (50-60). Fruit drooping berry, globose (depressed), usually 7-8 ribbed, green when young, changing orange-bright red and finally blackish when fully ripe, skin thin smooth. Flesh orange red, juicy acidic to sweet in taste, faintly gummy or resinous. Seed single, large, sometimes 2-3 small flattened seeds also found.

Species

Eugenia dombeyi (Sprengel) Skeels Syn. *E. brasiliensis* Lamk.

Small evergreen tree, trunk short branches sub erect, leaves opposite ovate to obovate, elliptic, glossy coriaceous, young shoots rosy, Flowers solitary, axillary, white slightly fragrant, sepals 4, green, petals 4, white, stamens 100. Fruit globose or oblate berry, sepals persistent, bright red (young) dark purple to black or white, flesh red or white, sweet, soft and juicy. Seed 1 to many, light brown to grey green, hard, 1cm diameter. It has three distinct botanical varieties categorized on the basis of fruit colour.

E. dombeyi var. iocarpus ---- ----- - Fruits deep violet

*E. dombeyi var.*erythocarpus-------- ---- Fruits red

*E. dombeyi var.*lencocarpus---- Trees short statured, fruits white

The genus Eugenia differs form syzygium in that cotyledons are usually united seed coat smooth, free from pericarp, inflorescence usually racemes of pedicelled flowers.

❑❑❑

15
Litchi

Genus : *Litchi*
Species : *chinensis Sonn.*
Family : *Sapindaceae*
(Sub family-Nepheleae)
Chromosome No. : 2n=30

A Chinese renowned statesman Chang Chow-ling of the eighth century in a poem on litchi indicated it as the most luscious of all fruits, and in quality surpasses the other two southern China's famous fruits i.e. orange and peach. Litchi cultivation has been restricted to few selected sites mainly for its somewhat exacting in its climatic and cultural requirements. In recent times as litchi has established in the tropical America, there appears no doubt that very soon the fruits shall be in abundance in different parts of the tropics and subtropics. The most limiting factor in early dissemination of litchi was the perishable nature of seed and prior to steam navigation it was very difficult to transport the seeds from one continent to another successfully.

The name of fruit is spelled in different ways litchi, lichee, lychee, leechee, litchi, laichi and many more. According to Yule and Burnell in northern China the litchi is pronounced as lee-chee, while in southern China it is ly-chee. Name litchi has been part of the botanical name for very long and for its extensive use this name has been retained.

The soapberry or Sapindaceae family includes some of the most prized fruits namely litchi, rambutan and longan and for better understanding of their relation i.e. differences and resemblance with each other they are discussed here together. According to one Chinese statesman 'litchi is the most luscious of all fruits'. Further orchard of litchi with few hundred trees and with ordinary care would certainly give handsome annual return for a hundred years. Litchi is famed as a long lived tree, with broad round topped crown and glossy light green foliage has an additional ornamental value.

Litchi is a very old fruit crop and the Chinese knew it as early as the 3rd century of present era and in 18th century it was introduced in India (Bengal). Litchi is tropical in its requirements i.e. likes moist atmosphere with plenty of rainfall and free from frost. Considering that litchi is more exacting in its climatic requirement it is not extensively grown in India as it is in southern China, but there are certain places where it is commercially grown i.e. Bengal, Bihar (Muzaffarpur), Saharanpur, Dhehra Dun, and few isolated places in Himachal Pradesh.

The major food value of delicious litchi fruit is sugar and acid and is a matter based upon climate and cultivar, also contain minerals like calcium, phosphorus, vitamin C, A, B_1, B_2 in fair amount and is low in protein and fat content.

Origin and Distribution

Litchi is native of southern China, northern Vietnam and Malaysia. The crop has a long history of cultivation in China and it is here that it has undergone intensive selection. Malayan people cultivated it as early as 2500 BC. On account of exact climatic requirements and short seed life its spread in last 400 years has been very slow. In south east Asia, it is commercially grown in Bali and in northern Thailand.

Uses

The delicious fruit is eaten fresh, much fruit is used as dried lychee nuts, canning, fruit processed into juice and wine. Honey of litchi is of very high

Flowering shoot and inflorescence

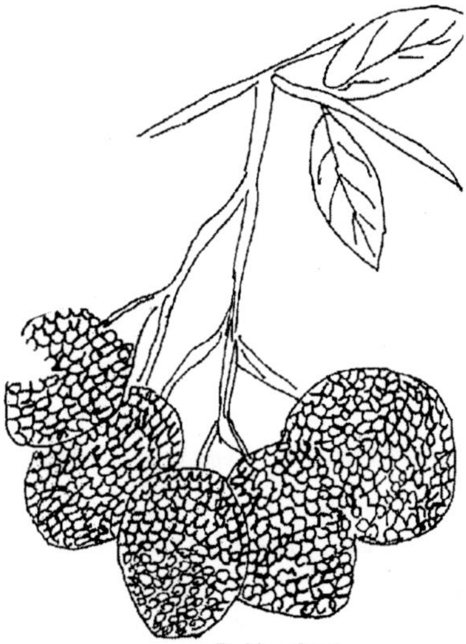

Fruiting shoot

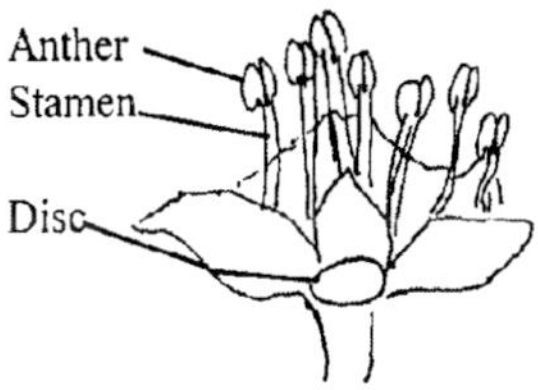

Male flower

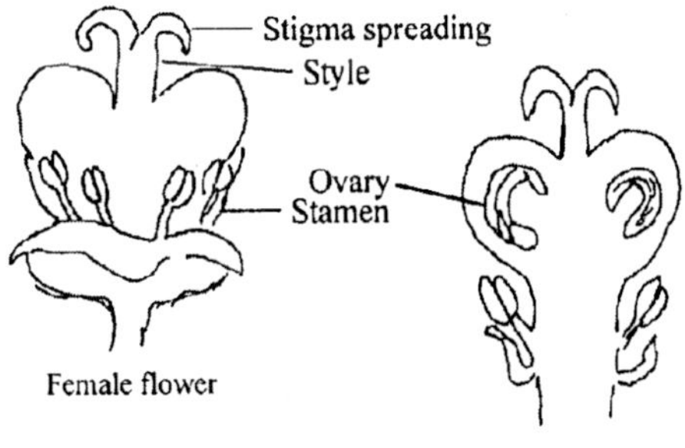

Female flower

VS of female flower

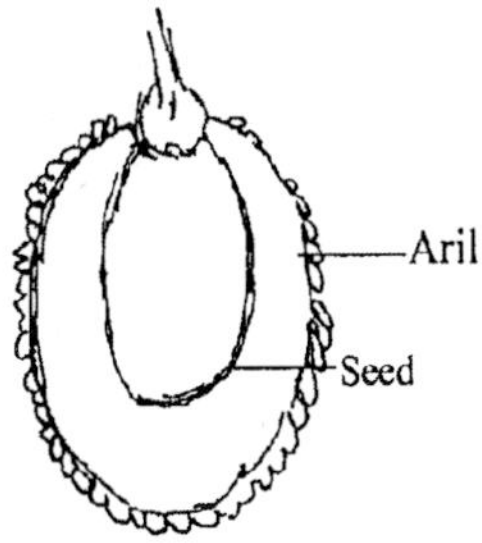

VS fruit

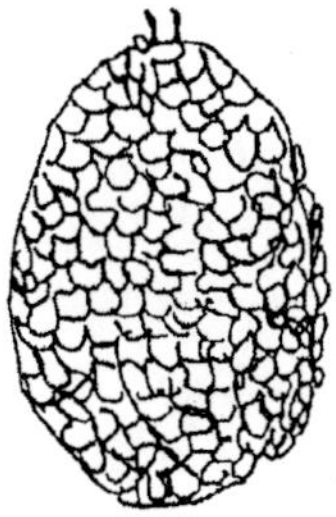

Fruit

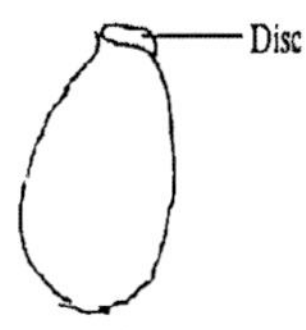

Seed

quality, fruit, peel and seed has medicinal properties while flowers, bark and roots are used for gargle.

Description

Litchi is a dense evergreen tree, long-lived, polygamous, tall (10m), round topped, trunk short, stocky. Leaves are coriaceous, paripinnate, glossy, deep green above, glaucous beneath, margins smooth, leaflets generally in pairs of 2-4, elliptic, oblong (8-15 x 3-4 cm), apex short pointed. The sex of the flowers on the same tree is different in various flushes, either functionally male or female, calyx 4, corolla absent, stamens 6-(10) filaments in male flower twice the length of the calyx, in female flowers it is very short. Flowers occurs in large terminal branched panicles mostly on new shoots of current season growth. Flower small, pale greenish–yellow. Flowers are of three types i.e. male, hermaphrodite and pseudo-hermaphrodite, corolla wanting, in male flowers anthers are functional, pistil grey or pinkish white and abortive. Anthers are developed with pollen grains, but fails to shed pollen in hermaphrodite flowers and acts as female. Although morphologically they are bisexual. In the third type, ovary is neither well developed (bisexual) or ill developed (male). Fruit usually globose, oblong to round-ovate, about 3cm in diameter, borne in bunches, pendent in loose clusters, epi/exocarp with short tubercles, usually red bright to purplish when ripe. Fleshy edible portion is the aril which is outgrowth of the seed, seed large dark brown, covered with white juicy fleshy translucent aril, has agreeable sweet acid taste.

Litchi at present comprises of only one species *L. chinensis* and can be grouped into 3 subspecies which are as following :

a) L . *chinensis spp. Chinensis :* Originated in northern Indo-China, Vietnam and Combodia, commonly cultivated in China. Twigs slender, leaves 2-4 jugate, flowers in cymose clusters, stamens 6-10, fruits smooth or with short 1mm projections. This botanical type has two cultivar groups

1. *Water litchi :* Adapted to lowlands, fruit almost smooth.
2. *Mountain litchi :* Adapted to high hill region. Fruit small, tubercles very sharp.

b) L. chinensis spp. Philippinensis : Very common and widely distributed in Philippines. Twigs slender, leaves 1-2 (3), jugate flowers in cymose clusters, stamens 6-7, fruits with pointed tubercles (3mm).

c) *L. chinensis spp. Javensis :* Adapted to more humid tropical conditions, common in southern Indo-China and west Java. Twigs thick, leaves 2-4 jugate, flowers in sessile clusters, stamens 7-11, fruits almost smooth or with slight projections (1 mm)

Cultivar

Detail of some important cultivars with pomological traits are detailed below.

Muzaffarpur : Very important and popular cultivar grown in India, plants are spreading type, tree top round, bears heavily, fruits pink to deep orange in colour, tubercules distinct, resists splitting. Average fruit weighs about 20 g, seed small pulp thick tough juicy sweet with pleasant flavour, pulp to seed ratio high (as aril is thick).

*Rose Scented :*Important cultivar from Indian perspective, plant growth vigorous moderately susceptible to cracking and sunburn (two major litchi problems). Fruit usually oval-oblong conical in outline each fruit is 20-25 g in weight. Skin deep rose coloured, pulp soft greyish white very sweet and juicy, seed small smooth and shiny pulp to seed ratio high.

Shahi : The cultivar bears consistently and heavily, suitable for canning, grown commercially in Muzaffarpur area of Bihar (India), fruit oval-oblong conical usually large 22-25 g fruit weight, tubercle bright red in colour at fruit ripening.

Tai So : Flesh recovery relatively high as seeds are small and fruits are large, each fruit weighs around 22-26 g., oval or egg shaped, apex rounded dull red when mature, with acidic sweet taste when mature and sweet when ripe. Trees are fast growing, spreading crown open rounded.

China : Fruits are less prone to cracking and sunburn, regarded as one of the most popular and excellent cultivar grown in India. Trees medium in growth, semi dwarf, leaves shiny pale green in colour somewhat smaller in size. Fruit usually globose yellow orange in colour, large 22-27 g. in weight, flesh soft juicy very sweet, flavour pleasant.

Bombai : Popular cultivar grown in west Bengal (India). Tree growth vigorous, spreading, bears in clusters. Fruits usually heart shaped, tubercles red when fruit is ripe, 15-20 g. in weight. Soft, juicy aril white sweet in taste with pleasant flavour, seed somewhat large shiny and smooth.

Brewester : Is of Chinese origin grown extensively in Florida, require cold winter for flower initiation. Trees are upright small, branches wide spreading rough (corky lenticels) foliage dense, fruit borne in clusters heart shaped, 20-24 g in weight, exocarp with attractive pinkish red colour, rough, flesh or aril sweet juicy and fragrant, seed small, large proportion undeveloped (seedless type) pulp to seed ratio high, 65-75 per cent.

❑❑❑

16
Rambutan

Genus : *Nephelium*
Species : *lappaceum L.*
Family : *Sapindaceae*
Chromosome No. : 2n = 22

Rambutan is one fruit from the Malay Archipelago which has become very popular in its place of origin and is well distributed here. Besides, rambutan a number of other valuable tropical fruits are present which are yet to be cultivated extensively in other tropical and subtropical parts of the world. In Singapore and Penang the fruit grows in abundance. Though very popular in Malay Archipelago, Sri Lanka, however it is unknown in India, Madagascar, Mauritius etc. The fruits common name has been derived from Malayan word rambut referring to hairs by which the fruit surface is covered with. The fruit is also known by other names like Rambustan, rambotang, ramboetan etc. The crop is strictly tropical in climate requirements and is very sensitive to frost. For best performance it likes a hot and moist climate and is a major factor why the crop has not succeeded in most other places.

The hairy litchi (*Nephelium lappaceum*) on account of soft red spines which are hair like that cover the entire fruit surface is regarded as an important fruit of family Sapindaceae, which has about 37 genera and 72 species. The fruit is indigenous to the Malay Archipelago were number of valuable tropical fruits not cultivated extensively are found and rambutan is one among them. From here it spread to other tropical regions of the world where the temperature and humidity round the year was relatively high, i.e. conducive for its growth. Later the crop moved westwards to countries like Burma, Thailand, India, Sri Lanka, Philippines, Indonesia, Hawaii and Vietnam. The fruit is close to litchi for its delicious juicy, flavoured aril which is translucent white and sub-acid-sweet in taste. The common name rambutan in taken from Malayan word Hambut meaning hair or rind. The entire fruit surface (pericarp) is covered with spines which are soft and yellow or red in colour. The outgrowth of the seed-aril the edible portion is pearl white translucent juicy with blend of sub acid sweet taste. The entire tree with its shiny attractive beautiful foliage, flower and colourful fruit (varying shades of yellow to red) makes it useful ornamental tree and adds to the aesthetic value in landscape. In climatic requirements rambutan is strictly a tropical fruit, but it is difficult to believe that it is not known/grown in India, Madagascar and Mauritius whereas it is plenty in Sri Lanka. Some of the species which yield edible fruit other than rambutan are :

1. *N. chryseum* (Mountain rambutan) grows wild.
2. *N. obovatum* (Vine longan)
3. *N. malaiense* (Malayan longan)
4. *N. topengii* (Hainan rambutan) grows wild in Hainan Province of China.

Origin and Distribution

The exact origin of this popular fruit is unknown. The specie is seen growing in provinces of Hainan and Yunnan (southern China), extends to Indo -China region including Indonesia, Malaysia, Sumatra Java, further to Philippines. Presently the crop is cultivated in the humid tropics of Asia (Sri Lanka, New Guinea) and marginally in Australia, America and Africa.

Flowering shoot and inflorescence

Fruiting shoot

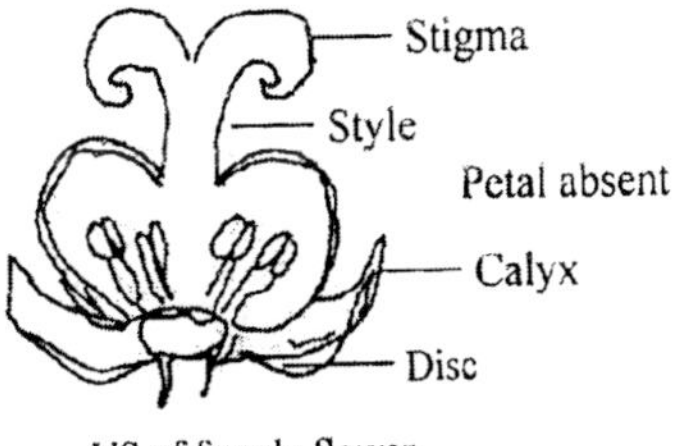

VS of female flower

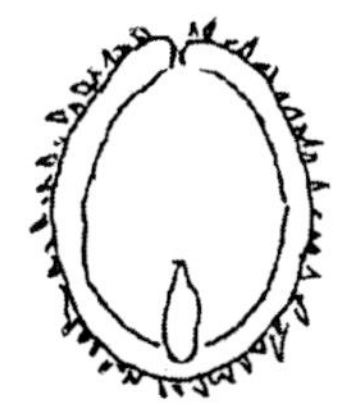

VS of fruit

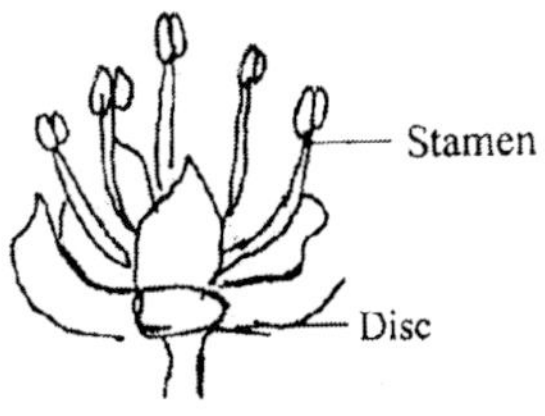

Male flower

Seed

Uses

Plants are grown for its fruit, the sweet juicy aril (sarcotesta) is generally eaten fresh used in preperation of jams or is canned, ornamental value. Leaves with mud forms a black dye, fruit rind yields toxic saponin, young shoots employed as green dye for silk. Also has number of medicinal properties viz. leaves in poultices to overcome headache, roots to treat fevers, bark for tongue ailments, fruits for stomache and anthelmintic.

Description

Plants are vigorous, spreading, large sized, (4-7m tall), leaves paripinnate, leaflets 6 in number, alternate, ovate to obovate, glabrous above, slightly pubescent below and along the midrib. Inflorescence appears terminally or is sub terminal, flowers may be either male, hermaphrodite or female, actinomorphic greenish yellowish or whitish, calyx usually 4-5 (may be upto 7), almost free or united for more than half the length, petals generally absent but when present are 4 in number and highly reduced (1.5mm), disc glabrous or hairy, complete, stamen number variable usually 4-5, but may be 8-9, filaments long with dense hairs at the base, anthers split lengthwise, pistil 2 but sometimes may be 3 with dense hairs, style fully developed, stigma spreading to recoiled. Fruit is an ellipsoid to sub globular schizocarp, measuring about 7x5 cm, weighing 30-100gm, has 1 nutlet. Fruit purplish red, glabrous, densely set with filiform curved appendages. Edible part is the sarcotesta, the seed envelope which is translucent, thick, juicy white to yellow.

Nutritionally rambutan is considered to be high is sugar, vitamin C and minerals like calcium and phosphorus and is fair in vitamin A, B, B2, B5 and fibre content.

Cultivars

In Malay Archipelago a number of potential seedlings with useful pomological traits have been selected and are presently grown on seedling stock. Description of some important rambutan cultivars is summarized below.

Seematjan : This cultivar is the most popular and commercially grown all over the tropics, mostly in Malayasia, Thailand and Burma. Tree medium tall, much branched, evergreen, crown open branches are long and spreading. The fruit when fully ripe the epicarp becomes intense red in colour, bright attractive in appearance, entire surface covered with tender curved spines which are long about 2 cm. Cultivar has two distinct strains.

(i) *Besar*, aril thick, white adhere to the seed more juicy, sweet in taste with very pleasant flavour.

(ii) *Koombarg* has small sized fruits, epicarp or rind is thin and the spines are also not too many and are present thinly on the surface, edible part aril is soft and adherence not very strong and less sweet in taste.

Seelengkeng : Plants are dwarf in nature propagation through air layering is difficult, bright green in colour glossy, flower appears both on terminal and axillary inflorescence. Fruits are somewhat smaller (3x2 cm) ovate, surface covered with bright red soft spines, aril is very sweet and in taste resembles very closely to litchi, fetches high price in the market.

Seetangkooweh : Plants are large tall growing, tree top broad, branches spreading, foliage bright green attractive. Fruits are large (5x4 cm) ovoid but somewhat compressed laterally, surface covered with short (1 cm) tender spines, flesh or aril yellowish white when fully ripe and adhere tightly to the seed coat, juicy, sweet in taste with pleasant flavour has good storage quality and withstand long transportation.

Lebakboeloos : Tree large, vigorous, crown broad, branches spreading foliage bright green. Fruit dark red in colour with short spines about 1-1.5 cm long, placed sparsely on the surface, very attractive when ripe. The edible part aril is whitish about 0.5 cm thick and adhere firmly to the seed coat, juicy sweet-acidic in taste, withstands long transportation.

See-Chohmpoo : Most widely cultivated and a popular cultivar of Thailand. Trees are medium in growth, crown broad and open, branches spreading. Fruit medium in size, ovate, pink red in colour, spines tender and curved. Most significant feature is that seed is small and aril is thick, flesh/aril white, firm tough in texture, juicy, sweet with characteristic pleasant flavour.

Rohng – Rian : Another very important rambutan cultivar grown extensively in Thailand. Trees are of medium size, crown broad, branches spreading. Fruit bright red in colour, shiny, attractive, ovate spines are many, red in colour while the tips remain green even upto maturity, flesh white, firm, texture tough, juicy, sweet in taste flavour pleasant.

B-R-1 (Boting Rambutan No. 1) : Is a new cultivar selected at Hainan, China. Tree is large fast growing, branches are profuse, long and spreading, crown therefore is broad and open. Foliage shiny evergreen fruits are large in size (30 gm or more), ovate in shape with dense and fine red spines, seed small, aril thick whitish hence edible portion is more, juicy sweet in taste.

❑❑❑

17
Longan

Genus : *Euphoria/Nephelium/Dimocarpus*
Species : *longana (Lour.) Stend. cinerea*
Family : *Sapindaceae*
Chromosome No. : 2n = 20

The longan (*Euphoria longana*) belongs to the Sapindaceae or soapberry family to which more important fruits like litchi and rambutan also belongs. The family contains more than 1000 spp. from 125 genera widely distributed in the tropical and warm subtropics. Most of the species are native to Asia though some are also found in south America, Africa and Australia. The genus has seven species but only *longana* is both horticulturally and commercially important as it yields/ produces edible fruits. For most of the growth, flowering and fruit characters

Flowering shoot with axillary paniculate inflorescence

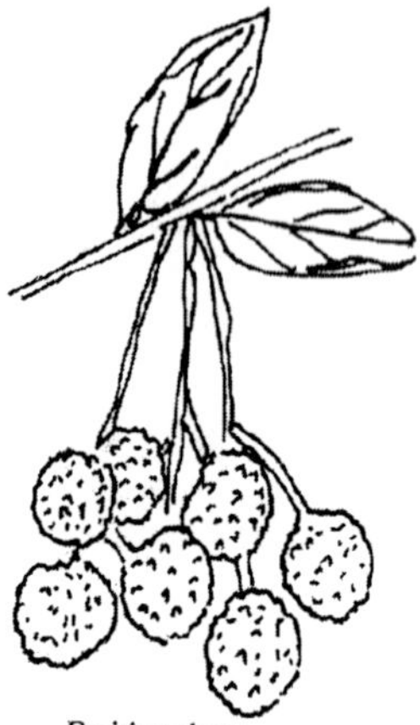

Fruiting shoot

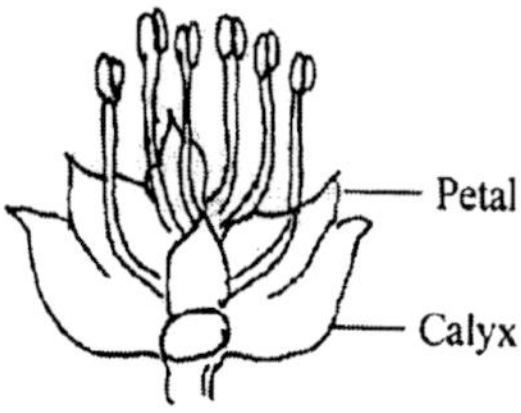

Flower

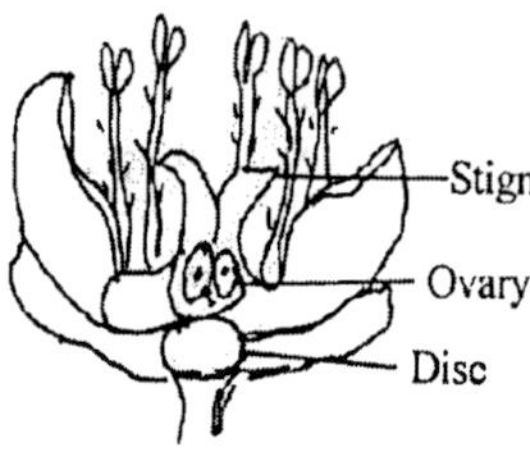

VS of flower

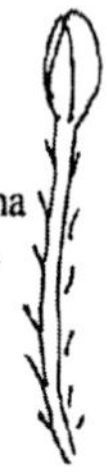

Stamen

Fruit

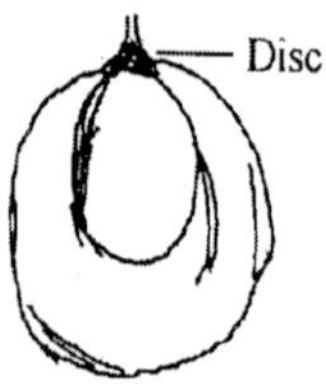

VS of fruit

it is similar to the two above horticultural crops mentioned. In respect of soil and climate it is less demanding and does adapt or tolerates drier and cooler condition better than litchi. The longan fruits are smaller, yellowish to tan brown, fruit surface is smooth (without tubercles) less acidic in taste and with mild flavour, flower has petals. In south east Asia especially Thailand and China where number of new plantation have come up in recent times the taste of longan fruit is rated to equal or sometimes even superior to that of litchi.

Longan fruits are fairly good in the fruit composition major constituent being carbohydrates, mineral matter mainly calcium phosphorus and iron and vitamins like $A_1B_1B_2$, ascorbic acid and niacin. The fruit is said to have originated in the subtropical China or in the area between India and Burma. From its natural habitat it moved east wards and reached Thailand in the 14th century, and by the early mid 20th century it was well established in number of places prime being Hawaii, Hong Kong, Singapore and as far as in Florida. At present Thailand produces almost as much longan as that are produced in Taiwan and China and compared to litchi production it is almost 12-15 times.

Origin and Distribution

Authors have expressed varied views about its origin, some say crop origin is in the mountain chain which runs from Burma to southern China. According to others, origin limit is extended further southwest India and Sri Lanka. Presently the crop is commercially grown in Taiwan, China (south) Thailand and limited acreages is in India, Florida, Queensland and south east Asia.

Uses

Longans are generally consumed fresh. In recent times longan canning industries have been established in China, Thailand and Taiwan, where large fruits with small seeds are canned, has high total soluble solids therefore retains its flavour better than litchi or rambutan, can be preserved dry either with or without pericarp, refreshing drink and liquor from its flesh can be made. Seeds have saponing, thereby used as shampoo. Has medicinal use, red hard wood is used as timber.

Description

Trees are vigorous (40 m. tall), scandent (climbing) shrub, branches usually terete with five faint grooves or lines, lenticels conspicuous dry corky, dense pubescence (tomentose),leaflets elliptical, coriaceous, tomentose below along midrib and veins, petiole long (1-20 cm) most of the axial parts with dense pubescence. Inflorescence generally terminal, long (about 40 cm long), densely pubescent (tomentose), flowers yellow-brown in clustered cymes (3-5 flowers), short pedicelled, bracts present. Calyx with 5 lobes (2-5 x 1-3 mm), petals 5, (2-6 x 1-2 mm), glabrous to densely pubescent, stamens 6-10 in number. Fruit drupaceous, yellow-brown, broad ellipsoidal to globular, smooth sometimes warty and granular. Seed blackish brown, surface shining, globular, covered with thin fleshy translucent white arilloid which forms the edible part.

Cultivars

Longans are easily propagated by seed but the resultant seedling have biennial bearing habit and fruit carried or borne invariably is of inferior quality. However from variability some very useful types have been selected and maintained through vegetative multiplication. According to one estimate there are about 300-400 longan cultivars grown in southern China. Characters of some important logan cultivar are summarized as.

Fu Yan (Lucky Eye) : Is an old Chinese cultivar, fruits are quite large (15-20 gm) skin thin with small seed, flesh thick firm and crisp, yields are higher than other cultivars, good for canning.

Wu Long Ling (Black Dragon): A new cultivar being high in sugar content (20-30%) is best suited for drying purposes. Major draw back is its alternate cropping, fruits are medium sized, skin thick flesh recovery is more.

Daw : Very popular cultivar grown in Thailand, early maturing type bears regularly, fruits are large with big seed, skin thin flesh firm crisp sweet and with pleasant flavour.

Biew Kiew : Grown commercially in Thailand where it is held in high esteem for its excellent quality fruits. Fruits are bigger in size, seed somewhat small hence better flesh recovery which is dry and crisp, creamish in colour, quality very good.

□□□

18
Fig

Genus : *Ficus*
Species : *carica* L.
Family : *Moraceae*
Chromosome No. : 2n = 26

Fig along with fruits like date palm, grape (vinifera) and olive was regarded as an important food crop of the ancient civilization that developed in the Mediterranean region. The fig description occurred prominently in number of songs as well as in the mythological and historical events that took place at that time. Fig is thought to be native of southern parts of Arabian peninsula, Italy, Balkan and USSR region. Fig was in cultivation as early as 300 BC and from Arabian peninsula it spread to Turkey, Syria, Iraq and later to many subtropical parts of the western hemisphere.

The edible fig (*Ficus carica*, L.) is a small deciduous subtropical fruit and mature tree can tolerate low temperature of –12^{0}C, but considerable damage is caused by autumn and spring frost when plant starts new growth. Under north Indian conditions the plants remain dormant and are not damaged by the freezing temperature. The plants perform well under temperature regime of 15-22^{0}C. Places with pronounced dry climate both during fruit development and maturation produce high quality figs, but high humidity and low temperature result in poor quality of figs with marked fruit splitting. Fig (common fig) cultivated in India has not drawn the due attention that it deserves despite the fact that both soil and climatic conditions desirable for fig cultivation are found in plenty, although wild figs have been growing for long time.

Figs, considered as food are nutritious and wholesome and is rich in sugars, lacks in vit C, moderate in minerals like Ca Fe Cu, protein content, can be dried and is highly valued for its laxative property.

The family Moraceae includes 60 genera and more than 2000 species mainly of trees, shrubs vines and some herbs. Important members of the family are Mulberry (*Morus alba, M. nigra M. rubra,* white, black and red mulberry), Jack fruit (*Artocarpus heterophyllus*), bread fruit (*A. altilis*) and Fig (*Carica*). The genus Ficus alone comprises of about 1000 species, a number of which have edible fruits and number of species yield fodder, natural rubber, bark, cloth and host plants for lac insects. Some figs are regarded as the wonders of the plant kingdom. The various types of figs available are used in a number of ways i.e. some planted for shade others as ornamental both for home and conservatory.

Origin and Distribution

Ficus carica L. is native of Asia Minor and reached Mediterranean at very early time. Ancient fig cultivation in Egypt dates back to 400 BC. Only two fig species are native to North America, the Florida strangler fig (*F. aurea)* and short leaf fig (*F. laevigata)*. The common cultivated fig does perform well in low wet tropics, but grows very well at higher elevations, provided rains are limited both at flowering and fruiting season.Fig is considered to be native of Indo-Malaysia region as this region is known to have large number of species. The specific epithet *Carica* refers to Caria, an ancient political division of Asia Minor bordering the Aegean sea. Botanical evidence points to the fertile region of southern Arabia, where wild fig and capri fig trees are still growing at its centre of origin. Fig has been domesticated millennia ago in Asia and Mediterranean region. During the time of exploration and colonization, fig was carried to all parts of the world including new world.

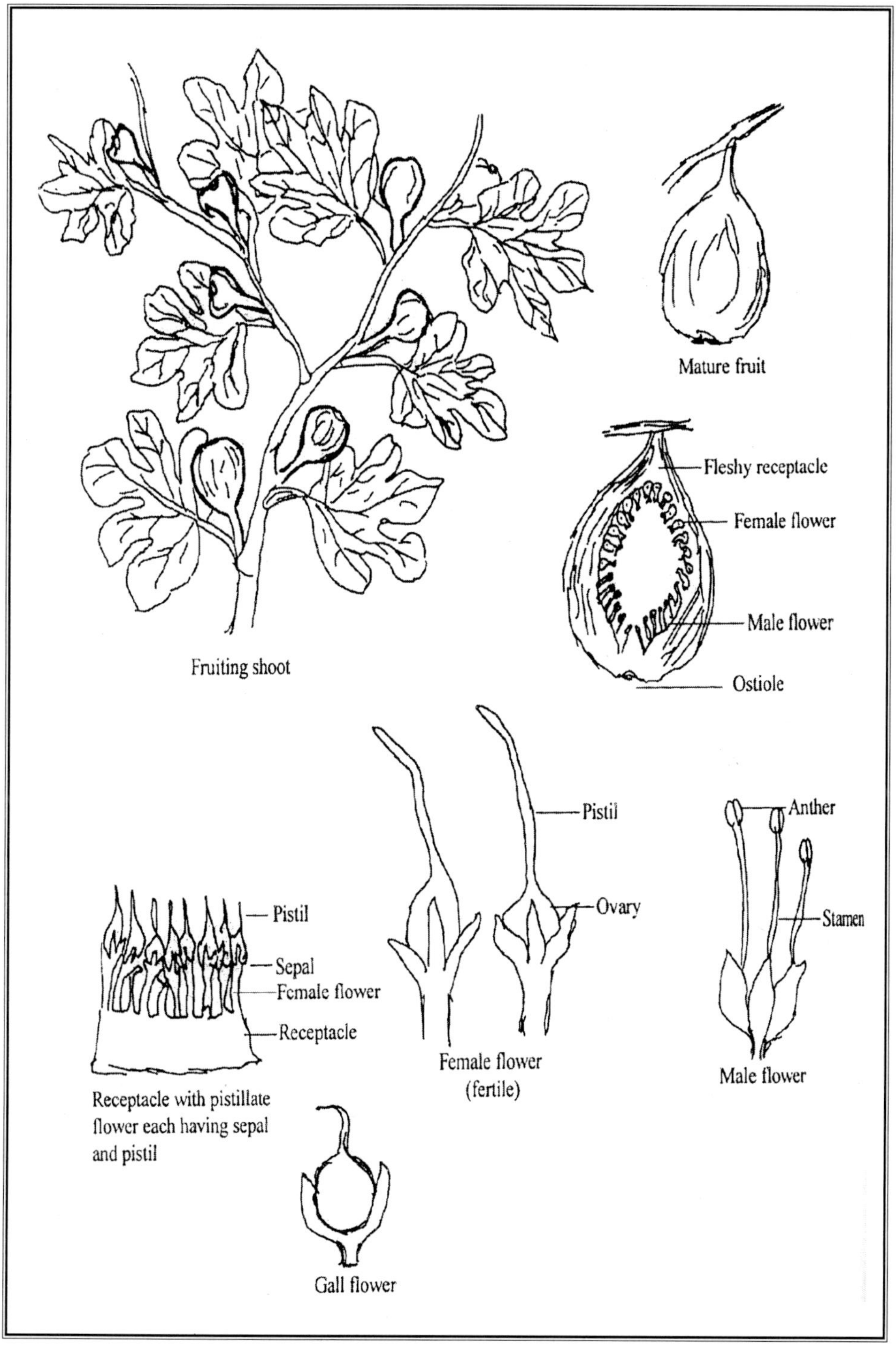

Fruiting shoot

Mature fruit

Receptacle with pistillate flower each having sepal and pistil

Female flower (fertile)

Male flower

Gall flower

Uses

Fig is a valuable and nutritious fruit. The sweet fruits are usually eaten fresh, may be dried or canned or used in variety of dishes, pastries and desserts, beside a laxative is also made. Wood is soft , light weight lacks durability, subject to attack of termites and damage therefore not commercially important as timber. Young leaves of some species are used as pot herbs, fodder for livestock.

Description

Plants are large shrub or low growing deciduous tree about 10 m. tall, with milky juice and smooth to shallow fissured bark, branches short, and stout, buds often large, pointed and naked. Leaves are simple, alternate persistent or evergreen (leaf fall in edible figs), large (10-20 cm. long) entire or 3-5 lobed, prominently palmately veined. The flowers are hollow and borne in the leaf axils, reduced to essential organs i.e. male and female and are produced inside globose-pear shaped receptacles with a narrow mouth (ostiole), interlocked by small green scale or leafy bracts. Flowers are borne axillary on previous growth and also on current growth. Near the ostiole male flowers are borne and towards the stalk female flowers are present. The flowers are of 3 types

i) Male flowers with 1-5 stamens

ii) Female flowers with ovary and along style each resulting one seed

iii) Gall flowers resembles female flower, short styled, ovary swollen, hibers fig wasp and does not set seed.

In the capri figs all three types of flower occur inside each fig, whereas in other types (dioecious), the male and gall flowers are produced in same inflorescence on one tree, and female inflorescence on another tree. The fleshy fruit or syconium is formed by the swelling of entire receptacle which encloses number of small male flowers and true fruits-achenes which develop from female flowers.

In fig cross pollination is affected by very small wasps *Blastophaga spp.*, are about 1mm long. The wasp specie is specific to a given species of *Ficus.* As the figs with gall flowers ripen, the adult wasp comes out of the ovaries by biting a hole in the ovary wall. The male wasps are wingless and almost blind copulate with larger females and soon die. The impregnated females moves between the scales and male flowers, the body is loaded with pollen grains. She then flies to another tree to oviposit. If enters a gall fig she oviposits an egg in each of the short styles, the process of egg laying is carried by the wasp and dies in the inflorescence. The eggs develop into larvae. However if female wasp enters a fig with female flowers her ovipositor is very small to penetrate the long styles. In an effort to find suitable place to lay egg, cross pollination is had as pollen is placed on stigmas by the wasp. The hatching overlaps with the dehiscence of the stamens and stigma receptivity. The fig fruit when ripe are eaten by birds and animals. The fruit is called Syconium, which develops from a hollow receptacle/hypanthium, which is pear shaped and fleshy in consistency and encloses number of small male flowers and true fruits i.e. achenes that develop from female flowers.

On the basis of pollination the figs are classified as,

1. *Common figs :* Have no male flowers and the seedless fruits that are formed, develop parthenocarpically eg. Adriatic
2. *Smyrna figs :* Pollination is required for both the crops eg. Calimyrna
3. *White San Pedro:* For spring crop no pollination is required, but is essential for main crop.
4. *Capri fig :* Such figs are used as pollinizers for other figs and they contain lots of insects.

Other important species are :

Ficus elastica Roxb. India-Rubber fig, wild in India and Malaya, rubber used for lining of receptacle in north eastern India, native methods of tapering like incising or slashing are wasteful and yield very low rubber, with high resin content.

Ficus vogelii Miq. Is potential resinous rubber source in west Africa.

Ficus benghalensis L. Indian banyan. An evergreen large spreading with very massive roots, enable it to extend laterally over a large area. This makes the plant to have largest crown amongst plants of various species.

Ficus benzamina L. Native to India and Malaya, known as Ceylon Willow, the drooping branches give tremendous grace to the plant.

Ficus religiosa L. Native of India, known as pipal tree, a religious/ sacred plant, can be raised by cutting and layering. Lac insect feed on it, tree suitable as avenue tree/for roadside plantation.

Ficus pumila L. Known as creeping fig, native to eastern Asia, roots flat against walls. The young leaves are small cordate, sessile, 2 cm. long, mature leaves 5-10 cm. long, large elliptic.

Species like *F. thonningii* Bl., *F. netalensis* Hochst. (East Africa) and *F. nekbuola* Warb. (Mozambique), yield bark cloth. The bark after removal is macerated, beaten by mallets, to cause stretching, potential shade plant for coffee and other plants.

Lac insect, *Laccifer spp.* feed upon *F. benghalensis, F. religiosa* and *F. cunia* to produce lac/shallac, which is a thick resinous fluid secreted by the larvae which coalesce and hardens to form an encrustation, dried and processed to yield lac of commerce. Today lac forms the basis of polishes and varnishes, gramophone records, and in surface coating industry.

Important species which are horticulturally and commercially important are detailed below

Ficus altissima Blume. The 'lofty tree' or Council tree is found in abundance in Assam to Malaya, Java and Philippines. A large spreading tree 75 feet or more, magnificent banyan tree with silvery grey bark, leaves thick leathery smooth, ovate to elliptic 6 inch long or more and veins ivory to reddish in colour. Fruit occur in axils, are about 2 cm long variable in colour yellow orange to scarlet red. Has tremendous value as a shade tree and has a stricking beauty with its red fruit on slender branches.

Ficus auriculate Lour. (F. *roxburghii* Wallich Miq). Ornamental fig. Plants are small trees or shrubs, young twigs hollow low spreading very showy for its large foliage on slender woody branches, leaves are ovate 16 by 13 inch in size, base cordate or rounded covered with fine pubescence with depressed veins, petioles covered with white hairs. Fruit appears both on branches and trunk which are big flat globose or pear shaped 2.5 inch or more across, white or rusty, flecked with silky hairs.

F. benghalensis L (*F. indica* L). Banyan Tree, East Indian Fig tree, Indian Banyan. Tree very large, top spreading, epiphytic when young. Leaves leathery broadly ovate to elliptic about 8 inch long. Fig fruit occur in leaf axils usually in pairs globose, orange red in colour about ½ inch in diameter. Regarded as very sacred plant in India and Pakistan. Has botanical variety Krishnae (*F. krishnae* C.DC) Krishan Bor, Krishna's Butter Cup, 'Sacred fig tree', leaves are curiously cup shaped deep green with raised ivory fine pubescece.

F. benjamina L (*F. nitida*, Thumb) Benjamin Tree, Weeping fig, Java Fig. Found in India, Asia, Malay Archipelago and tropical America. Small leaved rubber plant. Tree or shrub, branches are drooping, leaves thin leathery ovate-elliptic about 5 inch or more long. Fruit occur axillary usually in pairs, red globose, small ½ inch or less across. Has botanical types var. Exotica (*F. exotica*). Tree has thin drooping branches long tipped and twisted leaves.

var. nuda (Mi). *F. comosa* Roxb. Leaves narrow long more acuminate figs, whitish or red brown narrow base and no bracts when mature.

F. deltoidea Jack. (*F. diversifolia* Blume). Mistletoe Fig, Mistletoe rubber plant. Found growing in plenty in India, Java, Malaya etc. Plant is a shrub 6 feet sometime more tall, leaves generally hard broadly obovate dark green with brown spots above, olive green below. Fruits occur in leaf axils, yellow very small with greyish pubescence

F. elastica Roxb. ex Hornem (*F. rubra* Hort). Rubber Plant, Assam or Indian Rubber tree. Commonly found in India and Malaya more so from Nepal to Assam and Burma. On incision a latex /milky sap exudes. Tree tall about 30 meter high glabrous, young plants erect with brown thick stem. Leaves oblong to elliptic, thick about 12 inch or more in length. Fruit appear in leaf axils usually in pairs, ovate oblong in form, greenish - yellow about ½ inch across. Has number of botanical type major being.

F elestica cv. Decora (*F. decora* Hort.). Broad leaved Indian rubber plant. Leaves shiny dark green, midrib ivory and red below

...............cv. Doescheri (*F. doescheri* Hort.). Leaf margins green, marbled with grey green midrib creamy yellow, petioles are pink.

.... cv Variegate. Leaves are light green in colour with margins white or yellow.

F. palmate Forssk (*F. pseudocarica* Miq.). Is native of north west India, Nepal and Africa. Plants usually shrubs or small trees. Leaves round may be 3-5 lobed generally 5 inch long sometimes even more, margins may be entire or toothed almost scrabrous above and with short dense pubescence (tomentose) below. Fruits are with peduncles which are somewhat thick, pear to subglobose in shape small in size measuring about 1 inch across.

F. racemosa L. (*F. glomerata* Roxb.). Trimbal Cluster fig. The specie is wide spread in tropical Asia from India to north Australia. The specie is cultivated in India and it yields edible fruits. Trees are tall growing 60 feet or even more not epiphytic when young. Leaves somewhat variable ovate lanceolate to elliptic in shape, long (6 inch) base obtuse margin entire. Figs are borne on main branches and basal portion of trunk, green to red in colour about 1-2 inch in diameter, globose.

F. religiosa L. Peepul, Sacred fig or Bo tree. A common tree in India and most of the Asian countries. Is a scared tree for Hindus and Buddhists. The specie is distinct from others for its very long and thin leaf tip. Trees are very tall and fast growing deciduous initially epiphytic. Leaves are usually round ovate but tip is narrowly drawn out sometimes even more than half the blade length. Fruits are almost sessile usually purple to dark purple in colour and measures ½ inch in diameter.

F. subulata Blume (*F. salicifolia* Miq.). Specie is found growing in India, south China, Malaya Archipelago to the Soloman Island. Plants are shrubs semi scandent, dioecious. Leaves elliptic lanceolate 4-10 inch long, pinnately nerved 7-10 pairs, petiole short, fruit orange red about ½ inch in diameter.

Cultivar

Fig has been in cultivation for number of centuries but not much was done for cultivar improvement. Most of the cultivars are selections from the wild seedling made in the past by unknown people. However is recent past some improved types through systematic hybridization have been developed. Important pomological features are briefly described as below.

Conardia : Good for table and drying purpose, resists cracking. Fruit pyriform, yellowish green, pulp bright red, meat white in colour, eye very small individual fruit weighs about 50 g.

Flanders : Dries well, splitting occurs in unfavourable conditions. Fruit globose in form, neck thick yellowish green in colour, pulp amber white, eye very small and tight.

Excel : No splitting of fruit takes place, very good for fresh, drying and canning, fruit light yellow in colour, ovoid to globose, pulp amber, meat white, eye neither tight nor too loose.

Gulbun : Fruits with few splitting, large 65-80 g, good cropper, light yellow in colour shape round flat, pulp light pink, meat white, eye as that in Excel.

Evrem : Mid season cultivar high yielding, attractive with no splitting, fruit light yellow round flat somewhat small, average fruit weighs 35-50 g, pulp somewhat greenish meat white, eye medium tight.

❑❑❑

19
Jack Fruit

Genus : *Artocarpus*
Species : *heterophyllus L. (Lamk.)*
Family : *Moraceae*
Chromosome No. : 2n = 56

A famous traveller John de Marignolli has described jackfruit as 'there is again another wonderful tree, which produced fruit on the trunk, and was very pleasing to see and the fruit was as big as a lamb'. The rind is hard and has to be cut open by a sharp knife, the pulp inside is of pleasant flavour sweetness of which matches the honey and usually contains 500 or more chestnut like seeds which are eaten after roasting. In its place of origin (India) where it has been grown since ancient times it is held in great esteem. Theopharatus and Pliny who wrote about jackfruit also expressed similar opinion about the fruit. In India several vernacular names are given like Kanthal, Kathal, Panasa and Kantaka. The species being strictly tropical in its requirements, its early introduction in California and Hawaii did not prove successful as it failed to withstand the prevailing cold conditions. Breadfruit a close relative of jackfruit did not prove of any significance value when introduced to the West Indian colonies. In Polynesia (place of origin) it still occupies an important place, and is regarded as a staple food for its starchy characters and significant role that it has played in the dietary requirements of the local inhabitants. The name breadfruit given by English travellers is hence the most fitting one. In some of the Polynesian Islands, the tree is of ancient cultivation and has played important part in the life of the people that native inhabitants cannot see a place where it is not found.

The name is derived from 'arto' - bread and 'carpos' meaning fruit. Jack fruit originally is native to India and is presently cultivated throughout the tropical and low land of both the hemispheres and flourishes in the humid hilly slopes upto an elevation of 2400 m in south India. The tree is useful to mankind in many ways. The wild jackfruit are seen growing in abundance along the western Ghats of India. The tree is highly attractive for its dense dark green shiny foliage. Jack fruit is a multipurpose fruit plant and is the largest among the edible fruits. Young jackfruit appears in the market during spring and are available till summer, therefore enjoys high demand and premium prices as a vegetable as most of the vegetables during this period are either too costly or are scarce. Fruit is high in nutritive value and is a rich source of pectin, carotene, ascorbic acid, protein and also has mono, di and polysaccharides with substantial amount of fibre. The fruit latex has bacteriolytic activity and for its distinct flavour is sometimes not liked by many. Jackfruit (*Artocarpus heterophyllus*) and breadfruit (*A. altilis*) are the two fruit yielding species which are more important horticulturally. The genus Artocarpus has about 50 species comprising of evergreen or deciduous monoecoious tree. Wood is hard and durable not attacked by white ants. Seed is cooked and fruit is used in many culinary preparation, preserves like pickle, dehydrated leather, canning is successful, jack flakes are bottled and served with honey and sugar, pulp yields nectar.

Origin and Distribution

The jackfruit is considered to be indigenous to and in all likely hood has originated in the rain forest of western Ghats of India, as the crop has been cultivated since time immemorial. Later the crop has been introduced in many

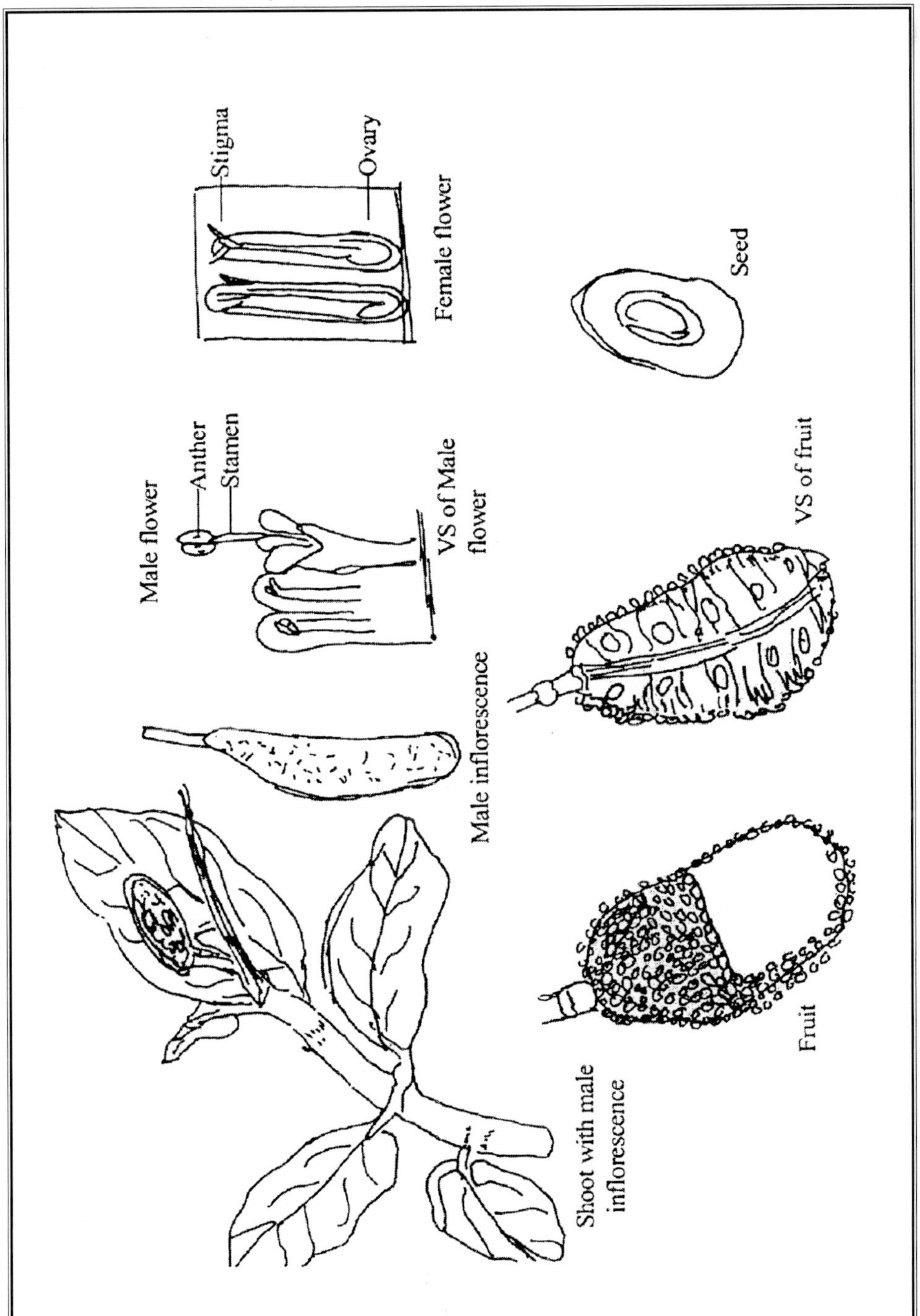
Stigma
Ovary
Female flower
Male flower
Anther
Stamen
VS of Male flower
Seed
Male inflorescence
VS of fruit
Shoot with male inflorescence
Fruit

parts of the world particularly the tropics and is grown all over the south east Asia.

Uses

Jackfruit when ripe is eaten fresh, or can be made into various delicacies i.e. jam, jelly, chutney, paste or preserved as candies by mixing or drying with honey, sugar or syrup, used in flavouring ice-cream and beverages, made into powder or concentrate to make various drinks. Seeds are consumed after roasting or boiling ,or may be salted to be eaten as nuts or made into flour and used in baking after adding it to wheat flour. Immatures fruits are cooked and taken as vegetable. Bark yields tannin, a yellow dye from wood is employed to dye cotton and silk, has number of medicinal values, pulp and seeds acts as cooling and nutritious tonic to overcome alcohol effects. Latex help in healing of snakebite and glandular swelling, roots against skin diseases and asthma. Heated leaves and ash heals ulcer. The wood/timber is classed as medium hard, easy to season, resistant to termite, fungal and bacterial decay, polishes beautifully.

Description

Plants are evergreen, monoecious, large handsome tall growing (20-30m), trunk round and smooth with dense crown and leafy branches, on injury white latex exudes from all plant parts. Crown dense, conical when young and round and spreading in old trees, bark rough, scaly, dark grey to greyish brown. The young twigs, shoots and leaves generally glabrous, some times short haired. Stipule, deciduous ovate–acute large sized (8 x 3cm), leaving behind a big scar. Leaves leathery, shinning thin dark green above and pale green below, obovate –elliptic to elliptic, base narrow, cuneate broadest at or above the middle, margin entire but may have 1-2 pairs of lobes, when young separately nerved and with reticulate veins, petiole long (4cm) faintly grooved and slightly pubescent. Inflorescence appear axillary on lateral short, leafy shoots that arise from old branches and main trunk. Male and female spikes are borne separately on same tree and are enclosed in a thick leathery deciduous spathe when young. During fruiting season 'foot stalk' the special fruiting branchlets appear on trunk on two or more year growth, also on scaffold limbs. Pistillate spikes occur only on foot stalks which appears on trunk, primary or secondary branches. Staminate spike appear on terminal leaf axils, main and lateral shoots or in the foot stalk. Male flower heads are barrel shaped or ellipsoid (3-8 x 1-3cm) having both fertile and sterile flowers, placed on receptacle (central core) which is dark green, stalk long (3x1cm) having ring near the distal end.

Male Flower

1. *Fertile male flower :* Tubular, with long perianth, bilobed, stamens 1-2 mm long.
2. *Sterile male flower :* Has solid perianth.

Female Flower

Female heads appear solitary or in pairs, distally placed compared to male, dark green, cylindrical or oblong (5-15 x 3-5cm),stalk thick, stout, distinct annulus at the top, subtended by a spathe like deciduous long bract (5-8cm)

Female flowers are tubular with pointed perianth, united at both ends, 3-7 angled, pointed or blunt, small projections, has spatula like styles and stigmas. Ovary is unicelled, style long and slender, stigma white and slender. Fruit is syncarous, oblong or pear shaped, with short protuberance, stalk long (5-10cm) and thick (1.5cm), rind thick (1cm), receptacle (central core) together are not separable from the fleshy soft or firm golden yellow covering (perianth) over the seed. Seeds are many, testa thin leathery, oblong–ellipsoid (4 x 2.5cm) which are covered by semi gelatinous exocarp and endocarp is horny, cotyledons unequal fleshy, endosperm either absent, or very small when present.

Artocarpus altilis 2n =27,28,56 (tetraploid)

Origin and Distribution

The exact region of its origin is not well established .The crop is considered to be native of Pacific and tropical Asia .The diversity of *altilis* has been seen to extend from Indonesia to Papua New Guinea. Presently the crop is grown all over the humid tropics of the world.

Key to economic spp.

A. Leaves deeply lobed 30-60cm long ------------------------- *A. altilis*

AA. Mature leaves entire 5-20cm long ---------------- *A. hetrophyllus*

Uses

The crop has wide uses, as indicated for heterophyllus, altilis is useful in varied ways.

Description

Plants are tall (30 m) trees, monoecious, evergreen under tropic and deciduous in sub tropical /or warm temperate weather conditions, trunk straight (5-8m tall and about 1.7 m in dia.), branches and twigs thick and spreading, buds are big (10-20 cm) covered with big stipule, lenticels, leaf and scar very prominent. Leaves dark green in colour alternate thick, leathery, shiny above and light green and rough beneath, ovate to elliptical, unlobed when young, but deeply lobed in 5-11 segments, petiole long 3-5 cm. Inflorescence axillary, stalk thick and long (8 cm) male inflorescence/catkin elongated/club shaped drooping (25 x 4 cm), flowers very small yellow, stamens 1, spongy. Female inflorescence short thick (10 x 7 cm) globose or cylindrical in shape usually upright stiff, flowers numerous embedded in receptacle, calyx tubular (tube like), ovary 2 celled, style narrow, stigma 2 lobed. Fruit is multiple fruit, formed from entire inflorescence, syncarpous, oblong cylindrical or globose, about 1 foot in dia., with 4-6 sides or faces, sometimes may have short spines. Rind initially green–later turns yellow, thick central cavity comprises of abortive flowers makes the edible part/pulp juicy light yellow in colour. Usually seedless and seeded forms are there and called as breadnuts. In these mostly the edible pulp is replaced by seeds which are long (2.5 cm) flat, or rounded brownish in colour.

Other important related species of importance from horticulture point of view are

A. lakoocha (Monkey Jack). This species yield edible fruits which are small in size

A. hirsuta semi wild species

A. champeden, bears small sized fruits with very strong odour.

A. hirsuta Lam. Common in western Ghats of India. Plants are tall growing about 80 feet or even more, leaves are broadly elliptic and measures about 6-9 inch in length, surface usually rough below, margin entire, male spike hanging down pendulous. Fruit 2-3 inch long upright erect, spines ¼ inch long and perforated seeds ovoid, hirsute up to ¾ inch long. Yields important timber which is teak like.

A. odoratissimus Blanco. Tree tall, leaves 7-9 lobed, villous beneath pilose along the veins above, lobes lanceolate, male flowers in a common conical receptacle, fruit small globose with many seeds.

A. hypargyraeus Hance ex Benth. Native of S. China and Hong Kong. Plants are trees, bracts rusty pubescent, leaves oblong 3-5 inch long, apex narrowly pointed, margin entire of sinuately toothed, glabrous shiny above, primary veins raised below, rough with white thick pubescence-tomentose. Flower heads obovoid single peduncles thickly pubescent, male head ½ inch long.

A. chaplasha Roxb. Deciduous tall growing plant, 100-120 feet even more, wood is yellowish brown changes quickly to brown or dark brown, somewhat strong and hard. The greyish brown bark exudes milky sap on incision. Yields durable timber, resists insect and white ant attack. The tree is found in abundance in the moist deciduous and evergreen forests in sub Himalayan regions of Bengal, Assam, Burma to Andaman and Nepal.

A. lakoocha Roxb. (Monkey Jack) The tree of this species is well distributed in the sub Himalayan region from hilly tract of Kumaon to Assam, Burma and Andamans. Is also present in abundance in Orissa, Bihar and along west coast. Plants are medium tall, deciduous trunk straight. Wood characters resemble that of *A. chaplasha*. The specie produces edible fruits which are roundish 2-4 inch in diameter, irregular in form, yellow with sweet sour taste. Full grown tree may produce as much as 7-11 quintals of fruit per year.

Jack fruit is highly cross pollinated fruit crop and since it is propagated mostly through seed, types or forms varying distinctly in fruit shape size, quality, projection on the rind, maturity period, taste flavour, acidity and sweetness are invariably spotted, or found. Such variability provides enormous scope for improvement through clonal selection method.

Cultivars

The crop is seed propagated and being cross pollinated crop the seedling are highly heterozygous. This heterozygosity has been made use of in the selection of different type, varying widely in number of characters like fruit shape and size bearing, quality, taste flavour sweetness, acidity and spikes on rind their density and form etc. Such variation provide suitable platform for further crop improvement. Two group are recognized soft flesh and firm flesh types.

Soft Flesh : The fruit yields to the pressure of fingers when ripe, pulp soft and very juicy, taste variable from insipid (flat taste) sweet acidic to very sweet.

Firm flesh : The flesh is firm, hard, and crisp, taste varying in sweetness.

Singapore or Ceylon Jack : Is early fruiting type (after 20 years), size as that of common jackfruit.

Rudrakshki : The fruits are round, rind smooth compared to jack fruit, fleshy, inferior in quality.

In recent times a number of jackfruit cultivars have been released for commercial cultivation mainly from south east Asia, particularly from

Malaysia, Thailand and Australia. Some which are more important commercially are described below.

Dang Rasimi : Originated in Thailand. Tree vigorous, spreading, must be pruned to check its height, very productive and high cropper 75-125 kg/tree. Deep orange fleshed cultivar. Fruit green to light yellow uniform in shape. Skin with sharp spines, which do not flatten or open at maturity, pulp deep orange firm mild in sweetness with pleasant flavour, suited for marginal areas due to its vigorous growth.

Tabouey : Originated in Indonesia. Tree top rounded, leaves dark green, shiny moderate in production 50-70 kg/ tree, a yellow fruited type. Fruit usually thin, long, tapers at stem end, often misshapen, bright yellow coloured, spines blunt and irregular, pulp pale yellow in colour, with no aroma, pleasant in taste.

Black Gold : Originated in Queensland, Australia. Tree medium in growth, spreading, with pruning tree can be maintained to a height of less than 3 meters. Regular and consistent heavy bearer 55-90 kg/tree. Fruit dark green, spines fleshy and do not flatten which makes harvest difficult. Mean fruit weight varies between 6-7 kg, pulp constitutes about 35%, seeds many 190-200. Pulp deep orange soft to somewhat firm, sweet with strong aroma, flesh separates easily from the ray compared to other cultivars.

Cheena : Originated in Malaysia, from cross between jackfruit and champedak (*A. integer*). Plant low growing, spreading, with annual pruning plant can be kept at 2.5 m. height, produces regularly, 50-70 kg/ tree. Fruit small 2.5 kg, uniform in shape, somewhat long and narrow, seed about 35-40, skin green, spines blunt, turns yellow on maturity. Pulp orange to deep orange fibrous, soft, flavour high and aromatic, separates easily from the rag. Pulp constitutes 33% of fruit.

Honey Gold : Originated in Queensland, Australia. Tree slow to moderate in growth, branches spreading, tree top open and spreading, can be maintained at a height of 2.5-3 m. by annual pruning. Deep yellow fleshed cultivar. Fruit dark green, blocky weighs about 4.5 kg, spines small sharp, usually open and turns yellow at maturity, pulp dark yellow to orange, comprises nearly 36% of fruit weight, seed about 40 or more (5% of total wt.) sweet aroma, flavour very rich. Produces 35-50 kg fruit per tree.

Apart from these some selection have also been made like J-30, J-31 and NS-1 (Malaysia) and are being cultivated in number of south east countries.

❑❑❑

20 Mulberry

Genus : *Morus*
Species : *alba, nigra, rubra L.*
Chromosome No. : 2n=28
2n=42 2n=28 (56, 84)

Plants are ornamental and fruit bearing trees have wide adaptability. Plant loves to grow on garden soils but also grow or rocky and marginal soils. The number of species is disputable but two species are native of North America. Mulberries are usually cultivated as food or for rearing of silkworms and for the edible fruits. The silkworm mulberry is *M. alba* and fruit bearing mulberry is *M. rubra*. In India and elsewhere mulberry is known primarily as a fruit bearing tree, although it is not yet planted extensively, though few

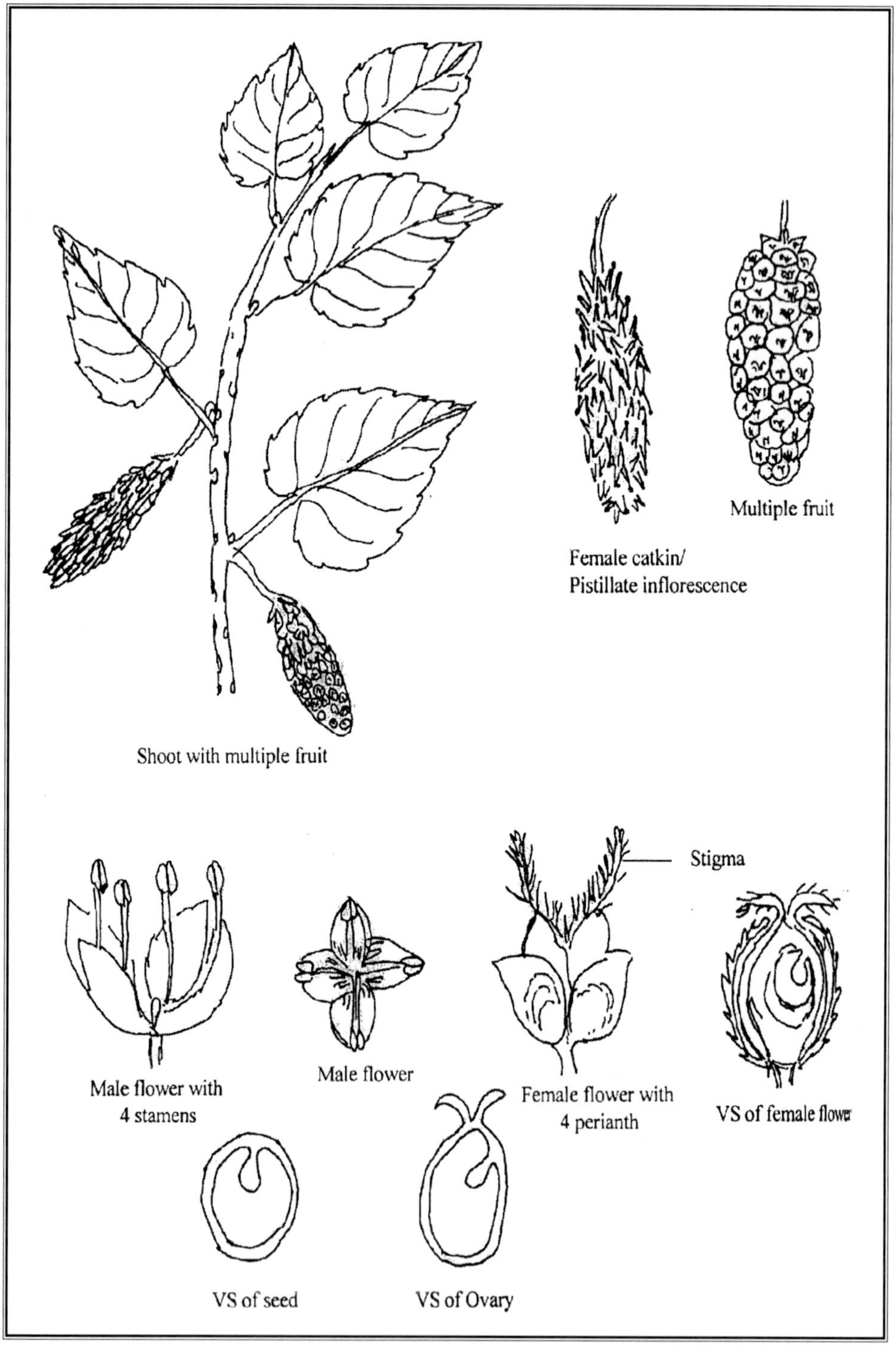
Shoot with multiple fruit
Female catkin/
Pistillate inflorescence
Multiple fruit
Stigma
Male flower with
4 stamens
Male flower
Female flower with
4 perianth
VS of female flower
VS of seed
VS of Ovary

trees around home ground are common. Mulberry is remarkable for its foliage and for variable forms. The extent of variability is so pronounced that several forms may be present of a single tree and different trees of the same species show distinct individual characters. The ripe mulberry fruits of different colours are also liked by birds, hogs and poultry.

Origin and Distribution

Mulberry has its origin in different parts of the world. White mulberry is considered to have originated in China and is grown mainly in tropical countries. The red mulberry is native of United States, whereas black mulberry is native to Iran and Asia Minor, was also cultivated in Mediterranean region for centuries. Mulberry cultivation is also mentioned in the Bible and so by Greek and Roman writers.

Uses

Ripe fruits are usually eaten raw or prepared into juice, jam and wine etc. Sometime it is also used as an ingredient in cough syrup, used in dye preparations etc. Leaves are fed to silk worm caterpillars.

White Mulberry
(*Morus alba* L.)
Origin and Distribution

White mulberry is native of China and has been cultivated for several thousand years there.

Uses

Fruits are generally eaten fresh or may be cooked. The leaves form the main food of the silkworm. In China bark has been used to make papers, wood used to manufacture sports goods like rackets and hockey sticks, light-yellowish brown wood is employed for decorative carving.

Description

Mulberry plants are medium sized, wide spreading trees and may reach 25 m. height monoecious and dioecious. The leaves are simple, alternate, smooth, light green in colour, broadest near the base, 6-16 cm long sharply toothed (serrate) along the margin, often deeply lobed. The female spike (catkin) is a syncarp, 2.5- 5 cm long, containing many drupes enclosed in fleshy perianth, may be white red or purple in colour.

Black Mulberry
(*Morus nigra* L.)
Origin and Distribution

Black mulberry is native of western Asia. The fruit crop specie of old world was introduced in eastern North America in the colonial days.

Description

The black mulberry plants are usually short, stout tree with wide spreading branches, the tree crown is rounded. The simple alternate leaves are dark green generally unlobed, but sometimes may be 2 or 3 lobed, about 4-14 cm long. Fruits often broadest at base to uniformly wide (1-1.5 cm) long cylindrical/ oblong, dark red to black coloured.

Red Mulberry
(*Morus rubra* L.)
Origin and Distribution

The red mulberry is considered to be native of North America and occurs occasionally extending from south-eastern Canada to southern Florida and Bermuda. Is a lowland tree primarily in river valley and flood plains and in rich soils of moist low hill sides.

Uses

Ripe fruits are eaten fresh, even squirrels, raccoons and birds compete for fruits.

Description

Red mulberry plants are medium sized (20 m tall), spreading, crown broad rounded, trunk, short stout straight, bark thin to thick, irregular fissures, ridges long, comprises of close scales, dark reddish brown. Branches thick strong, spreading, twigs, green smooth, later reddish brown. Leaves deciduous, simple alternate, thin dark bluish green, long (7.5-10 x 5-6 cm) generally broadest, almost rounded or heart shaped near the base, broad pointed at tip, entire or 1-3 lobed, margin serrated (sometimes doubly serrated). Leaf stalk and leaf (under surface) pubescent when young but at maturity becomes smooth. Flowers are small, greenish, bearing either male or female flowers, male flowers in stalked slender hanging cluster 3-5 cm. long, female flowers in broad elongated clusters, 2.5 cm. long. Fruits, cylindrical, fleshy, juicy, sweet, 2-3 cm. long, red, becoming dark purple to almost black. Seeds tiny, many, swollen at the base tapering to a point, light brown in colour.

Species of genus Morus also has number of botanical varieties and the name indicates specific plant leaf or fruit characters.

Morus alba var. pendula Dipp. Branches of this variety are characterized as slender and drooping.

Morus alba var. pyramidalis Ser. Trees are of pyramid form, that is lower branches are long and spreading which shortens towards the tip. Leaves are lobed and acute.

Morus alba var. laciniata Beissn (var skeletoniana, Schneid). Skeleton leaved mulberry. Leaves of this variety are deeply and continuously segmented into narrow lobes. The lobe at distal end (terminal) and two of the lateral are usually with long tips or point, which provides an attractive and distinct form.

Morus alba var. tatarica (Linn.). Russian Mulberry. Low growing with bushy top, small tree, leaves also small more lobed. Fruit variable in colour creamish white, violet or black very small, taste insipid.

– – – – – – – – – var venosa Del. This botanical variety has number of synonyms, fibrosa, nerrosa, urticafolia. Specie has ornamental value as veins are very conspicuous and white leaves are narrow rough edged and uneven outside. Fruits are about ½ inch long. Variety provides considerable distinct effects.

– – – – japonica Audib. Leaves are large thin long pointed, teeth deep and large, margin rough narrow, uneven and deeply lobed. Fruit small oblong red in colour.

– – – – – multicaulis Perr. This variety also has number of synonyms like latifolia, sinensis, and indica. Plants are small but strong growing, leaves are large and long pointed, dull green in colour, rough to feel rarely lobed. fruits are almost black, sweet in taste.

Botanical Types of Red Mulberry

Morus rubra var. tomentosa Bureau. Leaves with soft whitish pubescence on underside, shiny rough above.

– – celtidifoli HBK. Trees are smaller than red mulberry, leaves cordate ovate, somewhat lobed small and smooth. Fruit small almost globular black in colour and sour in taste.

M. indica Linn. The specie is found wild in the sub Himalayan region in the dry forests from the Sutlej to an elevation of 5000 feet. Cultivated to feed silkworm. A deciduous tree, leaves ovate, margin unevenly and coarsely serrate about 2-5 inch and petiole 1 inch long. Male spike occur on short slender stalks, female spikes (catkin) ovoid, short, styles connate pubescent. Fruit dark purple.

M. serrata Roxb. Found growing in Kashmir and in the Himalayan region. Used to feed cattle. A large deciduous tree, buds scaly, young shoots and under surface of leaves are tomentose, leaves cordate, apex short pointed margin coarsely toothed very often deeply lobed. Male spike hairy, female long and cylindrical, styles pubescent united below. Fruit purple in colour.

Cultivars

No standard cultivars of mulberry are known and grown commercially.

❑❑❑

21 Persimmon

Genus : *Diospyros*
Species : *kaki* L.
Family : *Ebenaceae*
Chromosome No. : 2n=60, 90

Persimmon-Kaki, though native of China, but it was in Japan where the fruit developed to its present status and much of the cultural practices were established or developed in Japan, hence the Japanese name by which it was popular in Japan was accepted all over the world. Japanese persimmon sometimes called as date plum and Chinese date-plum are the names generally used to describe persimmon in United States. The Chinese name is *Shitz* and in France it is called as *plaquemine*. In Japan Kaki is regarded to be their prime and best fruit and about 800 cultivars have been recorded and described. Likewise the Chinese also fancy persimmon and large areas have been put under its cultivation.

Persimmon is a tropical and subtropical fruit with its origin in China, but most of the development which has brought the fruit to its present cultivation status took place in Japan where it is regarded as a national fruit. The name Kaki is used for this fruit in Japan, other names by which the fruit is known are Japanese or Oriental persimmon, date plum and Chinese date plum. For simplicity, botanically the cultivated Kakis are grouped under *Diospyros kaki*. Persimmon is a subtropical fruit as it readily fails in most of the moist lowlands tropical areas. Despite long history of its cultivation in its native place the fruit failed to establish itself in the western countries reason being cultivation of astringent cultivars with poor quality were grown predominantly. The fruits of such astringent cultivars rendered the fruit inedible until it was complete ripe and soft, astringency being caused by soluble fruit tannin which decreased as the fruit ripens. With the findings of non astringent cultivars a renewed interest in persimmon cultivation all over the world has been initiated. Such types have gained greater popularity in very short period of time. Fruits of such cultivars are totally free of astringency and the fruit can be consumed while it is still firm and fruit colour has not changed to complete yellow, orange or orange red in colour besides they also have greater shell life and can withstand long transportation.

The persimmon of commerce belongs to the genus *Diospyros* family Ebenaceae and approximately has 200 species. Of the vast number of species the three most important fruit bearing species are *Dispyros kaki* (the Oriental or Japanese persimmon), *D. virginiana* the American persimmon and *D. lotus*, the date plum. The last two species are not so commercially important as is the Japanese persimmon because fruit in these is of inferior quality and hence are used as rootstock for the oriental persimmon.

Origin and Distribution

Diospyros is widely distributed in the tropical and sub-tropical regions of the world especially Asia, Africa and America. *D. kaki*, also called as kaki, the oriental persimmon is the classical fruit which originated in China and was later introduced to Japan in very early times. Persimmon's commercial centres has primarily remained clustered in Japan and China. In recent times California, Brazil, Italy, Israel have also started commercial cultivation of persimmon, whereas in Malaysia, Java and Sumatra it is grown on limited scale only.

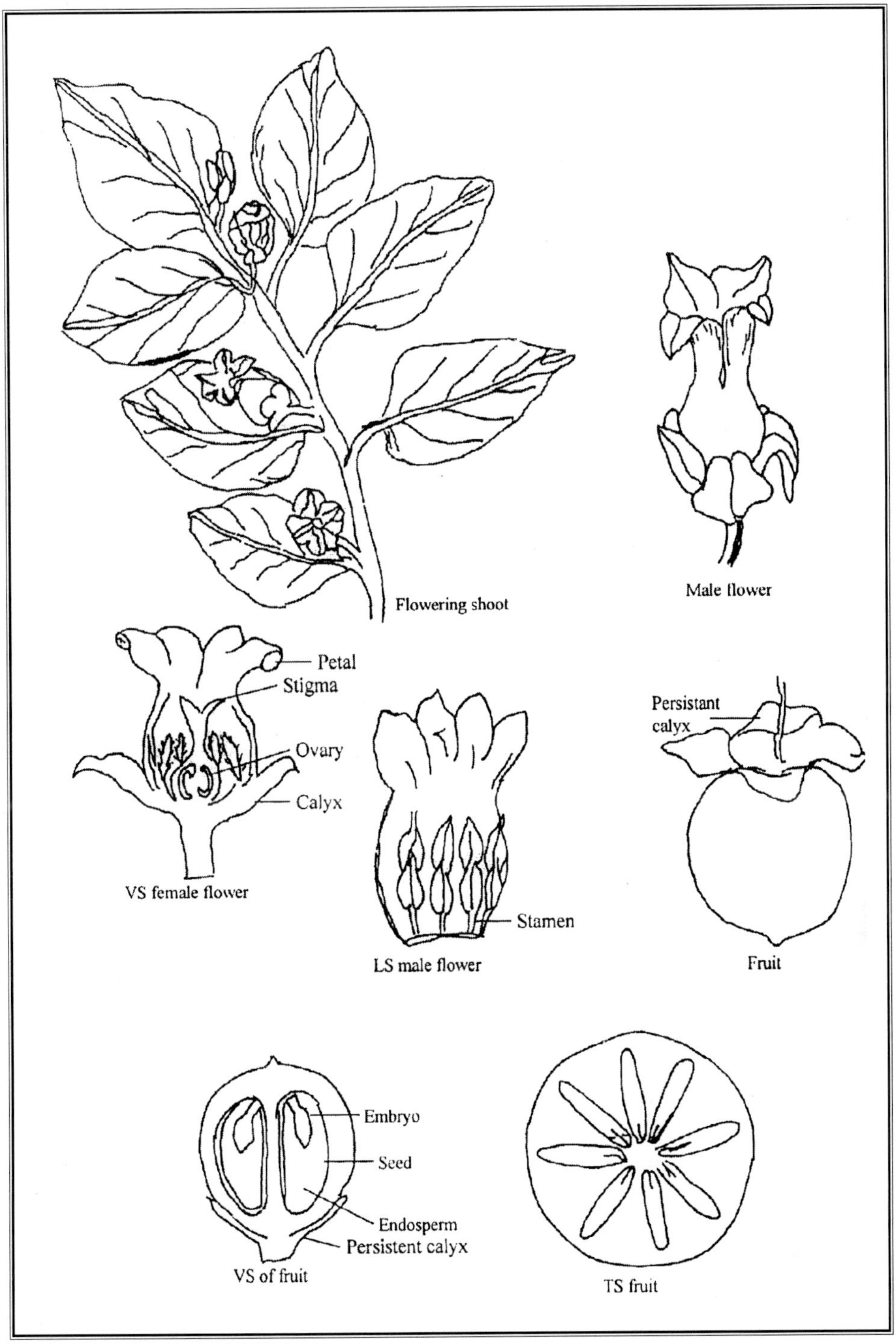
Flowering shoot
Male flower
Petal
Stigma
Ovary
Calyx
VS female flower
Stamen
LS male flower
Persistant calyx
Fruit
Embryo
Seed
Endosperm
Persistent calyx
VS of fruit
TS fruit

Uses

Persimmon yields quality wood Ebony wood, which is very hard, dark coloured (almost black) and wood takes fine polish hence commercially used in making cabinets and high quality furniture, decorative panels on interior decorations, Fruits without astringency are eaten fresh, pulp processed into jams, jelly, ice-cream and puree. Persimmon tannin makes paper and cloth durable after painting.

Description

Persimmon plants are slow growing trees, generally dioecious sometimes monoecious deciduous trees. Trunk short straight rough with age develops prominent dark grey to greyish brown square to irregular blocks. Crown round, branches profuse, spreading, branchlets slender light brown when young becomes darker with maturity. Young growth pubescent. Terminal bud absent, lateral buds 2-4 mm, with 2-3 reddish brown pubescent scales, apex pointed. Leaves are short and thick petioled, ovate orbicular to broad elliptic, margins entire, coriaceous, glossy, dark green above, pubescent, lighter beneath, and less pubescent. Pubescence more on mid ribs and along veins, apex round obtuse, some short pointed (acuminate). Male and female flowers are produced in leaf axils of current growth on separate or same tree. Male flowers produced in cluster of 2-3 sometimes 3-5 flowered cymes, in some cultivars the central flower is hermaphrodite in the male cluster. Calyx 4 lobed greenish yellow, corolla twice the length of calyx, stamens 14-24. Female flowers are stalked, solitary pendulous, 4 lobed campanulate, calyx, tubular-campanulate 4 lobed corolla, stamens 8 or with 16 staminodes. Ovary 8-10 celled with 1 ovule per cell. Fruit is a berry more so like tomato, quadrangular-globose, ovate, very variable in size shape and colour. Yellow green, orange to red in colour, with green 4 lobed persistent calyx, mesocarp thick juicy edible. Flesh orange to orange red purple. Seeds flattened, reddish brown, ovoid oblong in shape.

Pomological description of some useful species from horticultural usage is given below.

D. virgianana Linn. Common persimmon. Plants are tall 50 feet or more round topped head, with spreading branches often pendulous. Leaves ovate or elliptic, apex acuminate, shining above, pubescent below about 3-6 inch long. Flowers short stalked, greenish yellow, male in cluster of three's small 1/3 inch long, stamens 16, female flowers large single, style 4 lobed connate at the base. Fruit plum like globose or obovate, with large persistent calyx at stalk, pale orange often with red coloration on cheeks. Fruit edible variable in colour, size and flavour.

D. lotus. Medium tall tree 40 feet round topped, leaves oblong-elliptic apex acuminate, pubescent, very often glabrous about 3-5 inch long. Flowers reddish white, male/staminate in three's, stamens 16, female single. Fruit yellow when young turns black when fully ripe globular in shape ½ - ¾ inch in diameter, edible wide spread in western Asia and China.

D. texama Scheele (Syn *D maxicana*, Scheele). Tree small 40 feet tall, densely branched. Leaves oblong or obovate, pubescent below 1-2 inch long. Flowers appear with leaves, sepal and petal 5 segmented/lobed. Male flowers with 16 stamens, female with 4 connate/united pubescent styles at the base. Fruit small about one inch in diameter black in colour.

D. tessellaria Poir. (*D. reticulata* Willd). Plants are trees or shrubs, leaves leathery oblong or oval rounded at both the ends shiny above pubescent and reticulate below about 3-6 inch long. Flower occur in clusters, calyx tubular sessile, 4 segmented. Corolla 4 lobed stamens 12-13 glabrous. Fruit ovoid 1.5 inch long, edible.

D. discolor Willd. Tree medium tall 45 feet or more. Leaves are oblong about 10 inch long leathery/coriaceous. Flowers 4 segmented male flowers in 7 flowered cymes, petals usually white stamens 24 in number, glabrous, female flower single with 4 stamens, staminodes are found. Fruit globose in shape brown in color pubescent 4 inch in diameter, edible.

D. ebenum J. Konig ex Retz. Ebony East Indian Ebony. Specie is highly valued for its hard black marked with brown line ebony wood. Leaves are elliptic 4 inch long leathery, petioles short. Flowers fragrant mostly 4 merous, male usually in cluster with 16 stamens, female solitary. Fruit globose about 1 inch in diameter, small. Specie common in India and Srilanka.

Cultivars

Persimmon cultivars are classified as pollination constant or pollination variant and also as astringent or non astringent types. In pollination variant the fruit shape and colour changes according to pollination-flesh light coloured when seedless and dark reddish brown when seeded, whereas in pollination constant no change in flesh and fruit shape occurs.

Persimmon is rich in cultivar number in both astringent and non astringent types, Detail of important cultivars is given below.

Hachiya : Widely cultivated persimmon cultivar, fruit large, oblong glossy orange red astringent type, apex pointed calyx persistent. Astrigent till it is soft, usually seedless, quality very good, pollination constant.

Hykume : Fruit medium to large in size angular, four sided pale orange in colour russetted or mottled, pollination variant.

Izu : Non astringent, matures early, orange red flat in shape, appearance attractive susceptible to fruit fly attack. Plant with medium upright growth.

Jiro : Non astringent, tree upright and vigorous, fruit ripen in mid season, large orange red, oblate with eight conspicuous lines, calyx large, quality excellent.

Fuyu : Late ripening cultivar, plants medium in vigour with spreading growth habit, crown rounded fruit medium sized oblate, skin orange to deep red with four distinct lines, quality and storage excellent.

❑❑❑

22 Avocado

Genus : *Persea*
Species : *americana Mill.*
Family : *Lauraceae*
Chromosome No. : 2n =24

Horticulturally avocado (*Persea americana*) is a new fruit and is often considered to be one of the most nutritive of present day fruit crops. Genus *Persea,* name has been derived from ancient Greek name of an Egyptian tree with sweet fruits derived probably from *Perseus.* Avocado belongs to the family Lauraceae, has 47 genera and about 1900 species. Genus *Persea* has some 150 species of tropical evergreen trees. Its discovery in the new world has contributed much to the human diet. Fruit is a rich source of fat (27%), 2 per cent protein content is higher than in any other

Flowering shoot

Inflorescence

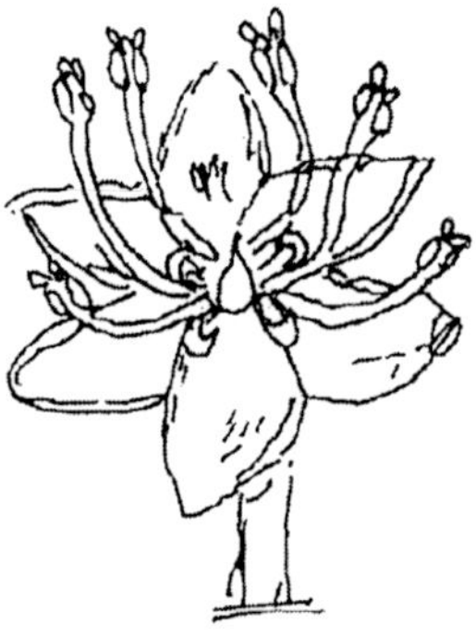

Flower

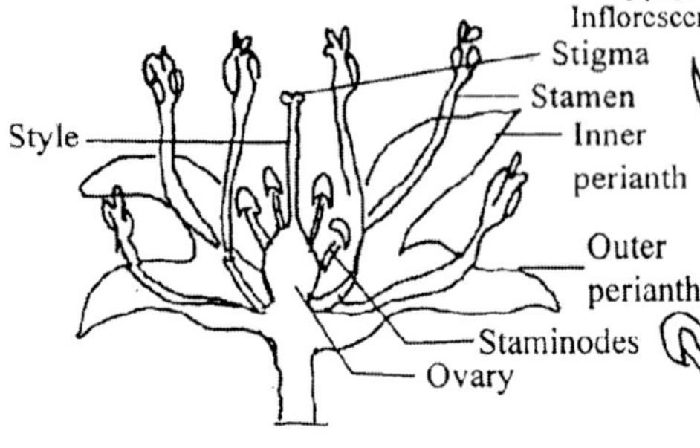

LS of flower

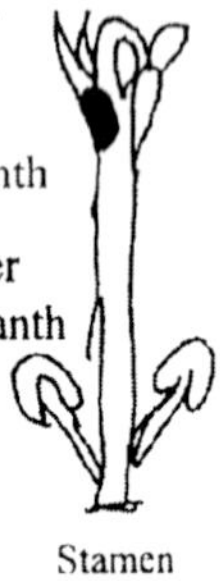

Stamen

VS fruit showing drupe

LS of Fruit

fresh fruit, mineral like Ca, Cl, Fe, Mg, P, K, Na and sulphur are also higher that makes avocado as a useful food which produces more of base forming elements rather than acid forming elements that are found more only in nut crops. Vitamins like A and B, ascorbic acid, niacin, riboflavin and thiamine are also in sufficient quantities, but lacks carbohydrates as they have no sugars. Pulp preserved by freezing, filling or spread of sandwich, milkshakes, ice cream etc. the buttery consistency of the pulp in general resembles that of cows butter, has flat taste and nutty flavour and lacks any smell. Fruit though has high fat content is harmless to human being but may cause some indigestion when taken in large quantities, further being low in sugar content it is very useful for diabetic patients, oil is used in preparation of number of cosmetics.

On the central Americans diet, avocado takes the place of meat as it is cheap nourishing, appetizing and above all the fruit is available throughout the year and in every likelihood it is presumed that avocado soon would become the staple food of people in United States.

Origin and Distribution

The avocado is said to have originated in the Chiapas- Guatemala-Honduras. The Spanish invaders also found avocado growing in Central America, Mexico to Peru and further down to Venezuela . From here the crop migrated to Caribbean islands and got established, and in the last century avocado reached south east Asia and today is grown successfully in number of tropical and subtropical countries of the world.

Uses

Avocado is used in a number of ways .The oil is utilized in cosmetic industry i.e. in soaps, skin moisturizer products, perfumes, also as traditional medicines. The central Americans have used this nutritious fruit in preparation of number of diets, dishes mixed with spices or herbs, ingredient of salads, dessert or vegetables etc.

Description

Avocado plants are tall (20m) vigorous , evergreen , round topped, trunk has fleshy bark, leaves simple, large, leathery, arranged spirally, shape variable, elliptic to lanceolate, ovate or obovate, young and opening leaves red tinged, later turning variable dark green, waxy, shiny above, glaucous beneath, midrib and veins depressed prominently below. Margins entire, young shoots/twigs often pubescent later becomes smooth. Oil glands are present both in leaves and bark. Inflorescence appear terminally, spreading panicle sometimes entire tree appears to be covered with the bloom. Usually indeterminate and ends with vegetative bud. Flowers in multiple of 3's, small half to 1 inch in diameter at anthesis, perfect, greenish yellow and fragrant. Perianth in 2 whorls 3 outer and 3 inner sepals, inner whorl longer than outer whorls, small (5mm long), tomentose, stamens, in 3 whorls of 3 each (9) and 1 whorl of 3 as staminodes. Stamen of inner whorl each has two orange nectaries at base, pistil with 1 celled superior ovary, style slender, hairy, stigma simple, pepillate. Fruit is single seeded large (7-20cm) fleshy berry, globose or pyriform in shape, skin of varying thickness, texture and colour yellow green to maroon or purple in colour, exocarp smooth or warty, thick (1-3mm), mesocarp yellow green, soft, consistency butter like. Pulp contains 2-30 percent oil or fat, seed large centrally placed, with 2 large fleshy creamish cotyledons encloses the small embryo which is present at the distal end of the stalk.

In avocado, one species is recognized with three (ecological) races, namely Mexican, Guatemalan and West Indian.

1. *Mexican:* Is native to the highlands of the Mexico, where wild forms and its progenitors are found, hardiest race, resistance to cold, commercially less important. Important characters are, leaves anise scented, fruits small (80g), skin smooth, seed large lie loose in the cavity, oil content high (30%) ripens in about 6 to 8 months after flowering.

2. *Guatemalan:* Native of highlands of central America, Guatemala in particular were wild forms are present in abundance, less resistance to cold than Mexican race. Fruits are larger (size/weight 1-2 lb), skin thick, hard brittle, very often warty, seed small held tightly in the cavity, oil content varies from 8-15 percent, takes 9-12 month to ripen after flowering, borne on long stalks.

3. *West Indian:* Native of lowlands of central America, least hardy of all races. Fruits are larger, but smaller than Guatemalan race, skin smooth and leathery, less thicker than Guatemalan, seeds large and lie loose in cavity, cotyledons rough, low in oil content 3-10 percent, borne on small stalks, leaves are light coloured (all races), fruits ripen 6-9 months after flowering.

Pomological description of some other important species is given below.

P. Barbonia (L) K. Spreng. (Red Bay, Sweet Bay Laured Tree Florida Mahagony). Tree 30-40 feet tall leaves 6 inch long lanceolate to lanceolate-oblong tapering at base, slightly pubescent, shiny, panicle with few or several flowers short pedicelled flowers. Fruit small ½ inch in diameter, dark blue or blackish. Planted as an ornamental in N. Amercia.

P. indica Spreng. The tree has attractive foliage with 3-8 inch long oblong elliptic or lanceolate oblong leaves Panicle 3-6 inch long, 3-5 flowered, flowers white about ¼ inch long. Fruit hardly with any flesh.

Cultivar

Several avocado cultivars after selection made from seedling population through vegetative multiplication are presently under cultivation. Brief description of more important one is detailed below.

Fuerte : A hybrid between Mexican and Guatemalan races, very popular cultivar grown all over the world. Fruits are large pyriform with 20-26 per cent oil, performs well under subtropics than in tropics although withstands some cold.

Hass : Is a seedlings selection from Guatemalan race, an early maturing cultivar than Fuerte in California. Suitable for cultivation in subtropical conditions. Fruits are rounded, medium in size, initially, green and turns to purple on ripening surface not smooth.

Lula : More suitable for cultivation under tropical climate conditions. The cultivar has originated in Guatemala, fruits are very large and weight ranges from 400-700 g, with about 15 per cent oil, pyriform in shape.

Adrith: Is a Californian cultivar produces regularly, yields are also high, fruit medium sized, average weight being around 250 g, green in colour flavour good.

Green : Belongs to Guatemalan race, fruit large oval, neck thick and stout, skin thin and warty surface rough yellowish green, flesh greenish yellow later turn green at skin, thick 2-2.5 cm., flavour nutty, seed lightly placed in the cavity.

Purple : Belongs to West Indian race, fruit large pear shaped, neck distinct and long, skin shiny smooth dark red to maroon in colour, leathery easily peels off from pulp, flesh thick 2 cm, smooth deep yellow at seed and yellowish green at skin (periphery), flavour rich and nutty, seed lie loose is cavity.

❑❑❑

23
Sapota

Genus : *Achras / Manilkara*
Species : *sapota* L. *zapota (Mill.) Forberg*
Family : *Sapotaceae*
Chromosome No. : 2n=26

Thomas Firmingaer, an English horticulturist while in India in praise of sapota wrote 'a more luscious, cool and agreeable fruit is not to be within this or perhaps any other country in the world'. A French botanist described sapota as 'the sweet perfumes of honey, jasmine and lily of the valley'. In view of these the fruit is considered to be one of the best fruits of tropical America, although cannot be compared with the other more popular fruits of the region i.e. pineapple and cherimoya. In many tropical parts of the world this delicious fruit is held in high esteem. The tree wood is hard and very durable and the branches usually extend horizontally and are low on trunk which help the plants to resist cyclones and hurricanes more effectively than many other tropical fruit trees. Sapota / Sapodilla usually is a dessert fruit and rarely is cooked or preserved in any way.

Acharas zapota, sapota or sapodilla is an important fruit introduction from tropical America (South America). This delicious fruit crop in short time has naturalized well to the Indian conditions and locally is called as Chiku. The unripe fruit and bark produces a milky white latex which quickly solidifies when comes in contact with the air. This latex (gutta parcha) forms the base for chewing gum industry for making of chicklets and is commercially used in number of countries like Guatemala, Honduras, Mexico. The fruits when fully ripe are delicious, sweet in taste and with distinct flavour, provides valuable raw material for the manufacture of industrial pectin, glucose and number of other products like natural fruit jellies. The immature fruits however are very astringent. Fruits are fair to good source of carbohydrate, protein, fats, mineral matter like calcium phosphorus and iron, ascorbic acid contents, is considered as one of the best fruits of tropical America although it fails to compete with more esteemed tropical fruits like the cherimoya or pineapple. The common name sapodilla is from Spanish word Zapotillo meaning small sapote.

Sapota is predominantly a dessert fruit and rarely it is preserved or cooked in any way Fruit is not rich in various component though has 1.1 g (per 100 g edible portion) of fat, adequate in mineral contents like calcium phosphorus and iron and is low in ascorbic acid (6 mg.). *Acharas zapota* is the only specie important horticulturally as it produces edible fruits.

Origin and Distribution

The edible sapota is placed under sapodilla a group of fruits which have their origin in central America, Mexico and West Indies. The crop after its migration to the south east Asia, has naturalized to prevailing weather conditions so much so that it is extensively grown all over south east Asia. The crop has adapted to the tropical conditions of both the hemispheres.

Uses

Sapodilla fruits with dry and gritty texture are eaten fresh, or processed into jams, preserve or used for topping of ice-creams, juice fermented to make vinegar or wine. White latex from injury to plant is the base for chewing gum

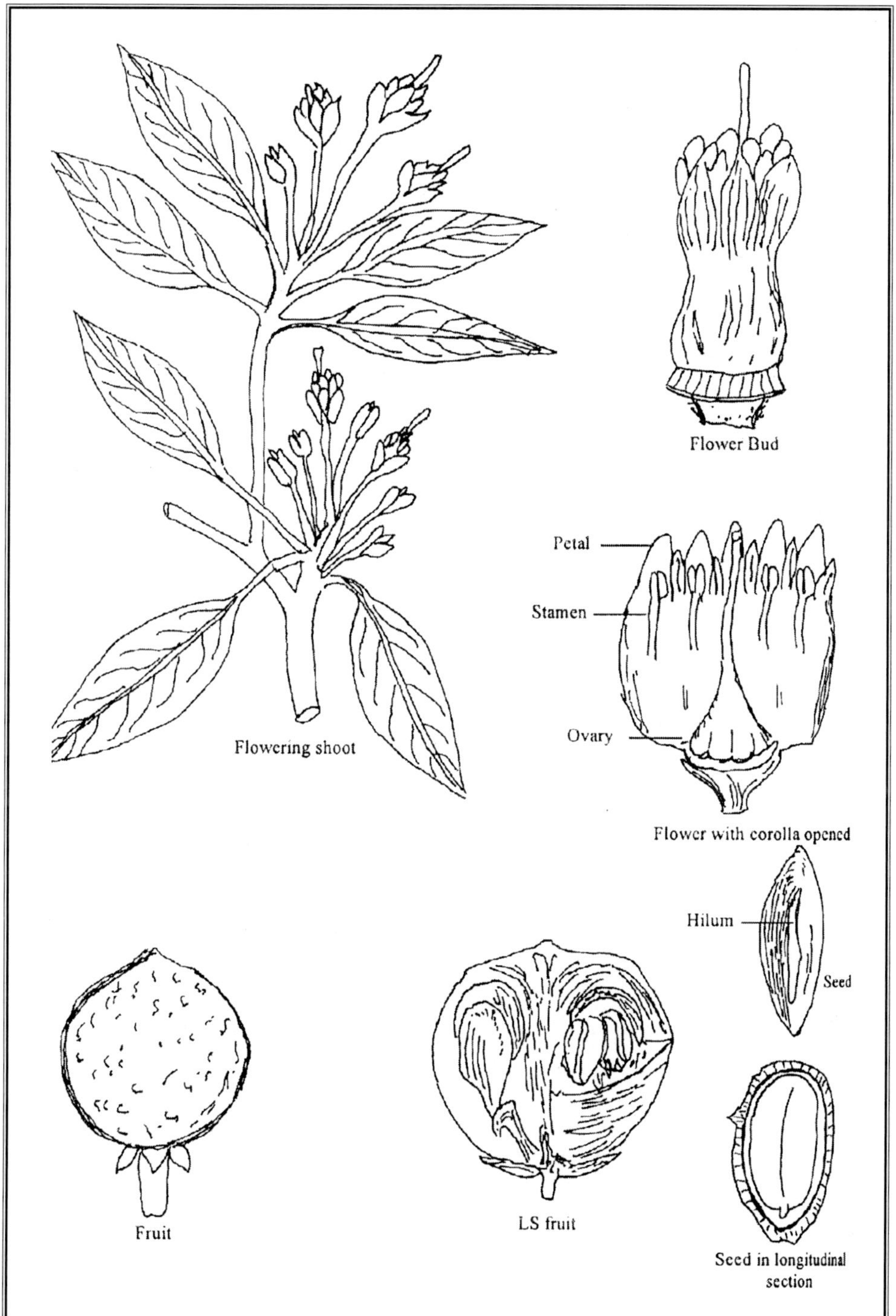
Flowering shoot
Flower Bud
Petal
Stamen
Ovary
Flower with corolla opened
Hilum
Seed
Fruit
LS fruit
Seed in longitudinal section

industry, and for dental surgery. Tannin from bark is employed to cure fever and diarrhoea. Wood being hard and heavy form excellent timber for furniture.

Description

Plants are tall trees, spreading/ upright, slow growing, evergreen in nature bark rough contains milky latex with whorls of horizontal branches produced at regular intervals, dark brown in colour, tree top round/ globose or pyramidal in shape, trunk low branched. All plant part on injury exudes white latex. Leaves simple, alternate, dark green, glossy above, lighter with dense pubescence below, ovate elliptic to oblong lanceolate, wedge shaped (cuneate) to broadly acuminate at both the apices, margin entire, mid rib depressed below, thick and prominent below, lateral veins parallel and many, petiole medium thick and long. Flowers axillary and solitary in upper leaf axils and in clusters of 2-3 or more in lower leaf axils, pendulous, about 1.5cm in diameter, regular and perfect, pubescence hair like thick and brown, pedicel short (1-2cm). Calyx brown 6 present in two whorls, outer three united at base and pubescent, inner three are light green, rounded and leathery. Corolla 6, gamopetalous lobed in single whorl, tubular, longer than calyx, white companulate about one third of its length divided into 12 segments which are equal oblong erect and glabrous. Stamens, six perfect and six staminodes, usually placed between petals, filaments short, erect obliquely placed on calyx tube, yellowish brown, four celled, ovary 10-12 celled superior, syncarpous style, oval shaped (subulate) distinctly exerted from the flower before opening, stigma oblique, disc or cushion like growth prominent on ovary. Fruit is round, globose ellipsoid hanging berry, apex with style remains at stalk end, skin rough, thin, yellow -brown (rusty brown) to reddish overlaid by fine scaly matter. Epicarp thin mesocarp (flesh) is yellowish brown, granular at skin/ fruit periphery and yellowish and soft at seed, unripe fruit astringent but on ripening develops sweetness flavour and juicy pulp. Seed number variable 0-12 usually 2-4,oblong, laterally compressed black and brown coloured, separate easily from pulp, hilum distinct.

Cultivars

On the basis of variation in tree growth habit foliage and nature of branches two district varietal group has been established namely,

1. Tree with erect branches : Branches whorled, foliage intense green, broad oval, fruit yellow smooth large, pulp sweet butter like.
2. Tree with drooping branches. Foliage light green narrow elliptic fruit round brownish taste inferior.

Important cultivar grown in India are described briefly.

Kalipatti : Commercially cultivated in Karnataka. Gujarat and Maharashtra. Tree medium tall round topped, branches spreading, leaves are dark green broad ovate, thick, margin entire. Fruit oval occur singly, flesh soft, sweet quality excellent, seeds 1-4, heavy cropper.

Chhatri : An important cultivar of Maharashtra. The tree form resembles an umbrella hence the name, branches appear in whorls on the trunk in all directions and are horizontal and drooping, leaves ovate-lanceolate light green in colour margin entire. Fruits resemble Kalipatti, lacks in quality less sweet but good yielder and produces regularly.

Cricket Ball : Known as Calcutta Large, grown mostly in West Bengal, Andhra Pradesh, Karnataka, Tamil Nadu, Maharashtra etc. Performs well in drier areas. Tree round topped, branches are spreading, leaves light green shiny, fruits are large round (resembles cricket ball) surface rusty brown, pulp very granular/gritty moderately sweet, does not bear heavily.

Oval : A shy bearing cultivar like Cricket Ball, fruits small to medium in size oval in form, flesh gritty or granular, less sweet,

Thagarampudi : Grown mainly in Tamil Nadu, is regarded as an export cultivar. Tree has round top, branches are spreading, leaves green ovate to ovate lanceolate, fruit round oval in shape medium sized, skin thin and rough flesh dull coloured very sweet in taste melting and juicy, seeds 1-4.

❑❑❑

24
Ber

Genus : *Zizyphus / Rhamus*
Species : *mauritiana Lam. jujube Lam.*
Family : *Rhamnaceae*
Chromosome No. : 2n = 24, 48, 96

According to famous writer David Fairchild it is the jujube, the large fruited Chinese type which is at its best compared to the inferior type which are in abundance in Arabia, northern India and southern Europe. The resultant inferior fruits are either from a different species compared to those of China or from varieties which have not been much influenced and improved by selection and cultivation. Pliny in his work has indicated that jujube was introduced from Syria to Rome by the end of the reign of Augustus, showing its presence in southern Europe for over 2000 years. In most of the case both the species are cultivated in China, Philippines, from Malaya to India, Africa, westward through Afghanistan, Arabia, Persia, Asia Minor to Mediterrranean coast of Spain, France and north Africa. In India *Z. mauritiana* is called as Indian jujube, ber or bor and *Z. jujube* is known as common Jujube anab or unnab, while in China usual name is *tsao*. As regards climate conditions Fairchild has indicated that 'no weather is said to be too hot for it and can withstand low temperature of 22°F without injury'. The successful cultivation of ber is determined more by the length or duration of the summer season and not by the extent of winter. The jujube is said to be very precocious and fruits bear heavily and on very rare occasion it fails to produce a good crop. The crop has a productive life of about twenty five to thirty years and does not live above forty years.

Amongst the several tropical and subtropical fruit crops ber is perhaps the one which is most suitable for cultivation under arid and semi arid where most other fruits fail to grow for one or the other reason.

Originated in India and is found growing in wild, semi wild and cultivated forms in all parts of India extending from foot hill of Himalayas to southern tip.

Origin and Distribution

Indian ber is considered to have its origin likely in the Middle East or the Indian subcontinent. Commercially the crop is cultivated in India and China and on a limited scale in south east Asia and other tropical parts of the world.

Uses

Almost all parts of the plant are used in one or the other ways, fruits are generally eaten fresh, unripe fruits are eaten with salt, leaves serves as fodder, also cooked as vegetable. It has medicinal properties especially against digestion and to poultice wounds. In India, the species is employed to rear lac insects, besides shellac is also made. The wood is very fine in texture, used to make implements and utensils. Dye is extracted from fruit and tree bark.

Ber is often called as poor mans fruit but is rich in its nutritive value as it contains good amount of sugars, vit C and B, ber roots, bark and seed also contain alkaloids, mostly eaten fresh but products like the squash, juice, ber butter, candier pickle are also made from ber extract, In Ayurvedic and Yunani, medicines are made that help in digestion and blood purification, bark overcomes diarrhea, root decoctions and powder are useful to overcome ulcers, fever and healing of old wounds.

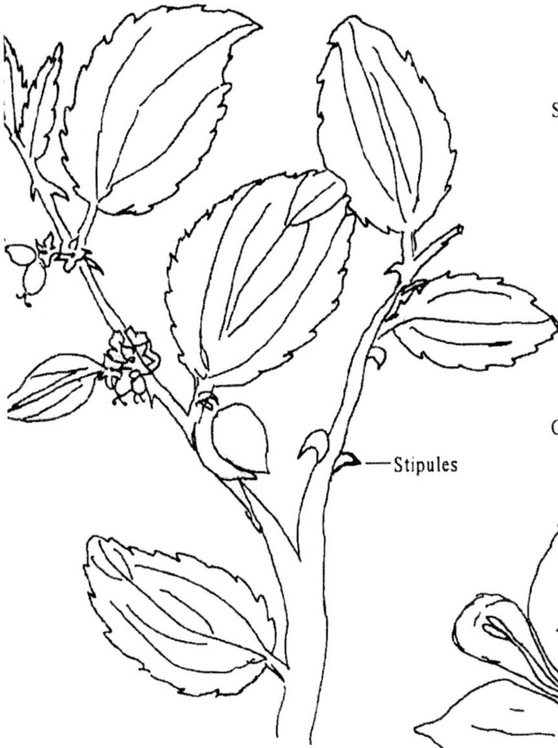

Flowering and fruiting shoot

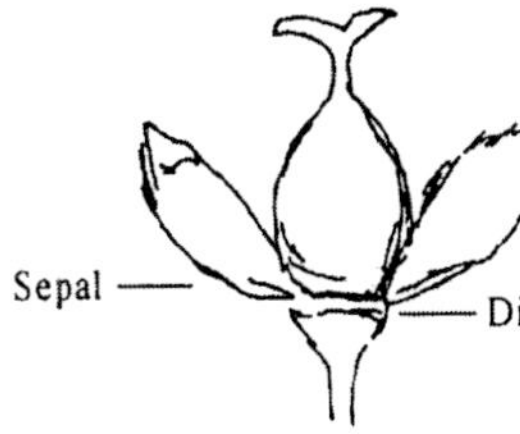

Ovary with sepals

LS of flower

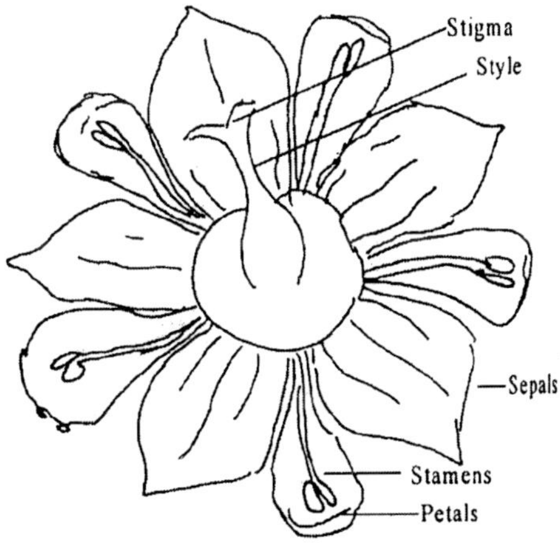

Cut section of flower

VS of fruit

VS of fruit

Description

Plants are small trees or bushy shrubs (25-40 feet tall), somewhat spiny, erect or spreading in nature with evergreen or semi deciduous branches either drooping or in a zigzag fashion, highly tomentose in Indian jujube, whereas Chinese ber are glabrous beneath. Stipules spinous, single and straight, or may be in dimorphic pairs, the second being shorter and curved, spines sometimes missing. Leaves simple, short petioled, alternate, round, broad elliptic ovate to oblong elliptic (2-10 cm x 1.5-5 cm), margins slightly crenate or entire, dark green glossy and glabrous above, with thick dense pubescence (tomentose) below, marked with 3 distinct longitudinal veins. Flowers are borne in axillary cymes both on one year old as well as current season growth, measuring 1-2 cm in length with about 7-20 flowers. Flowers are small (2-3 mm across), pale yellow, mild fragrance, short pedicelled (3-8 mm), calyx green 5 lobed star shaped (deltoid) distinctly hairy outside and glabrous inside, petals 5, pale yellow to yellow, curved, reflexed, subspathulate in shape, stamens 5, pale yellow, small, placed alternate to sepals, styles usually divided, bifid, ovary 2 celled, suppressed, placed on a disc which is distinctly lobed (10) or grooved. Fruit is a typical drupe, round, ovoid or globose in shape, skin smooth, thin, generally tough, glossy. Fruit colour variable (yellowish to reddish or sometimes blackish), flesh crisp, juicy, white, sweet to sub acid in taste, becomes mealy when fully or over ripe, seed irregularly furrowed each having 1-2 elliptic brown kernel.

The genus Zizyphus approximately has 40 species and belongs to the family Rhamnaceae. Two species *Z. mauritiana* (Indian ber) and *Z. jujuba* (Chinese ber) are the most important as they produce edible fruits and are of commercial significance.

Z. mauritiana (Indian or cottony jujuba) is more adapted to tropical and subtropical regions. Plants are vigorous in growth, branches are drooping evergreen (may partially shed leaves is summer) leaves highly pubescent (tomentose, almost silvery) on the undersurface, flowers in autumn, fruit ripen is spring are round to oval greenish yellow to reddish brown on ripening.

Z. jujube. Common jujube, Chinese Jujube, Chinese Date. Mostly found in temperate parts of the world, sheds leave in winter, plants are small somewhat spiny tree upright with zig zag branches, flowers is spring on slender deciduous branchlets sometimes on old wood also and fruits ripen in summer or autumn, are oblong elliptic to ovoid in shape 2 inches in length, skin thin, dark brown flesh white crisp textured, sweet in taste, seed hard two celled elliptic or oblong in shape surface is rough.

Beside these two species other species which are of significance from horticulture perspective are described below.

Z. joazerio Mart. Leaves broadly ovate 2-3 inches long apex acute base cordate glabrous margin serrulate. Flowers occur in many flowered cymes. Fruit about the size of cherry yellow, native of Brazil.

Z. lotus Lam. Native of South Ecuador and N. Africa. Small shrub 3-4 feet high very thorny or prickly. Leaves ovate oblong, margin crenulate glabrous. Flowers appear in few flowered axillary cymes. Fruit subglobose yellow.

Z. mistol Griseb. Native of Argentina. Small tree 30 feet tall spiny, leaves coriaceous/leathery oval or obtuse, base subcordate, petiole about 1 inch long, short, margin finely serrated, pubescence dense. Fruit black about 1/3 inch in diameter.

Z. nummularia DC. Found wild in temperate Himalayas, the spines are about ½ inch long one slender and other shorter hooked and bent downward. Leaves ovate in form, apex acute, margins finely serrate-crenate. Fuits are small ½ inch long, black elliptic, with acid taste edible, stone rugose shell bony to hard.

Z. rotundifolia Lam. Native of Persia and area eastwards. Plants are shrub, spines dense, leaves ovate to rounded orbicular highly pubescent (tomentose) on both leaf surfaces. Fruit globose and black in colour.

Z. rugosa Lam. Found in abundance in foot hills of Himalayas. Large evergreen shrub or a small tree sometimes climbing types are also found, spines prominent. Leaves elliptic glabrous above 3-6 inch long. Fruit fleshy usually one seeded globose or obovoid.

Z. spina christi Willd. Specie is native of N. Africa and western Asia. Plants are small spiny trees. Leaves oval to oblong margin crenulate surface shiny or pubescent more on the veins below. Flower occur in leaf axils in cluster, pedicels very pubescent. Fruit ovoid globose in shape, red in colour.

Z. oenoplia Mill. The plants are found growing in plenty in the sub Himalayan region. A climbing shrub type, spines sharp and small, single. Branchlets, inflorescence and undersurface of leaves are tomentose. Leaves oblique ovate or ovate lanceolate, shortly dentate or entire with two nerves, lateral veins many. Flower occur on short axillary sessile cymes. Fruit small black edible, seed tuberculate.

Z. xylophrus Willd. Found growing wild in the foot hills of north west Himalaya, central India and western Peninsula. Shrub or small tree straggling in nature, leaves, inflorescence and branchlets all tomentose, leaves broad ovate, base cordate with 3 sometimes 4-5 distinct nerves. Fruit globose with grey pubescence about 1 inch across, totally dry stone 3 celled 3 seeded and furrowed.

Z. sativa Gaertn (*Z. vulgaris* Lam). Common Jujube native to south Europe, east Asia. Shrub or a small tree, prickly or unarmed shiny, spine usually more than 1 inch, branchlets slender and usually fascicled and resembles pinnate leaves. Leaves ovate or ovate lanceolate, apex round or sharp, base oblique margin finally serrated. Flowers in axillary cymes, yellowish, fruit oblong or ovoid with long stalk dark red in colour.

Cultivars

On account of variability present in its natural habitat (India) a number of types have been selected and on account of distinct characters they are presently being cultivated as cultivars. The nomenclature is somewhat loose as one cultivar has been assigned different names. Some important cultivars grown in India with their pomological traits are given below :

Kaithli : Native of Kaithal. Tree has bushy growth medium tall, branches semi spreading thorny, thorns straight, leaf ovate margin narrowly serrate, upper surface dark green less pubescent with brownish colouration under surface smooth, fruit ovate oblong with short pointed apex, stone oblong with pointed apex, surface furrowed, sweet quality good.

Umran : Medium sized tree, bushy growth, branches drooping with few thorns, each thorn curved, leaf oblong ovate, margin serrated, pubescence slight on upper surface smooth below. Fruit elliptic, fairly large with faint lines, stone oblong furrowed, sweet in taste with good keeping quality.

Nazuk : Tree with round top medium tall, branches with few thorns, spreading, leaf oblong shiny green on upper surface dense white pubescence below, apex rounded, fruit elliptic oblong surface warty (not smooth) seed oblong furrowed.

Chhuhara : Medium tall tree, branches spreading, leaf ovate oblong margin serrate, under surface smooth upper with thick pubescence, fruit ovate oblong sweet in taste, yellowish green in colour, skin smooth, stone sub globose with furrowed surface, apex pointed.

Ilaichi : Plant medium tall, branches spreading leaves obovate, base narrow. Fruit round flat to rounded, green shiny surface later turn golden yellow to chocolate brown in colour as fruit ripens, flesh absent.

Sandhura Narnaul : Selected from Narnaul (place in Haryana). Plant with more upright growth, leaves broad ovate, base and apex both rounded light green in colour. Fruit ovate-oblong, apex short pointed, greenish yellow to golden yellow when ripe, sweet in taste.

❑❑❑

25 Pomegranate

Genus : *Punica*
Species : *granatum* L.
Family : *Punicaceae*
Chromosome No. : 2n = 16,18

Pomegranate has originated in Afghanistan and Iran the abode of prophet Mohammad and for very long has been associated with grape and fig and is still held in higher esteem. Acknowledging the multi usefulness of pomegranate Prophet Mohammad said 'Eat the pomegranate sententiously for it progress the system of envy and hatred'. The crop was very much in demand and liked by the inhabitants prior to the days of Mohammad in the Orient, as pomegranate has something special in the form of refreshing which the people of hot arid region like and adore

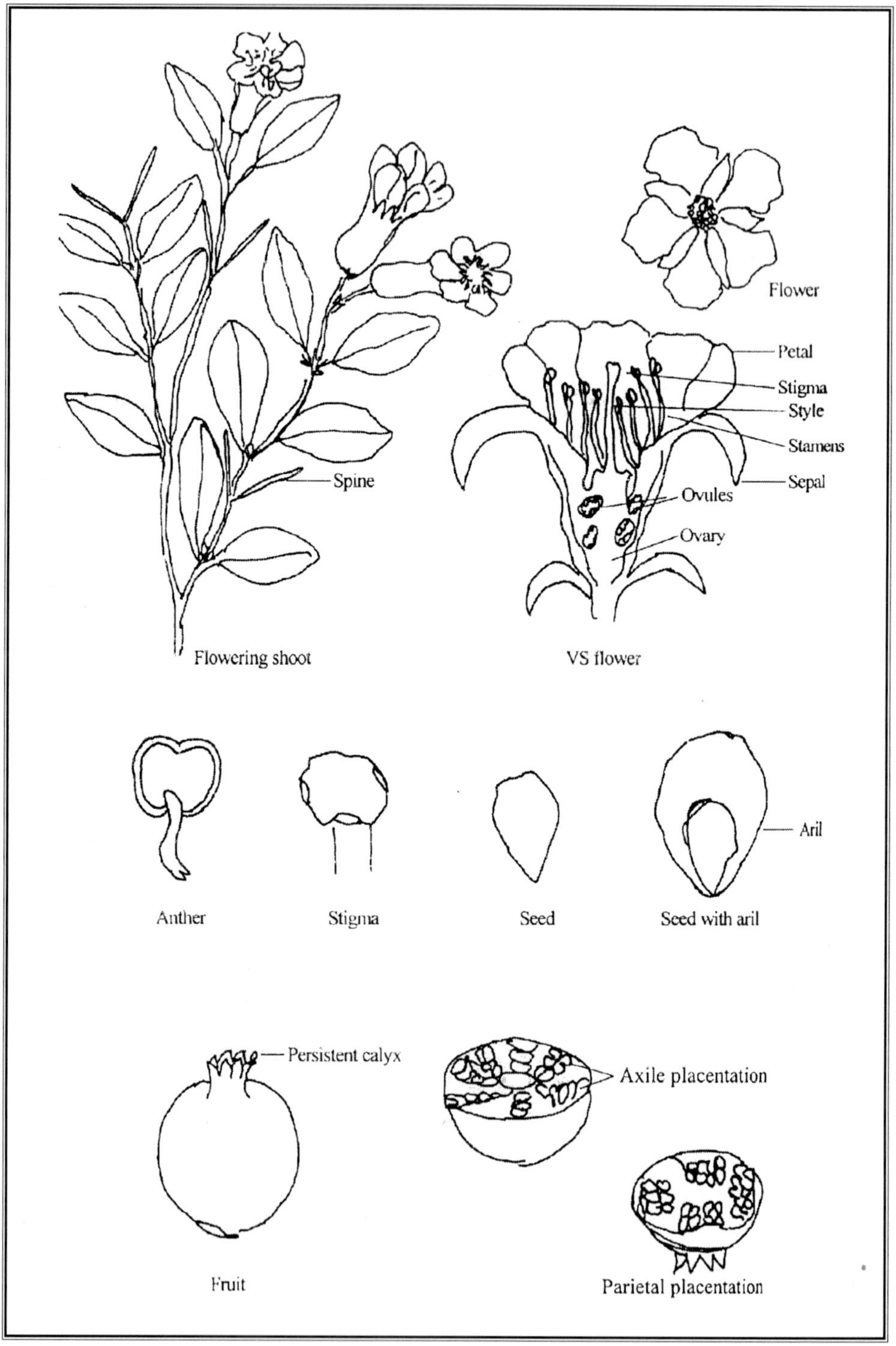
Flower
Petal
Stigma
Style
Stamens
Sepal
Ovules
Ovary
Spine
Flowering shoot
VS flower
Aril
Anther
Stigma
Seed
Seed with aril
Persistent calyx
Axile placentation
Fruit
Parietal placentation

enormously. The ancient Semitic name *rimmon* was adopted by Arabs as *rumman*. The Potruguese *roma* or *roman* was formed from *rumman*. The present botanical name *Punica granatum* has been derived from Roman name *malum punicum* (apple of Carthage) and *granatum*. Of the various names given to the fruit, the Indian name anar is considered to be the most common.

Pomegranate (*Punica granatum*) a subtropical plant is winter hardy drought tolerant and can even withstand desert conditions and therefore has been rightly called as plant with versatile adaptability to varied climatic conditions which is evidenced from the fact that the tree is deciduous under low temperature condition and evergreen or partially so under tropical or subtropical conditions. The plant performs best under semi arid climate where cool winter and hot summer prevails for prolonged time especially during ripening, hence resembles date palm, though is less exacting as regards heat than the latter and more frost resistant. Pomegranate is commercially grown for its sweet acidic fruit, also as an ornamental type on account of its green shiny foliage and large bell shaped orange red flowers. Pomegranate is a good source of protein, carbohydrate mineral matter (Ca Mg P and Fe) and appreciable quantities of oxalic and nicotinic acid, thiamine and riboflavin, seed yield oil which has scope in industrial use.

Origin and Distribution

The crop is native to Asia especially to Afghanistan, Iran and the Himalayan region. Later it moved to Mediterranean and become adapted and got naturalized to this region. At present pomegranate is grown all over the world in the tropic and subtropic conditions. The plant has high adaptability, under tropical conditions there are usually three flushes and each may also result in flowering. In the subtropics, trees flower in spring along with a flush of vegetative growth. Fruits of this flush ripens late in summer, whereas the vegetative growth during same period flowers in the next spring.

Uses

The fruit is subjected to number of uses, is eaten fresh, from extracted juice refreshing drink is prepared, has medicinal uses too. From dry fruit rind gargles are prepared, fruit juice controls fever. The bark , leaves and fruits are good source of tannin and on account of this is effective against diarrhoea and dysentery, tapeworms and intestinal worms.

Description

Small deciduous shrub or small trees (6-10 m) largely crooked. The plants are deciduous under subtropical and evergreen under tropical conditions.

Trunk thin, much branched at the base. The branches and lateral shoots end with spines. Leaves may be clustered, opposite or sub opposite on small axillary brachlets, leaves shiny, very short petiole broadly furrowed, dark green and shining above and light green below, oblong- lanceolate, base acute or obtuse, apex obtuse or acute, midrib prominent. Flower buds in evergreen varieties are borne on mature wood of previous season's growth, while those flowering in rainy season, flowers are borne on current season's growth whereas in deciduous varieties flowers are borne only on current growth. The flowers are found either terminally or in leaf axils and on short side shoots in clusters of 1-5, campanulate (bell shaped), about 4-5cm long and wide, calyx fleshy gamosepalous and persistent, toothed, 5-8 in number, red, pale yellow in colour, petals red white variegated in single or double whorl, generally crinkled, stamens numerous/infinite, placed irregularly on the calyx tube, unequal in length, filament slightly curved at tip. Ovary 3-7 celled many ovules in each cell, style solitary with swelled base, yellowish red stigma depressed, faintly grooved and suppresses the stamens. Fruit is a round/globose berry with persistent calyx also with reminent of style and stamens, rind leathery, varies in colour, yellow-green to black-violet. Fruit when cut is separated by membranous walls and into different compartments by white spongy tissues. Aril the outgrowth of seed forms the edible part, is juicy and of variable colours, red, pink or yellow-white. The fruit has two types of placentations i.e. axile and parietal, for which the punica berry is designated as Balusta, about 6-12 cm in diameter, brownish yellow to red. Seeds are angular, hard to soft.

The genus *Punica* has two species, *Punica granatum* and *Punica protopunica* the latter is wild type and in found growing wild in the Socotra island while the former specie is a cultivated type and is grown in almost all tropical and subtropical regions of the world. *Chlorocarpa* and *Porphyrocarpa* are the two subspecies of specie *granatum*, the former is indigenous to Trans-Caucasus region and latter specie is native of central Asia.

P. granatum var nana. Dwarf pomegranate. Not very tall growing usually has ornamental value light green shiny foliage with double flower, specie is hardy, fruit yields acidic aril. Pomegranate has limited species variability. Apart from those described no other specie is known from cultivation point of view.

Specie *granatum* has number of botanical forms or varieties which have tremendous ornamental value as the flowers are mostly in double whorls, variegated and are beautifully coloured but do not fruit. Some well established forms are described.

P. granatum var. Double Red. The calyx tube much larger than the other forms from which emerges a number of large bright scarlet petals. Flowers profusely more so in summer and fall.

.... var. Double Dwarf or Punica nana racemosa. Has dwarf growth habit, flowers occur in cluster usually double flowered brilliant scarlet. Flowers profusely and over a long time.

.... var. Double Variegated or Legrellei. Is considered to be a mutant of double red. A very attractive form with large handsome flowers, flowers variegated mottled or striped with red and yellow.

.... var. Double White. Flowers are white in colour and in other characters is similar to Double Red.

.... var. Double Yellow. Flowers are pale yellow in colour otherwise quite similar to double variegated.

Cultivar

The cultivars presently grown in India show distinct variation in fruit shape, colour, taste, colour of aril, rind thickness etc. Pomological detail of few important cultivar is given below.

Ganesh : A very prolific yielding cultivar, fruit usually medium sized, soft seeded type, aril or flesh light to pinkish in colour, very juicy sweet in taste with pleasant agreeable flavour and taste. Plants are medium tall.

Kandhari : Plants medium tall with attractive foliage, flowers orange red in colour. Fruits larger in size, angular in outline, rind thin, crimson to deep red in colour, aril or fleshy testa blood red or deep pink coloured, flavour pleasant with sweet acidic taste, juicy, seed hard.

Dholka : Cultivar produces consistently large sized fruits rind somewhat greenish overlaid by light yellow and very faint pink blush, aril / flesh white to pinkish white, juice some what acidic, flavour mild, soft seeded type.

Alandi : Seeds are hard otherwise very good cultivar, bears regularly and heavily, fruits medium sized, aril deep pink to intense red coloured (blood red), sweet in taste, juice slightly acidic.

Muskat Red : Seeds medium hard, small to medium size fruits, rind quite thick (compared to other cultivars) aril red coloured, sweet in taste.

Spanish Ruby : Seeds are soft, produces fruit which are small to medium in size, rind smooth glossy thin, aril red or rose coloured, good cropper.

Kabul : Fruits are large (as that of Dholka) rind quite thick smooth yellowish overlaid with deep bright red colour, somewhat squarish, flesh/ aril big, red in colour, juice somewhat bitter in taste.

❑❑❑

26
Passion Fruit

Genus : *Passiflora*
Species : *edulis* Sims.
Family : *Passifloraceae*
Chromosome No. : 2n = 18

The passion fruit (*Passiflora edulis*) is another fascinating tropical fruit besides having both dessert and processing qualities, the crop also has a big multi coloured very attractive flower and fruit with district taste and flavour. Passion fruit is very popular in number of tropical and subtropical countries. Despite having variability at the species level the commercial development has not progressed as desired which is evident from the fact that the available commercial types are highly susceptible to diseases, lack of adaptation to soil and climate extremes, high establishment costs growing management problems harvesting and market costs have forced the grower to focus on other remunerative crops.

Passion fruit is said to be good source of minerals especially calcium, phosphorus and iron, vitamin A, niacin and riboflavin. Maximum proportion of various nutrient constituent of sugar, ascorbic acid, juice and flavour are in mature fruits compared to those present in immature fruits. Most common processed products of passion fruit are jam, jelly, squash, ice cream, fruit nectar, passion fruit concentrate and syrup.

The family passifloraceae is very large and diversified comprises of 12 genera and about 550 species, perhaps the only family with the largest number of fruiting species than in any other fruit family. Genus *Passiflora* has 400 species, and 60 species bear edible fruits. The purple passion fruit, *P. edulis* is adapted to the colder subtropics or prefers higher altitudes in tropics and is commercially grown in New Zealand, Kenya and South Africa, whereas the golden passion fruit (*P. edulis* var. flavicarpa) is more tropical in its adaptation or adapts well to tropical lowland conditions. In many of the countries where the fruit has naturalized (other than place of origin) commercial production of passion fruit is obtained from cultivars of golden passion fruit (*P. edulis* var flavicarpa) Some cultivars are self incompatible hence require pollination source for adequate fruit set.

In Australia hybrid between purple and yellow forms are more in use, characters are intermediate, usually high yielding more resistant to wilt and nematodes with purple gold skin, yellow orange pulp, more flavour and reduced acidity.

Origin and Distribution

The fruit in native of southern Brazil, more precisely the outskirts of rain forests. Passion fruit has two distinct forms one is edulis-called as purple passion fruit is adapted to higher altitudes and requires cool weather conditions, while the other type/form is flavicarpa-yellow passion fruit is more tropical in adaption. From its natural habitat, the crop travelled through Europe , Australia and reached southern east Asia in 19th century. Presently crop is grown in throughout the tropical and subtropical region of the world.

Uses

Usually the fruit is eaten fresh, used for processed products like ice-cream, jams, squash, nectar, jellies etc., makes the natural concentrate on account of high acidity and flavour, pulp preserved after cooling or heating to blend with other fruit juices.

Flowering and fruiting vine

Fruit

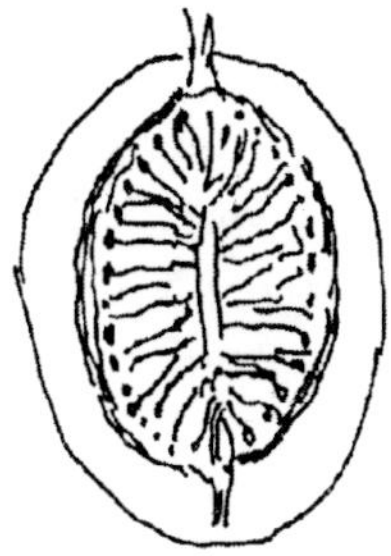

VS of fruit

Flower from above

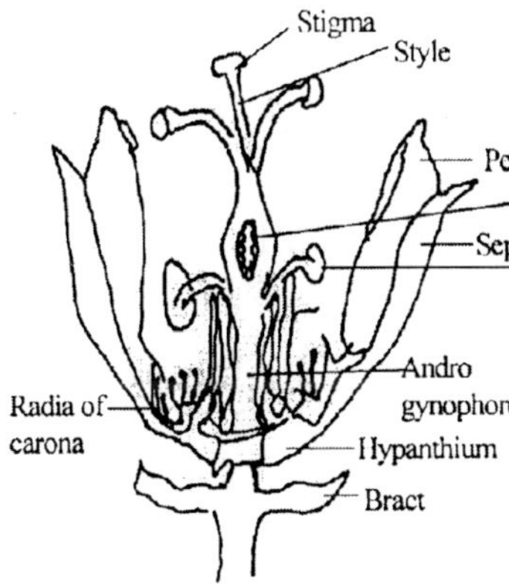

VS of flower

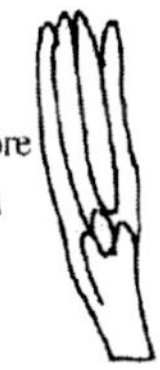

Stamen

Style with capitate stigma

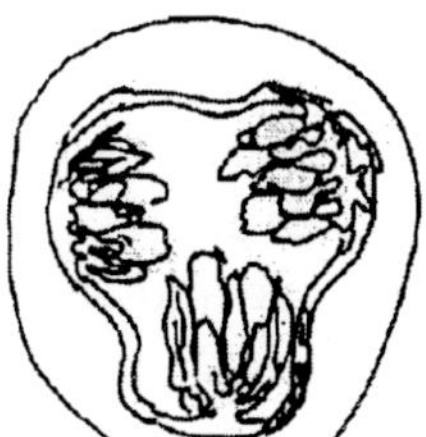

TS of ovary

TS of ovary seed and flesh

Description

Passion fruit vines are semi woody climber upto 15m long, lives for short time, quick growing, perennial, stem green grooved and glabrous, tendrils long spirally coiled, axillary, stipules and petiole present, stipule about 1cm long lanceolate, petiole thin, long (2-5cm), grooved on upper side, glabrous, with two circular glands at top. Leaves alternate, unlobed when young, but palmately 3 lobed later, cordate at base, lobes ovate in shape, margin with curved teeth and are glandular tipped. Flowers are borne in leaf axils/axillary, solitary large (7-10 cm in diameter) showy and fragrant, peduncle long (2-5 cm) triangular, with 3 leafy bracts, ovate to lanceolate in shape, calyx, base tubular with 5 lobes, reflexed spreading, white above, yellow-green below, margin has 4 glands, apex has thorn like appendages, petal 5, usually free, thin, white, alternating with calyx lobes, white and thin, corona with 2 outer rows of wavy radiating threads 2-3cm long, white, base purple and 3 rows of short purple tipped papille. Stamens 5, filaments united to form a tube round the gynophore for about 1cm and then widely parted for 1cm, anthers large, two celled, versatile transverse, pale yellow, drooping down, ovary on gynophore, ovoid, locular with 3 parietal placentae, style (3 parted) has longitudinal furrow, stigmas reni or cordiform. Fruit ovoid or globose berry (4-12 x 4-7cm) yellow or dark purple, exocarp thin, hard, mesocarp greenish, endocarp white. Seeds numerous, attached with hook-like funiculi on the ovary wall, surrounded by yellow aromatic juicy/pulpy arils, with tart but pleasant flavour, Testa, black, hard, 3 toothed at base and flattened.

Passion fruit has large number of well recognized species that constitutes a large family or a group of evergreen climbers that are very interesting herbs or shrubs mostly of the cultivated types and climb by means of tendrils. *Passflora edulis* and *P edulis* var flavicarpa are important as they yield edible fruit. Other species which have successfully adapted to the local conditions of many countries need to be systematically evaluated for fruit qualities, source of resistance or as rootstock species. Some other species useful as ornamental or that yield fruit are described briefly.

P. lutea Linn. Wild Yellow passion flower. A perennial herb 5-10 feet tall shiny or nearly so, leaves 3 lobed, broader than long, heart shaped at base, petiole glandless. Flowers occur singly or in pairs ¾ inch across, green yellow, fruit is a berry globular, ½ inch across when ripe purple and smooth.

P racemosa Brot (*P princeps*, Hort. *P amabilis,* Hort.). Native of Brazil. One of the best red flowered passifloras and parent of number of garden hybrids. Leaves truncate at base, shiny usually 3 lobed, margin entire flower large 4 inch or more across, petals deep red, spreading, upright crown purple appear singly. Fruit oblong ovate greenish yellow when mature.

P. ligularis Juss. Native to tropical America. Young foliage has metallic colouration. Woody in nature tall growing distinctly branched, leaves ovate, cordate, large with lobes or teeth on margin. Flower single both sepals and petals green, corona white with distinct red purple areas. Fruit oblong larger than *P. edulis* brown in colour, shell thick.

P. quardangularsis Linn. True or giant Grandilla, Native of tropical America. Robust climber with quadangular winged stem, glabrous leaves round ovate to ovate, base heart shaped margin entire, 2-3 pairs of glands are present on petiole. Flower large 3-5 inch across, fragrant. Flower need hand pollination if fruit is to be had. Calyx and corolla both are ovate, calyx white from inside and corolla are reddish. The filaments of the crown present in five series, the outer extend beyond the floral parts. Fruit 5-9 inch long, oblong in shape, yellowish green in colour, flesh pulpy and edible. Has one botanical form var. variegata, in which the foliage is bloched with yellow.

P. incarnate Linn. Wild passion flower, May pop. Vine is perennial high climbing 20-30 feet long, leaves broad cordate 3 lobed to about half the leaf, serrate, two glands present on bearing petiole. Flower single, 2 inch across and axilllary, white, corona purple. Fruit about 2 inch long oblong, yellow with three district sutures, edible.

P. alba Link and Otto. Found in Mexico and S America. Stipules very large and leaf like, stem terete, leaves cordate broad/ovate, 3 lobed shallow shiny below margin usually entire, peduncles larger than leaves, one fid. Flower slightly more than 2 inch across generally white. Fruit obovoid green later turns yellow in colour.

P. alta Dry. Wide spread in South America. Leaves, shiny ovate or oval, base cordate, stem winged margin entire and uneven, petiole with 2 pairs of glands. Flower quite large 3-4 inch across fragrant. Inside of calyx and corolla red, numerous filaments of various colour, red white or purple Fruit big, 5 inch or more in length ovoid yellow in colour very fragrant and most common edible passion.

P. mixta Linn. The specie is seen growing from Venezuela to Bolivia, Indigenous to S. America. Stems angled, leaves 3 lobed ovate oblong in shape margin serrate, petioles with 4-8 stalked glands, flower large 3-4 inch across sepals and petals pink to orange red, corona deep lavender to purple, fruit ovoid 2-3 inch long.

P. mollissima Bailey. Curuba, Banana passion fruit. Native to tropical America as *P. mixta*. Stem cylindrical with red to yellow pubescence, leaves 3 lobed margin serrate-dentate, dense reddish pubescence below, flower 2-3 inch in diameter, sepals white petals pink both from inside. Fruit 2-3 inch long ovoid oblong with pubescence, edible, yellow in colour.

P. maculifolia Mast. Indigenous to Venezuela. Foliage is variegated, branches thin and wiry with soft pubescence, leaves cordate round, 3 shallow petiole glandless. Flower, single or in pairs small in size ¾ inch across greenish yellow. Fruit globular berry small sized ½ inch in diameter smooth purple when ripe. Though inferior in fruit quality to purple passion fruit but has vigorous vines, larger leaves, flowers and fruits, with pink, red or purple pigmentation in the leaves and stems, flower corona is also deep purple in colour, prolonged flowering duration, rind yellow and thick pulp yellow, seed dark brown has more resistance to leaf spot, Fusarium wilt, phytophthora blight and to nematodes.

Cultivars

The passion fruit industries all over the world is entirely based upon seedling lines of the yellow or purple passion fruit. Many of the species of the genus *Passiflora* can be easily crossed and nearly nine species that have edible fruit, have been extensively employed in inter specific hybridization. Through this approach number of drawback have been overcome especially passion fruit woodiness virus (PWV) etc. Passion fruit cultivars along with their important pomological traits are described as below.

Noel's Special : Cultivar originated in Hawaii, and named after a student who discovered it. Selected from open-pollinated seedling. Vines are vigorous, produces crop one year after plantation, leaves dark green. Fruit round berry filled up with dark orange pulp, weighs about 80-90gm, total soluble solids range between 15-20 percent, self incompatible and requirea cross pollination for good fruit set. Tolerant to alternaria brown spot.

❑❑❑

27 Loquat

Genus : *Eriobotrya*
Species : *japonica* Thunb./Lindl.
Family : *Rosaceae*
Chromosome No. : 2n = 34

In Orient this beautiful and excellent fruit has failed to acquire the commercial significance which the crop deserves, further no concerted efforts have been made to improve it through selection and cultivation as has been done for many other Asiatic fruits. In contrast to major tropical and subtropical fruits which are said to be too sweet, loquat is one fruit that has subacid, sprightly flavoured fruits that match in taste and flavour to those of temperate fruits. The loquat has been in cultivation since antiquity in Japan and under present fruit culture scenario is one of the important fruits of that country.

Loquat (*Eriobotrya japonica)* also called as Japan plum, or biwa of Japanese is an important evergreen plant of family Rosaceae, which is extensively and commercially grown throughout the subtropical regions of the world for its small trees with large handsome foliage and acid fruits. Presently it is grown in Australia, South America, south Africa, India, California including China and Japan where the crop originated and has been cultivated since antiquity. The genus word *Eribotrya* is derived from the Greek words 'erion' meaning wool and 'botrys' meaning cluster which is implied to the woolly inflorescence.

The pubescent, pyriform fruits are mostly consumed fresh, also used to make jam, jelly, preserves and for canning purposes. Being a good source of acid and pectin, fruit makes very good jelly products. Fruit has substantial amount of carbohydrates and minerals (Ca, P and Fe).

Loquat tree is very productive and is regular bearer. Except for crop failure on account of frost injury to the blossom, plants rarely fail to produce well every year.

Loquat is a close relative of apple and pear and in its sub acidic flavour it resembles to cherry. It is ironical that in its native place this excellent fruit has not achieved the desired commercial status which is due to it, nor has been substantially improved through cultivation and selection. For its large, dark green dented leaves highly pubescent and dense foliage, loquat has an excellent ornamental appearance and is often planted in parks and gardens for additional attraction in landscape. Fruit has only one cultivated specie *japonica* which is important from horticulture point of view.

Origin and Distribution

South eastern China is most likely considered to be the place of loquat's origin. Records indicate its cultivation in China and Japan at very early times.Presently the crop is grown commercially all over tropics and sub tropics, is also being cultivated in South America, India, Australia, Mediterranean region South Africa and to a limited scale in United States (California).

Uses

Fruits are generally consumed as fresh or may be preserved as jam or jelly, alcoholic drink made from juice. Seeds possess almond like flavour therefore used to flavour cakes and drinks. The tannin rich leaves are astringent and has anti-diarrhoeic properties, oil from flower is employed in cosmetic, as a insect repellent, fruits good to quench thirst and stops vomiting.

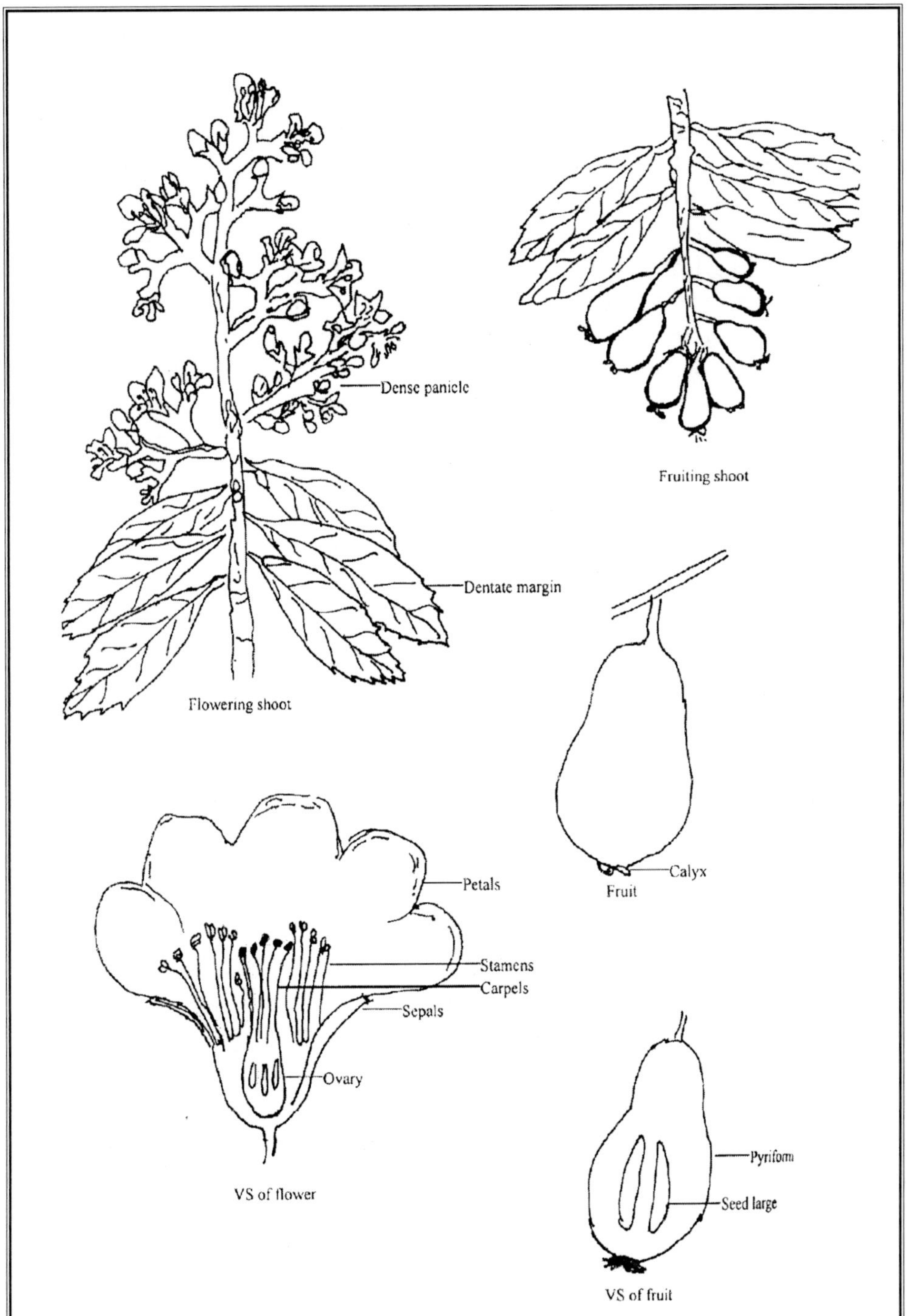
Dense panicle
Dentate margin
Flowering shoot
Fruiting shoot
Calyx
Fruit
Petals
Stamens
Carpels
Sepals
Ovary
VS of flower
Pyriform
Seed large
VS of fruit

Description

The only evergreen plant in family Rosaceae. Plants are shrubs or small trees (10 m tall), trunk usually straight with low branches, crown dense, rounded, branchlets densely pubescent, rusty brown in colour. Leaves simple, alternate, stiff, dark green, glossy above, with thick rusty brown pubescence below, margins dentate, oblong lanceolate, midrib and veins surface wavy, petiole about ½ cm long stipule persistent. Inflorescence densely branched panicles measuring 12-20 cm in length containing 70-150 flowers densely pubescent, rusty brown. Flowers fragrant, sessile about 2cm in diameter, calyx 5, highly pubescent, ovate, persistent even on fruit, petals 5, white to creamish, ovate, stamens 20, styles 2-5 usually 5, ovary inferior, 2-5 celled. Fruit is ovoid, globose or pyriform densely pubescent, pale yellow to deep orange, flesh yellow or orange with sweet-sour taste, seeds variable 1-5, dark brown in colour.

Detail of some other species is given below.

E. Hookerana Decne. Plants are small trees, leaves very large 8-12 inch long 2-4 inch wide rusty, thickly pubescent, tomentose when young margins serrated with round or subacute base. Veins thick conspicuous in 20-30 pairs. Flower white ¼ inch across, fruit small ¾ inch long ellipsoid yellow. Found in eastern Himalayas.

E. grandiflora (Rehds and E.H. Wils). Found in Taiwan. Leaves are oblong rounded at apex, cuneate at base, 4-6 inch long, 1-2 inch wide, margin crenate serrate. Fruit small 1 inch in diameter. Has one botanical var. Koshunensis Kanch and Sasaki, in which leaves are obovate oblong 3-6 inch long 1-3 inch wide apex obtuse, base acute.

Cultivars

On the basis of variation in maturity time, fruit colour and shape, as well as on the origin basis loquat cultivars have been grouped as

a) *Chinese group :* Fruit pyriform deep orange and large

b) *Japanese group :* Fruit small rounded yellow and slender.

Description of some cultivar which are grown in India and world wide is given below.

Golden Yellow : Plants are typically tall growing, upright-spreading in nature, quite fast growing, compact with roundish-oval crown. Leaves are typical large lanceolate with few sharp teeth, dark green above and densely pubescent below. Fruit medium in size, oval to oblong. Calyx persistent, pubescence thin velvety golden yellow, pulp pale yellow to orange, soft, smooth, sub acidic in taste, flavour mild with few black smooth seeds.

Improved Golden Yellow : Tree and fruit characters resemble very closely to Golden Yellow except that crown is more open fruit shape is oval to pyriform, pulp thick and more seedy in nature.

Large Round : Tree vigorous, spreading, branches wide crotched crown compact and uneven, leaves dark green on upper surface pubescent below more so along the veins and mid rib lanceolate-elliptic, margin serrated at base and dented at top. Fruit medium sized, ovate globose in form, yellow in colour flesh creamy, whitish at seed, thin, firm and somewhat coarse.

Safeda : Tree growth vigorous, upright, medium to tall in stature, tree top rounded, leaves typical elliptic-lanceolate, margin serrated pubescence conspicuous. Fruit large, oblong-pyriform, surface pubescent, pale yellow in colour, flesh creamish white, whitish at seed, smooth tender, melts rapidly, sub acid, juicy, taste excellent with few seeds.

Thames Pride : Plants are tall, fast growing, upright, top round or dome-shaped. Leaves elliptic-lanceolate, margin dentate at apex pubescence dense. Fruit medium in size, shape pyriform, deep yellow surface velvety, pulp pale orange, somewhat course, medium thick, sub acid in taste, flavour mild, seeds moderate in number.

Tanaka : Tree dwarf in nature, growth medium, upright, crown compact and semi rounded. Leaves elliptic-lanceolate. Fruit small orange in colour oval-ovate, pulp medium thick, yellow, firm, sub-acidic in taste, moderately seedy.

Centenaria : Evolved from a cross between Mizuho x Mogui. Trees are vigorous in nature, produces regularly and consistently. Leaves oblanceolate, dark green, pubescence dense. Fruit large yellow-orange, attractive uniform with some purple spots, flesh yellow firm, taste pleasant sub acid with 5-6 medium sized seeds.

California Advance: Tree growth moderate, upright dwarf in nature, crown more open and round. Leaves oblanceolate, margin dentate. Fruit oblong-pyriform medium in size, surface pale yellow, pulp yellow, firm, thick, taste sub-acidic, pleasant with many seeds.

❐❐❐

28
Durian

Genus : *Durio*
Species : *zibethinus* Murray.
Family : *Bombacaceae*
Chromosome No. : 2n= 56

The exact place to which durian is indigenous is not established with certainity. The specie is said to be native to Borneo and other adjoining islands of Malay Archipelago. The species is seen growing in Sri Lanka and some other tropical countries. Outside the Malayan region durian cultivation is greatly restricted and is usually confined to the botanic gardens from study point of view. The crop has failed to move away from its native place primarily due to the fact that seeds are highly perishable which has restricted durian movement within tropical places of the world. Further due to the disagreeable flavour this fruit has not met with the favour from the

Europeans. The banks of the Sarawak River are covered with many fruit trees that basically fulfill the food requirements of the local inhabitants called as Dyaks. Of the different kinds of fruit tree, durian is the most important found in abundance and is rated very highly. The fruit is of hot and humid places and at first instance the fruit smells like a rotten onions -disagreeable flavour which has limited its cultivation. However when fully ripe some prefer it over all the fruits. The fruit rind is very hard and thick and does not crack even when thrown from great heights. On the fruit surface there are about five faint lines that run from base to the fruit apex, each line represents a carpel and it is through these lines that the fruit is cut open with a knife to reach the edible part which is cream-coloured pulp, in which are embedded two or three seeds. Eating durian is altogether a new experience for its very rich butter-like custard, its consistency and rich flavour are not easy to describe. On account of sharp stout spines on fruit surface the durian are said to be very dangerous especially when fruit starts to ripen and ripe fruit fall regularly. At this stage one has to be very careful as falling durian result is very painful wound, as the spines usually tear open the flesh.

Durian (*Durio zibethinus*) belongs to the family Bombaceae, has 6 genera and about 27 species of which six are important horticulturally as they yield edible fruit. Durian is an important fruit of the tropics, is a widely cultivated fruit tree of Asian tropics for its malodorous aril and of the seeds being highly esteemed by some and not liked by the others. The genus Durio is a name from Malay language from the word 'dury' thorn alluding to the prickly fruit. The specific name *zibethinus* is said to be derived from the practice of using the decomposed fruit as a bait for the civet cat or zibet, name zibetto (Italian) meaning strong smelling. In the market, fruit is recognized by hard greenish thick thorny rind, yellowish sweet flesh with very strong disagreeable odour. Fruit is a good source of carbohydrate(s), protein, minerals like calcium, phosphorus, and iron, vitamin A and number of essential amino acids like oleic, palmitic, linoleic, linolenic, stearic etc.

In quality durian is said to be the best of all fruits because the fruit cannot supplement the place of the sub acid, juicy types, like the grape mango or orange and in exquisite flavour it is unmatched. The major drawback however is the spiny nature of fruit which very often is dangerous and may cause dreadful wounds.

Has only one species which is important from fruit point of view.

Origin and Distribution

The fruit is commonly known by its English name the 'Durian', though it has number of vernacular names in the south east Asia which is its place of

Shoot bearing flower and fruit

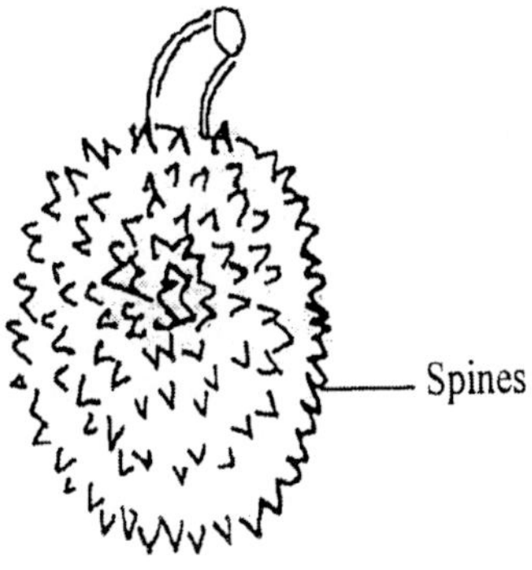

Fruit

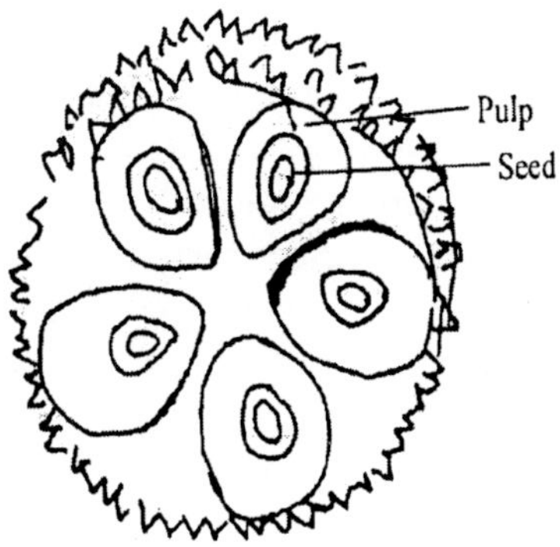

origin and its wild forms are in abundance here. The geographical distribution ranges from south India, Sri Lanka to far in the south east to New Guinea. Besides, isolated durian plantations are also found in Hawaii, Australia and Zanzibar. The crop is very important in Burma, Vietnam, Philippines and large orchards are found in Thailand.

Uses

The fruit is recognised by its distinct smell or flavour. Aril the edible part is eaten as fresh. Indian satisfy their appetite with mangoes, it is the durian doing same work in places of its origin. The durian flesh after drying is preserved as '*Durian cake*', used to flavour cookies and ice creams, arils are deep frozen for export trade. The dry fruit rind is used to smoke the fish, roasted seeds are consumed as snack, also yet to mature fruits and new young shoots are also cooked, fruits also used as medicine to treat a number of animal and human disease. Both the flesh and seed are considered to be very nutritious.

Description

Has six species, other species are of lesser importance from fruits point of view but are regarded as important source of timber. Has very long juvenile phase of 7- 12 years. Tree very large grows upto 40 m tall, bark dark red brown in colour which peels off irregularly. Leaves large coriaceous, alternate, elliptical to lanceolate with acute apex, upper surface glossy, glabrous, pinnately veined, under surface thickly covered with hairs and coloured scales which are either silvery or golden coloured. Flowering takes on terminal panicle.

Flower buds globose, ovoid in shape each measuring about 2 cm in diameter, epi-calyx splits into 2-3 ovate lobes (1-5 cm length). Inflorescence from fascicles of corymbs of 3-30 flowers measure upto 15 cm length. Flowers, large (5-6 cm long & 2 cm diameter) greenish white or white, calyx tubular (3 cm long) have 5-6 prominent teeth (triangular), petals 5, almost twice the length of calyx, at base terminate into a prominent claw. Stamens many, distinctly present in 5 separate bundles, style long, pubescent with capitate/ capitellate stigma. Ovary 5 celled, each cell with many ovules. Fruit ovoid, globose or ellipsoidal (25 cm long 20 cm diameter), green or brownish, very spinous in nature, each spine upto 1 cm long, mostly indehiscent, carpels with seeds (4 cm long), covered / surrounded by firm white or cream yellowish, very sweet, juicy, edible pulp 'the aril'.

In south east Asia, especially in Thailand number of seedling selections from heterozygous population obtained through seed propagation have been made. On the basis of variability some suitable types have been selected and are being grown commercially, Pomological description of some durian cultivars is given below.

Cultivars

Deception : In south-east Asia and especially in Thailand is one of the most popular and outstanding durian cultivar. Fruits are large and initially provides a feeling that edible portion too in large, but in fact it is not so, quality is very good.

Mawn-Tawng : The fruits of this cultivar are easily recognized by its large size. Fruits are somewhat more elongated, base with a pronounced beak, on maturity the fruit colour is golden to yellowish brown in colour, flesh creamish yellow sweet, juicy, flavour pleasant characteristic of durian. Individual fruit may weigh 2-5 kg.

Gahn-Yao : Cultivar is characterized by its large fruits, among various known cultivars is rated to be one of the best cultivar. Fruits are large 2-5 kg on average basis, globular in shape, surface greenish brown in colour at maturity, flesh/pulp is creamish yellow sweet in taste, firm, becomes soft, or mellows as fruit ripens.

Golden Pillow : A very important and commercially grown cultivar throughout Thailand. Rated very highly for its quality characters but is not a very prolific bearer. Fruit medium-large elongated-globular, surface greenish with light brownish tinge, pulp creamy yellow, sweetest of all known cultivars.

Gra-Doom-Tawng : Cultivar produces somewhat small sized fruits each weighing around 1.5-3kg, globular in shape, fruit surface greenish brown at maturity, flesh or pulp bright dark yellow firm, sweet, quality very good, firm, but softens later.

❑❑❑

29
Phalsa

Genus : *Grewia*
Species : *subinequalis DC., asiatica* L.
Family : *Tiliaceae*
Chromosome No. : 2n = 36

Phalsa (*Grewia asiatica*) belongs to the family Tilaceae and some members are native of India too and is mentioned in Vedic writings and various plant parts are employed to cure number of ailments. The family comprises of 18 genera and some 400 species distributed mostly in the tropical and subtropical areas of the world. Of the prevailing 140 species, 40 are indigenous to India of which *Grewia asiatica* is of significance horticulturally. Phalsa is commercially cultivated in Uttar Pradesh, Punjab, Haryana, Rajasthan, Madhya Pradesh and on limited scale in Bihar, West Bengal, Gujarat, Maharashtra and Andhra Pradesh.

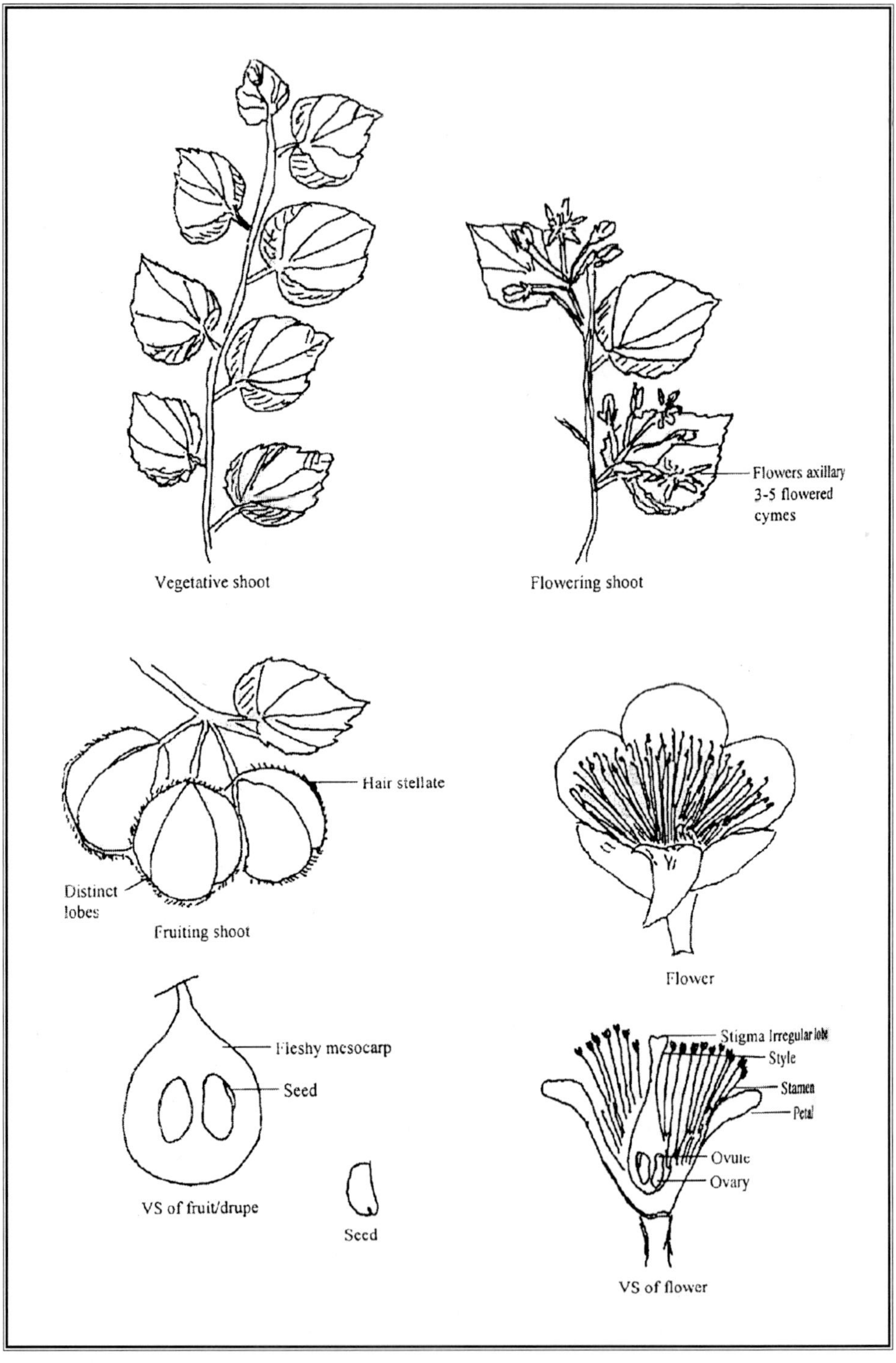
Vegetative shoot
Flowering shoot
Flowers axillary 3-5 flowered cymes
Hair stellate
Distinct lobes
Fruiting shoot
Flower
Fleshy mesocarp
Seed
VS of fruit/drupe
Seed
Stigma Irregular lobe
Style
Stamen
Petal
Ovule
Ovary
VS of flower

Origin and Distribution

The phalsa has its origin in the Himalayan region. The crop is primarily grown in India ,Bangladesh and Pakistan. Further in south east Asia, the crop is grown in Philippines in Luzon and in north and central Thailand.

Uses

The fruits when ripe are eaten fresh, juice is used to make a refreshing drink much liked in India during summer days as it has cooling, tonic and aphrodisiac effects which overcomes thirst and burning sensation. Cures fevers, inflammation, heart and blood disorders. Besides fruit other plant parts are also used for one or other purposes. Wood is strong, shoots are thin and flexible, therefore is a good raw material to make baskets, fibre, bark is employed to make ropes and gunny substance. Bark when soaked in water is used in sugarcane industry for clarification of gur.

Phalsa fruit when ripe are rich source of vitamin A and C with fair amount of minerals major being phosphorus and iron, fruits are sub acidic with fair blend of sugar and acid, mostly consumed as fresh, but also make high grade squash and juice.

Description

Plants are deciduous, large spreading, shrub, or small tree about 5m in height, bark grey, rough to feel, branches usually long and thin (slender), drooping, thick pubescence (hair like) on young shoots. Leaves deciduous,with 5 distinct veins, young leaves are brownish red, simple, alternate, petiole short generally broadly cordate (heart shaped) to ovate, base slanting (oblique), apex acute to acuminate ,margins coarsely toothed, pubescence light above and thick on under surface. Flowers appear in leaf axils in clusters of 2-8 or 3-5 flowered cymes on current season growth. Flower parts in multiple of 5's,flower yellow, hermaphrodite, peduncle about 2.5cm, pedicle measures about 1 cm, calyx oblong, corolla, obovate, stamens numerous (60-70), stigma lobed, ovary segmented. Fruit is a drupe, 1-4 celled, round to globose, nearly 2 cm in diameter (some times bigger), lobes distinct, usually red or purple in colour with star shaped tufts of hair. Flesh greenish white with purplish red tinge near the skin, soft fibrous with pleasant acid taste. Seeds small (5mm wide) 1-2, hemispherical.

Some important species of genus Grewia with brief pomological description are detailed below.

G. asiatica L. A small shrub about 4-5 feet tall with sub-herbaceous arching branches, leaves rounded or sub-orbicular, cordate 5 inch long with 5-7 veins at base densely tomentose on the under surface. Flowers appear axillary small in size flower stalk about the size of petiole. Fruit small red in colour.

G. Damine Gaertn (*G. salviifolia* B Hayne/ex Roth). Native of India and Sri Lanka. Small tree or a shrub, leaves elliptic oblong about 3 inch long apex acute or obtuse, margin finely serrated, base 3 nerved, pubescence dense grey below, flower appear in axillary cymes in clusters of 2-3, fruit shiny and two lobed.

G. biloba G. Don. syn. *G. parviflora.* Shrub 8 feet tall, leaves 5 inch long, ovate to rhombic ovate, doubly serrate sometimes 3 lobed 3 nerved from the base, glabrous or tomentose. Flower yellow small 1/3 inch across, occur in umbellate clusters at extremities of branches. Fruit red, usually 2 lobed. Common in China and Taiwan.

G. tiliifolia Vahl. Small to moderate tree, leaves ovate to rounded 4 inch long, 5-7 nerved from the oblique base, stipules leafy and curved. Bark grey or dark brown. Flowers are small borne on thick axillary peduncles. Fruit/ drupe globose about pea size, black in colour and edible, 2-4 lobed or not lobed. Found in sub Himalayan tract.

G. orientalis L. Native to India. Erect or straggling shrub, leaves ovate to oblong about 3 inch long, apex short pointed, base usually three nerved, pubescent below. Flowers few in axillary or terminal cymes, calyx yellow, corolla white or yellow, small fruit purplish shallowly 4 lobed about ½ inch in diameter.

Cultivar

The fruit is commercially grown through seeds, the seedling thus produced are variable and do not resemble mother trees. On the basis of variability expressed two distinct group on the basis of growth have been made i.e. tall and dwarf and some selections in India have been given names like 'Local' and 'Sharbati' etc.

30
Mangosteen

Genus : *Garcinia*
Species : *mangostana* L.
Family : *Guttiferae*
Chromosome No. : 2n = 56-76, 88-96, 120-130

In family Guttiferae to which mangosteen belongs, very few plants are fruits and yield drugs, gums and resins. It is a tropical fruit and occupies an important place in Garcinia family. The European voyagers on their return indicated wonderful stories of the East and mention of mangosteen received huge praise. Hacobus Bontius termed the fruit as 'Queen of fruits' the finest fruit in the world. Because of beautiful colouring of fruit alongwith pleasant delicate flavour makes mangosteen the most esteemed fruit of Asiatic tropics and no other tropical fruit equals to that of mangosteen. Despite being designated as the finest fruit of the tropics, the fruit is mainly confined to the limits of East Indies. The species is found wild in the forests of Sunda Islands and of the Malay peninsula. Writers have indicated that mangosteen is usual dooryard tree in the East Indies, more in Java and Sumatra. Some small orchards of mangosteen have come up in Sri Lanka, but they are most abundant only in Malay Archipelago. In India only a few scattered trees are found growing and more so in and around Madras. The mangosteen tree introduced either as plants or raised through seeds have started fruiting in Jamaica, Dominica, Trinidad and in Hawaii. This has raised the hope that very soon commercial production of this delectable fruit would take place in different tropical parts of the world and shall not remain confined to the interior regions of the eastern tropics.

Mangosteen (*Garcinia mangostana*) is a prominant unexploited tropical fruit tree native of Malay Archipelago, Malacca and Sunda Island, therefore requires strong tropical conditions which are characterized with high temperature, rainfall and humidity. Plants are very sensitive to low temperature (less than 5^0C). Tree is slow growing and may takes long time (10-12 years) to bear fruit. The tree may flower from February to May and fruit ripen from May to September depending on plant growth care and site. The plant has added attraction for its coriaceous leaves purple fruit, also used in landscapes.

Mangosteen belongs to the family Guttiferae, which comprises of about 9 genera and 86 species, further 68 species are important horticulturally producing edible fruits and some are even important as rootstock. Mangosteen fruit is recognized world over for its delicious sweet soft and distinct flavour. Round to subglobose, fruit has thick dark purple rind with persistent calyx at stalk end, 5-6 stigmatic lobes at the distil end of the fruit, which indicates the internal fruit segments through which fruit can be cut open, exposing the white or creamish edible part the aril, along with seeds which are edible. Fruit is a rich source of carbohydrates, vitamins, carotene, thiamine, niacin and riboflavin, important minerals being phosphorus, calcium, iron and protein (0.5g.). The thick rind is astringent and contains mangostin, xanthones, gartanin which are instrumental against control of cholera, diarrhoea and dysentery and other ailments.

Origin and Distribution

Out of the various species present in the genus *Garcinia, mangostana* is considered to be the only cultivated one. The specie is considered to have originated in south east Asia as some wild types /specimens have been found

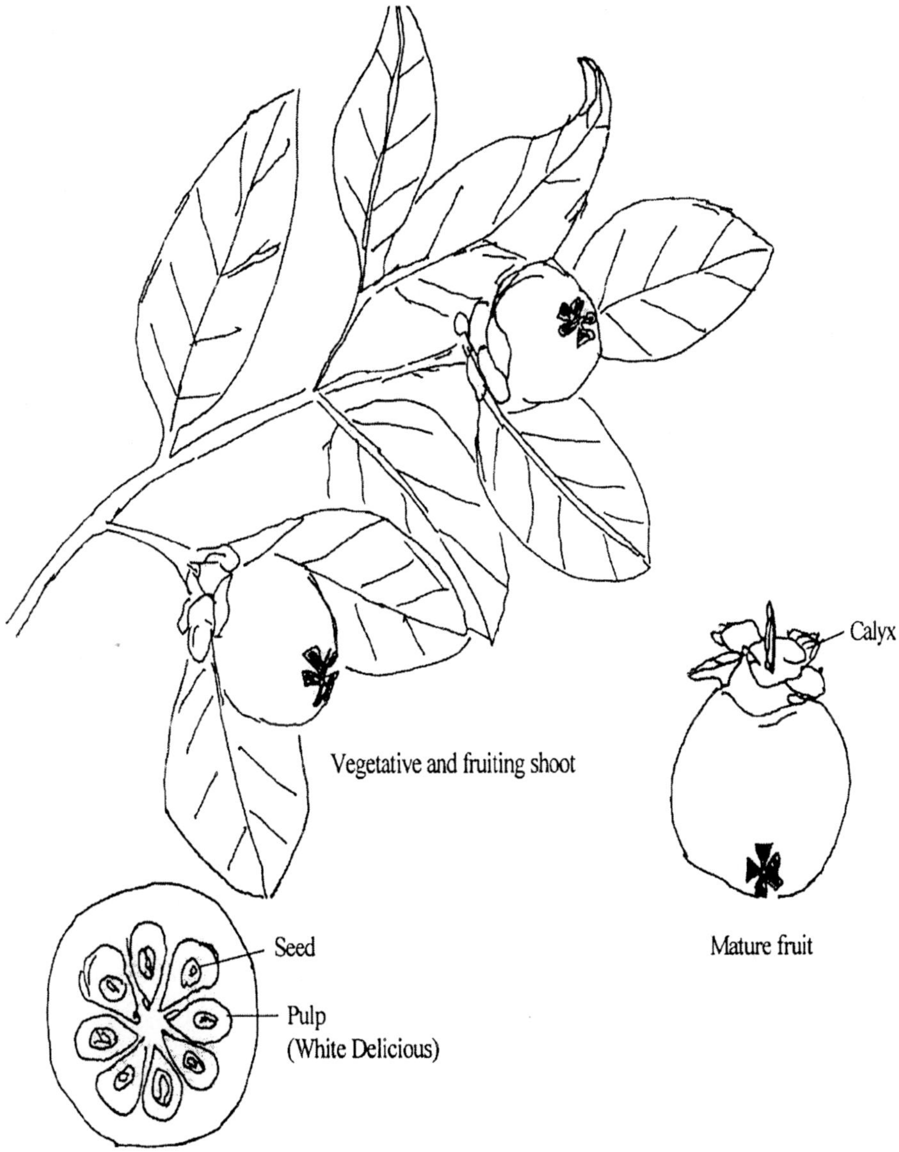

Vegetative and fruiting shoot

Mature fruit

TS of fruit

in Malaysia. Mangosteen is also regarded to be an allotetraploid hybrid resulting from hybridization between *G. hombroniana* and *G. malaccensis* (both the species are indigenous to Malaysia and *G. hombroniana* to Nicobar Islands). The mangosteen cultivated is primarily limited to south east Asia i.e. Indonesia, New Guinea, Philippines and in southern parts of Burma, Vietnam, Thailand and Cambodia. In recent times the crop has moved to other tropical areas of the world namely Sri Lanka, south India, Queensland, Brazil and central America.

Uses

On the basis of very high fruit quality the crop is rated as one of the most delicious tropical fruit, mostly eaten fresh, fruit rind and bark yields number of local medicines, for dye purposes and to tan leather. Wood is heavy, strong, coarse, red, highly prised for carpentry purposes.

Description

A characteristic yellow latex exudes from all plant parts. Slow growing and takes 8-15 years for bearing, deep green foliage, shines brightly in sunlight. The mangosteen plants are dioecious, tall tree(6-25 m), trunk usually straight, branches symmetrical resulting in near regular pyramidal crown. Leaves usually opposite, thick, coriaceous (leathery) with short petioles, oblong or elliptical, margin entire, dark green (olive) above, yellow green beneath, mid rib pale green in colour. Flowers are borne terminally on branchlets either singly or in pairs, polygamous i.e. male and bisexual flowers occurs together. Male flowers are borne in 3-9 flowered terminals fascicles while hermaphrodite are 2 inch across and appears singly or in pairs terminally on young branches, pedicel usually thick and short, calyx 4, in 2 pairs, thick, red, free and round, corolla 4, thick and fleshy, yellow green coloured, broadly ovate with reddish colouration on edges. Stamens many radiating from the centre, staminodes generally present and many, stigma 4-8 lobed and sessile, ovary superior, sessile or sub globose and is 4-8 celled each having one apomictic seed. Fruit is a typical round, globose berry (4-7 cm), smooth, rind thick(0.6-0.8 cm), dark red purple, sepals persistent so are the stigma lobes, edible part is white aril/ arillode the envelope of the developed seeds. At maturity rind becomes tough and thick, a circular cut with a sharp knife is required to get to the white pulp. Flavour delicious, seeds partially or completely lacking and when present are small thin and soft. Fruit usually develops parthenocarpically.

Some important species of genus Garcinia along with their horticultural traits are referred below as

G. morella Derr. Gamboge tree. Found growing in abundance from Bengal to Thailand. Plants are medium tall 30-50 feet, leaves usually more tapering or pointed at both the apices 4-6 inch long, veins usually indistinct. Flower yellow male about 3 inch long in the leaf axils, sepals very small female flower single large, staminodes about 12 in number. Fruits are about the size of cherry (Morello), four lobed

G. dulcis (Roxb) Kurz. Trees are about 20 feet tall, leaves long tapering at apex to a slender point, 6 inch long, margin entire, pale beneath. Flower creamish, globular. Fruit about the size of apple, seeds enclosed by yellow palatable aril, eaten raw or cooked, makes very good jam.

G. tinctoria (DC) W.F. Wight. syn. *G. xanthochymus*. Plants are tall about 40 feet or more, leaves oblong or elliptic oblong 18 inch long thick and coriaceous with many lateral veins. Flowers are white ¾ inch across, fruit dark yellow globose 2-3 inch in diameter. Common in western Himalayas in India.

G. spicata Hook. F. (syn. *G. ovalifolia*). Plants are medium sized with wide spreading branches, leaves broadly elliptic, 8 inch long. Flower, small, fruit globose walnut size, deep green, surface smooth, 1-3 seeded. Common in India and Srilanka.

G. Livingstonei T. Anderson. Plants are columnar, upright densely branched, leaves oblong rounded at apex, 5 inch long. Flowers pale yellow or white, ¼ inch long. Fruit orange yellow to reddish, globose 1 inch in diameter, produces edible fruits.

G. cambogie Desrouss, Roxb. Found in evergreen forests along the western ghats in India. Plants are medium sized tree, leaves broadly-lanceolate 4-6 inch long and narrow towards the petiole which is small about ½ inch long. Flowers usually yellow or orange, male flowers many, anthers two celled, filaments short and are placed upon centrally placed receptacle which is short and thick. Bisexual or female flower has 6-20 stamens usually sterile may be free or united in bundles that enclose the ovary which is 6-10 celled, stigma 6-10 lobed. Fruit shape variable, colour yellow or reddish about 3 inch across has 6-10 furrows that extends from base to the apex.

G. indica Choisy, Pierre. (Syn. *G. purpurea*, Roxb.). Grows in abundance along western Ghats in India. Plants thin, branches are pendulous and black coloured. Leaves, lanceolate, membraneous, red when young, almost sessile, veins in 6-10 pairs. Flowers are small, male in 3-7 flowered cymes, pedicelled, stamens many, 2 celled clustered around central receptacle, bisexual solitary stamens in 4 bundle 10-18 in number placed alternatively to the petals . Ovary

5-7 celled, stigma 6-7 lined or radiate, each with 2 tubercles. Fruit smooth, globose, purple 1-2 inch across, pulp red, acidic in taste with seeds embedded in it.

G. paniculate Roxb. Indigenous to Bhutan, Assam and Khasi hills. Medium sized trees, leaves elliptic lanceolate thin and membranous, petiole small about 1 inch long, secondary veins 8-10 pairs. Calyx small green shorter than petals, petals white, imbricate. Male flower in pyramidal panicles, anthers numerous around the central receptacle. Bisexual flower occur in branching spikes, stigma entire, some staminodes surround the ovary. Fruit small 1 inch in diameter, seeds immersed in pulp, resembles in taste to that of mangosteen.

Cultivars

As the crop is propagated through seeds the seedlings exhibit great variability and the resulting seedlings are not true to type more so as the seeds are not zygotic but develop from nucellar tissues of the carpel walls of the fruit. Some selections have been made which are recommended for local cultivation. No standard cultivars are known.

❑❑❑

31 Carambola

Genus : *Averrhoa*
Species : *carambola L.*
Family : *Oxalidaceae*
Chromosome No. : 2n= 22, 24

According to a Dutch traveller Linschoten (1598) has described carambola as a 'fruit with eight distinct corners as big as a small apple, sour in taste, resembles an unripe plum. The Chinese and Hindus eat this fruit as vegetable when green and as a fruit (dessert) when ripe. Carambola is esteemed in the Orient and is wide spread in the tropics. The *bilimbi* is a close relative to Carambola and in every livelihood is native of Malayan region and is considered as the only cultivated species. The fruit is very acidic in taste, and in appearance resemble cucumber.

Averrhoa carambola is regarded to be an important fruit of the genus Averrhoa which is native of warm tropical and subtropical regions of the world, Singapore, Malaysia, Hawaii, Japan, Taiwan, Australia, China, Thailand, India and Indonesia. The genus name Averrhoa is after Averrhoes an Arabian physician. The name carambola is also said to have come from Malabar which the Portuguese adopted it at very early times. In India it is, called as Kamrakh and its reference in early written accounts and Sanskrit name 'Karmara' indicates the presence of carambola in India much before the arrival of Europeans to India. The crop is also recognized as five cornered fruit. In recent times a number of fruits have been added to the basket of table fruits and carambola is one such interesting fruit, extensively used as fresh for its juice and above all for its use to decorate fruit salads, cakes etc. In India and China, the fruit is eaten as vegetable when mature (green) and as dessert when ripe.

Carambola fruit is a rich source of ascorbic acid, minerals like calcium, phosphorus, potassium and magnesium, predominant acids like oxalic and malic (sweet types of carambola) and only oxalic acid in sour carambola. Glucose, fructose and sucrose are the main sugar types major being glucose. Bilimbi (*Averrhoa, bilimbi,* L.), like carambola is indigenous to Malayan region, and is known as a cultivated type. In form it resembles cucumber and hence the name cucumber tree.

Origin and Distribution

The genus has two important species from fruit point of view ie *A. bilimbi* also known as cucumber tree or bilimbi and *A. carambola* as star fruit or carambola has eight prominent corners, sourish in taste just as an unripe plum. The fruit plant is widely distributed in tropics. The Indian and the Chinese consume the fruit as vegetable when green and as dessert when ripe, is believed to be native to the Malayan region, later the cultivation spread to southern China, India, Philippines and Hawai. The plant is long lived and crops regularly, is very susceptible to frost and young plants are damaged at low temperatures. Performs best under warm moist climates.

Uses

The unripe and ripe fruits are used for different purposes. Unripe fruits are preferred for making pickles and for jelly. Ripe fruits are eaten as dessert or juice preserves, in salads etc, as traditional medicine to overcome certain fevers and skin disorders.

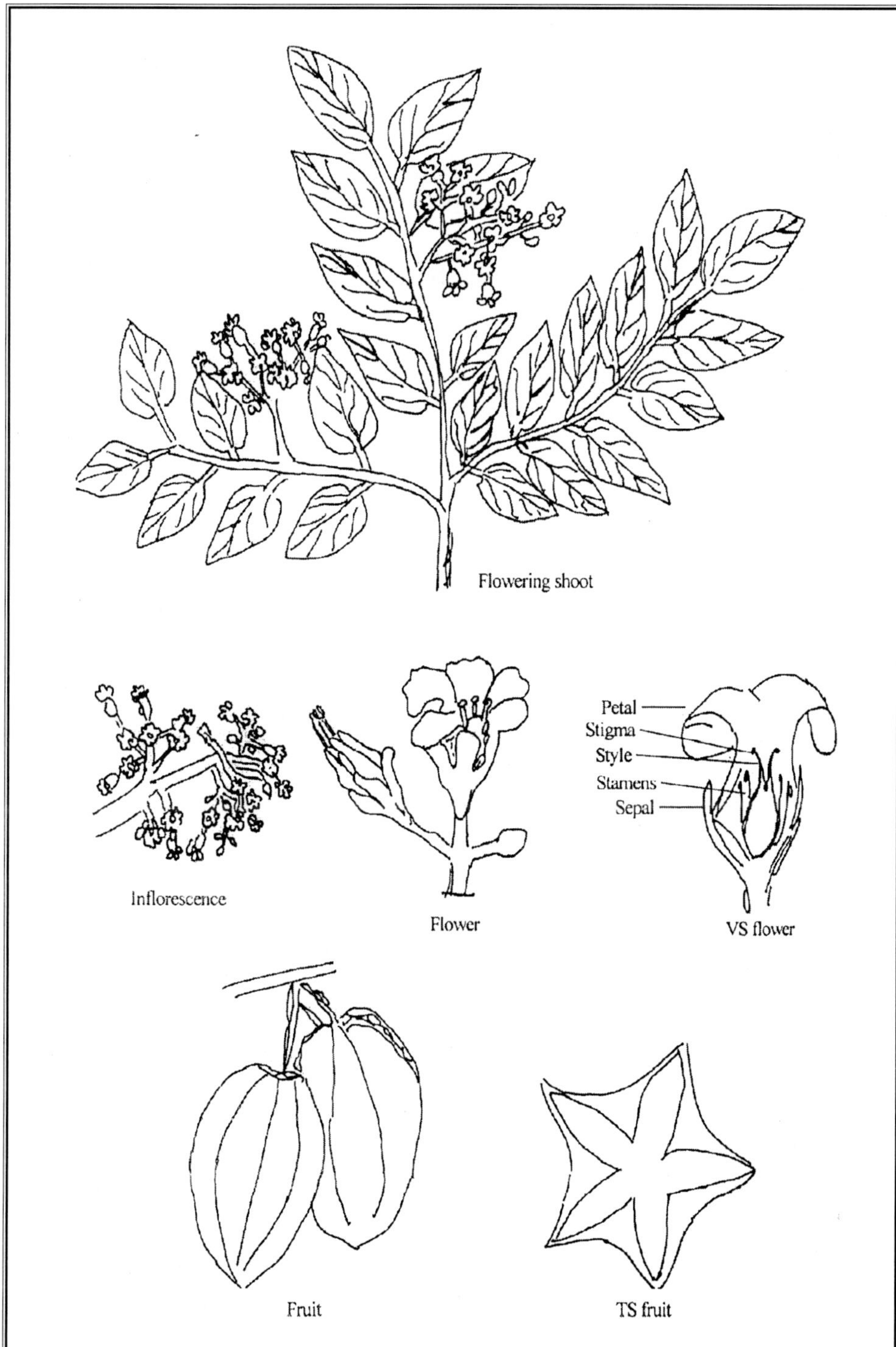
Flowering shoot
Inflorescence
Flower
Petal
Stigma
Style
Stamens
Sepal
VS flower
Fruit
TS fruit

Description

A small tree (6-9m) may attain a height upto 15m. Leaves usually ovate, leaflets jugate 7-10 in paripinnate, margin entire. Flowers pentamerous in axillary panicles, petals much longer than sepals, stamens 10 of which sometimes 5 are staminodes (rudimentary), ovary superior, five celled, usually with 5 styles. Fruit is a large berry ovoid to ellipsoid, seeds have endosperm, ellipsoid in shape.

Averrhoa carambola :A bushy tree, branches many, drooping. Leaves 3-6 jugate, flowers generally in axillary panicles, heterodistylous, petals attached, light red with purplish in the centre, measures about 8mm in length, stamens usually staminodes/rudimentary, anthers missing. Fruit has five prominent projections/ribs (star shaped), longer than broad (11-12.5 to 5.5-6cm) seeds usually covered with fleshy edible aril.

Averrhoa bilimbi: Plant is a tree and grows to about 30 feet in height. The leaves are larger than carambola, leaflet pairs are five to seventeen (2-5 in carambola). The red/crimson flowers have ten stamens, all are perfect. Fruit is acutely five angled or cylindrical, 2 to 4 inches long, greenish yellow when ripe, flesh soft juicy and translucent, contains few small flat seeds which are embedded in the flesh.

Very few standard carambola cultivars are available, and what so ever cultivars are grown have been raised through seed. Being heterozygous the seedlings producing fruit do not resemble the mother tree. On the basis of variability present, two cultivar groups are well known i.e. sour and sweet type. Brief description of some of the cultivars is given below:

Arkin : Is a seedling selection, fruit edges are slightly rounded at maturity, fruit colour is yellow to golden yellow, texture crisp, flavour pleasant, TSS content in ripe fruit is in the range of 10-12° Brix and average fruit weight is about 150-200 gm.

Fwong Tung : Found growing in Florida, fruit edges are slightly round yellow or orange coloured at maturity, texture crisp, flavour pleasant a blend of sweet-sub acid, fruit bigger in size and average weight is about 150-300 gm., with TSS content of 10-12.5° Brix.

Besides, these there are several other seedling selections which are being grown namely Starking, Thaiknight, Wheeler (Florida), Leng Bak and Jurong (Singapore), $B_1B_2B_4B_8B_{10}$ (Malaysia).

□□□

32
Bael

Genus : *Aegle*
Species : *marmelos* (L) Correa
Family : *Rutaceae*
Chromosome No. : 2n = 18(36)

Bael (*Aegle marmelos* Correa) besides being indigenous to N. India it is also one of the most useful medicinal plant of India, as it has number of curative properties. In Ayurveda the Indian medicine system use of unripe fruit is advised for acute dysentery and diarrhoea whereas ripe fruit acts as an effective laxative. Amongst the fruits native of India, bael is said to be the most nutritious fruit, is rich in protein minerals, carbohydrates, vitamin C as well as in riboflavin content. Despite these quality statistics the fruit failed to become popular (as dessert) primarily it being very

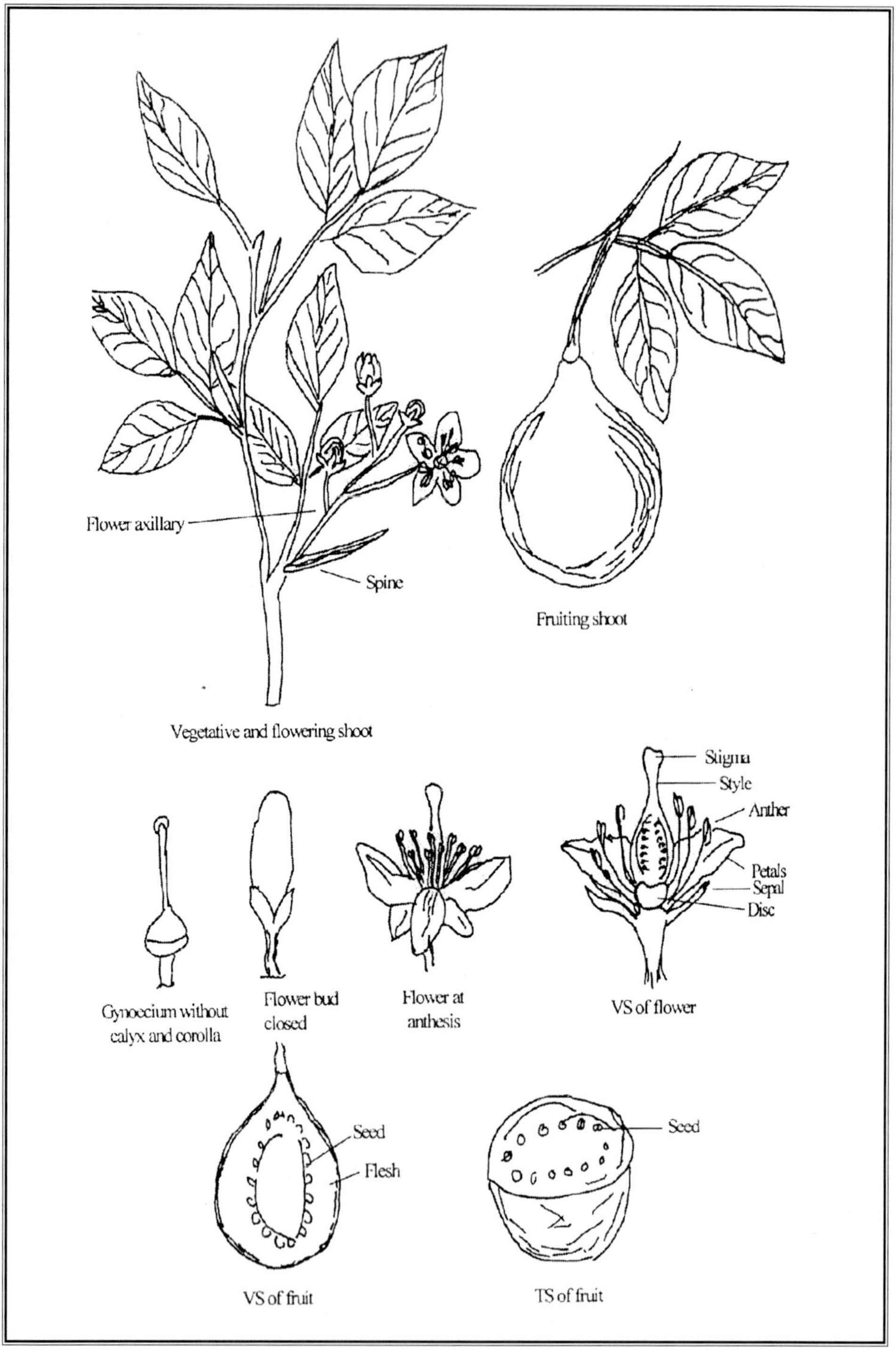

Vegetative and flowering shoot

Fruiting shoot

Gynoecium without calyx and corolla

Flower bud closed

Flower at anthesis

VS of flower

VS of fruit

TS of fruit

seedy with thick hard shell and mucilaginous flesh but the processing qualities needs to be exploited judiciously.

Bael is a prehistoric fruit plant, as it is mentioned in Vedas. According to traditional Hindu prevalent custom, trifoliate bael leaves are offered to Lord Shiva as a sacred offering. At the time of Ramayan, fruit was known to the Hindus and bael trees were found growing in Chitrakut hill and Panchvati, further crop is mentioned in Jain and Buddhist literature, too.

The crop is grown throughout Indian subcontinent i.e. Burma, Bangladesh, Sri Lanka, Pakistan and in number of south east Asian countries including Thailand, Indonesia, Malaysia, Fiji etc.

Origin and Distribution

Bael or bael fruit is common in different south east Asian countries by its vernacular name. The fruit is found growing wild in different countries in Indian peninsular mainly in Sri Lanka, Bangladesh and Pakistan. It is from here the crop spread to far south east Asia i.e. Thailand, Java, Malaysia and later in the other tropical parts of the world.

Uses

Almost all plant parts are directly or indirectly put to some use. The leaves and bark is used to overcome intermittent fever, leaves at early stages cause abortion and sterility in women, a paste made of leaves mixed with lime and betel pepper is used to overcome itching sensation, roots are employed to check heart palpitation, bowel inflammation and indigestion. Fruits have much wider application i.e. are eaten fresh or syrup, fruit nectar, sharbat and marmalade are extensively prepared. A yellow dye is prepared from unripe fruit rind and employed in tanning purposes, fruit is also dried, sliced and used against diarrhoea, constipation and dysentery.

Description

Plant is a small deciduous tree (10-15m tall) with trunk about 30-50 cm in diameter. Bael is a hardy deciduous and can tolerate harsh climate condition, e.g. very high summer temperature of 47-49^{0}C to as low as –5 to –7 ^{0}C in winters when grown at an elevation of about 1200 meters. Fruit is rich source of tannins and also contains number of important volatile oils, alkaloids and steroids, spiny, spines solitary or in pairs (1-2cm long). The older branches are more spiny. Leaves generally alternate, trifoliate, petioles 2-4cm long, terminal petiolules long (15 mm), terminal leaflets obovate and somewhat large (7.5 x 5 cm), lateral petiolules small (3 mm), leaflet shape ovate to elliptic

(7 x 4 cm) coupled with densely minute glandular punctuate. Flower, the inflorescence are axillary racemes. The bisexual flowers are about 2 cm in diameter and borne in clusters, sweet scented and greenish white. Calyx is small, shallow with five short broad teeth pubescent outside. Corolla 5, oblong oval, thick, pale greenish white dotted with glands. Stamens are numerous, sometimes coherent in bundles, ovary oblong ovoid, slightly tapering, cells are numerous 8-20, small and arranged in circle with numerous ovules in each cell. Fruit usually globose, pericarp nearly smooth, greyish yellow about 3 mm thick, hard and filled with soft yellow and orange flesh, very fragrant with pleasant flavour of pulp. Seeds are many compressed and arranged in closely packed tiers in the cell surrounded by a very tenacious slimy transparent mucilage that becomes hard on drying. The white testa has wooly hairs, embryo has large cotyledons.

The genus *Aegle* is of Greek origin and specific name *marmelos* is a Portuguese name, belongs of the family Rutaceae and has 2-3 species.

Cultivars

Bael in usually propagated by seeds and shows variation in seedlings. The crop does not have any recommended cultivar, and the ones which are grown further have no standard name(s). Most of the types are named after the place where they are found in abundance e.g. cultivars are named as Delhi, Agra, Varanasi, Calcutta. Fruits exhibit variation in shape may be oblong, cylindrical, flat spherical or pear shaped and also on fruit weight basis.

□□□

33
Karonda

Genus : *Carissa*
Species : *grandiflora A.D.C. carandas L.*
Family : *Apocynaceae*
Chromosome No. : 2n=22

The carissa (*Carissa grandiflora*) is regarded not only for its ornamental value but also for its edible fruits. The plant is not very demanding in its climatic requirements. It grows successfully in the warm moist tropical regions and also in the dry subtropics where the temperature never falls below 26-28°C. In United States, especially in California, Carissa is injured by frost, whereas in Florida frost has no influence on the crop. This crop species is valued highly as a hedge plant, as it resists severe pruning and resultant growth

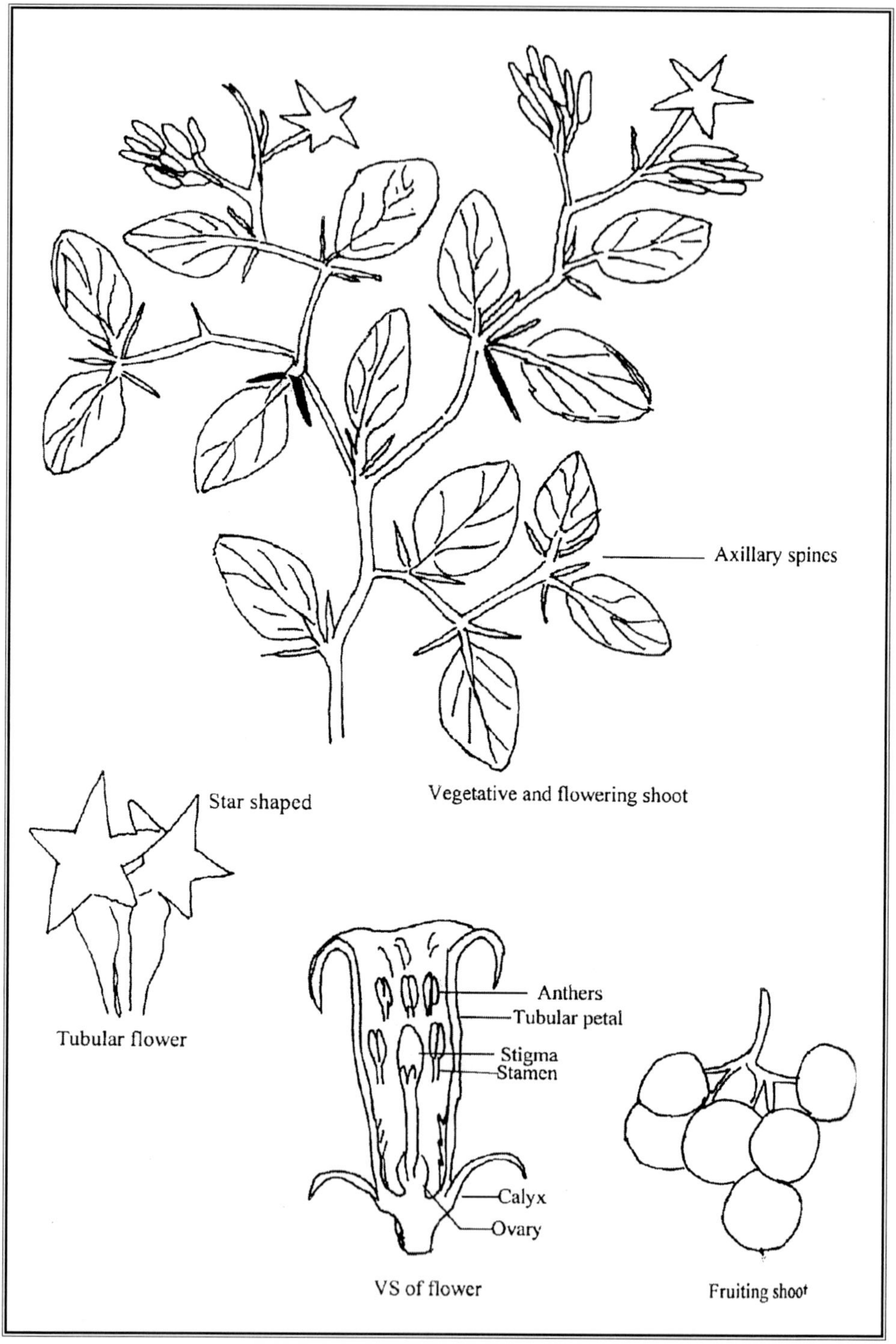
Axillary spines
Star shaped
Vegetative and flowering shoot
Tubular flower
Anthers
Tubular petal
Stigma
Stamen
Calyx
Ovary
VS of flower
Fruiting shoot

is usually low and compact, which are desired traits for hedge plant. The thorny, stout branches make an impenetrable barrier for animals. When in full bloom the tubular fragrant jasmine like flowers, red fruits with a dark green shiny foliage background provides a very attractive appearance. Another species very popular and common in India is Karonda (*Carissa carandas*). The species is smaller in all respect to *C. grandiflora* and in India, the fruits are extensively used to make pickles and preserves.

Genus *Carissa* has some 20 species most of which are evergreen usually spiny shrubs or small trees from the dry open woodlands in tropical and subtropical region of Africa and Asia.

Carissa plants have the inherent potential to bloom profusely in spring and with few flowers throughout the year. With its dense deep green and shiny foliage, brilliant red fruits with no definite climatic requirements needed for its cultivation they need to be cultivated more extensively in tropics than its present status. All carissa are highly valued as a hedge plant and it withstands sever cutting.

Origin and Distribution

Carissa is native of south-Africa and India.

Uses

The fruit pulp makes a fine jelly and also used as a substitute for cranberry sauce. Beside, gradiflora, the other species *C. edulis* and *C.carandas* are used for hedge plantation and also makes an attractive ornamental shrub.

Description

Plants are large, evergreen, erect shrub with dense dark green foliage grows to a height of 3-6 meters, sometime herb, generally with a milky latex, heavily armed with sharp bifid thorn. Leaves dark green, simple opposite or whorled, rarely alternate, thick glossy, ovate to elliptic in shape, tan red at opening later leaves turn dark green in colour. In the leaf axils slender to stout spines are present. Flowers white fragrant, tubular regular bisexual, complete and epigynous, solitary, or in cymose clusters. Sepals 4-5, green in colour almost free to the base, imbricate persistent ovate in shape. Petals are as many as sepals, lobes contorted in bud, the corolla often has hairy or scaly appendages or outgrowth of various kinds. Stamen as many as petals and alternate with them in one whorl, filaments short, anthers yellow and inserted on the corolla tube interpose, arrow shaped, often free or united in a cone. Disc often present, entire lobed or scaly. Style long simple, stigma thickened two lobed yellow coloured and hairy below. Carpels usually 2, very rarely more than 2, united (connate) or united by styles or stigma's only, but free at ovary. Ovary superior 2 celled. Fruit is a drupe, ovoid to ellipsoid, 5 cm long, coloured with reddish pulp at skin ,and lighter near the seed. Calyx persistent, latex present, flavour pleasant. Seeds 3-4 dark brown in colour and hard, flat, sometimes with endosperm and sometimes without endosperm.

Brief description of other Carissa species is given below.

C. *grandiflora* A.DC. Natal plum (syn. C. *macrocarpa*) are usually spreading shrubs or small trees (15-20 ft tall) somewhat drought resistant much branched and with stout or sturdy branches armed with bifurcated thorns. The foliage is dense deep green in colour and glossy. Leaves are thick, coriaceous, 1-2 inches long, ovate in shape, apex mucronate (short spiny tip). The fragrant white or slightly pink tinged flowers are borne in small terminal cymes and measures 2 inch across and are star shaped in form. Flowers more in spring, but few flowers can be seen year around. Most of the fruit is harvested in summer. Fruits are elliptic or ovoid in shape 1-2 inches in length skin thin with bright red colour, flesh firm granular and reddish with number of small rounded seed in centre. In other species the flowers are smaller ½ inch across or less, seeds 15-20 small in size. Mostly used as ornamental plant in Capetown, South Africa.

C. *bispinosa* Desf (C. *arduina* Lam.) Amatungula. An evergreen shrub hardy with thick glossy leaves, ovate or subcordate, glabrous and entire. Flower white in colour twisted to the right is the bud very large and fragrant, bears regularly and abundantly. Fruit dark red as the size of cherry.

C. *acuminata* DC. Resembles closely to C. *bispinosa*, spines not stout, or rather weak, leaves smaller ovate, sub cordate, peduncles short, forked and axillary. Calyx lobes lance-acuminate in form, corolla twisted to right in the bud.

C. *edulis* Forssk. Native of tropical Africa straggling bush about 10 ft tall, spines simple or branched, leaves ovate to sub lanceolate, about 2 inch long, apex acute or mucronate. Flower white inside and purple to red outside small about half inch across, lobes acute, corolla tube about ¾ inch long. Fruit is a berry globose small in size, red to purple in colour smooth 2-4 seeded fruit is edible.

C. *macrophylla* Wall. (C. *Dalzellii* Bedd.). Native to India found growing in plenty in western Ghats. A large shrub spines large about 1.5 inch long and leaves glossy on drying become dark brown, 2-4 inch long, apex sharp pointed. Corolla tube nearly 1 inch and lobes ½ inch long. Berry about 1 inch long, ellipsoid.

C. edulis (Forrek). Native of tropical Africa, straggling shrub, about 10 feet tall, spines simple or branched, leaves ovate to sublanceolate, about 2 inch long, apex acute or mucronate. Flowers white inside and purple to red outside, small about half inch across, lobes acute, corolla tube about ¾ inch long. Fruit is a berry globose, small in size, red to purple in colour, smooth, 2-4 seeded, fruit is edible.

C. spinarum DC. Plant is a small shrub which produces edible fruits. A common specie found in India thrives well on very poor and rocky soils, hence considered useful in reforestation. Is a very important hedge and the fruits are readily eaten by goats, sheeps etc.

The genus Carissa does not have many cultivars, few which are important and grown commercially are detailed below.

Frank : Originated in Riverside, California. Chance seedling, very old cultivar. Fruit large, quality good, self unfruitful, no more being cultivated.

Torrey :Originated in La Jolla, California. A spontaneous tetraploid of *Carissa grandiflora*. Plant large so are flowers and fruits, with thick growth as the usual types.

Chesley : Originated in California. Parentage unknown, fruit size large, flavour and texture not too good.

Serema : Originated in California, parentage unknown. Plants flower throughout the year under favourable condition and fruit is available for 8 months of the year, fruit stalk long helps in easy picking, storage potential good.

❏❏❏

34
Tree Tomato

Genus : *Cyphomendra*
Species : *betacea* (Cav.) Sendtner
Family : *Solanaceae*
Chromosome No. : 2n = 24

The tree tomato is a bushy fruit plant which in early time was planted in the gardens high upon the mountain slopes. Presently in India, Sri Lanka and number of other countries it is grown in the hilly region. Indians of Peru at very early stages gave the world a number of food plants of which potato is of great value, and tree tomato in the only woody species of family Solanaceae. The tree is fast growing and starts to bear in about two to three years times. The egg shaped and usually smooth-skinned fruits are produced in large number in hanging clusters. The sub acidic and

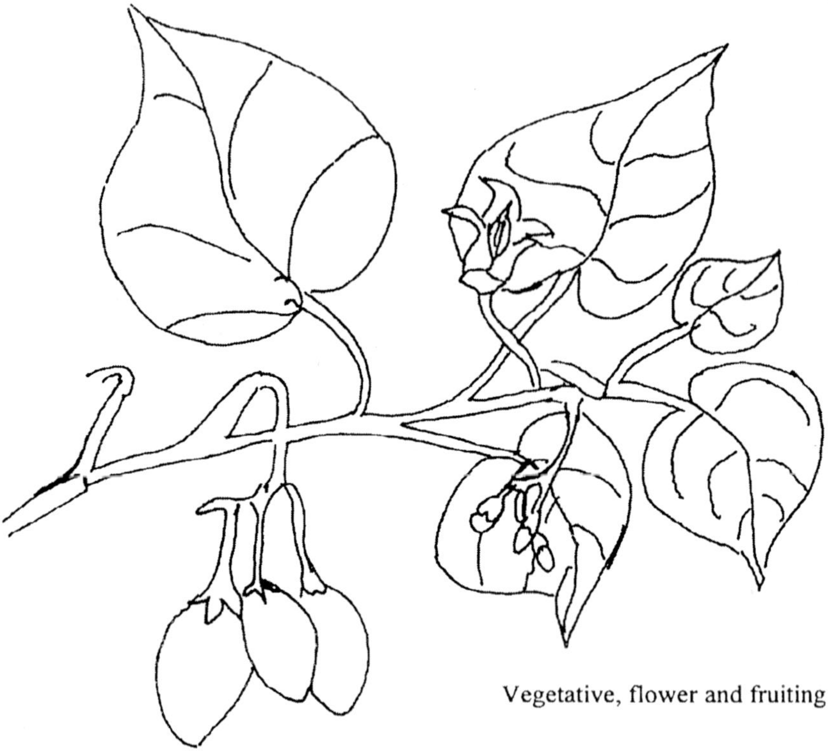

Vegetative, flower and fruiting shoot

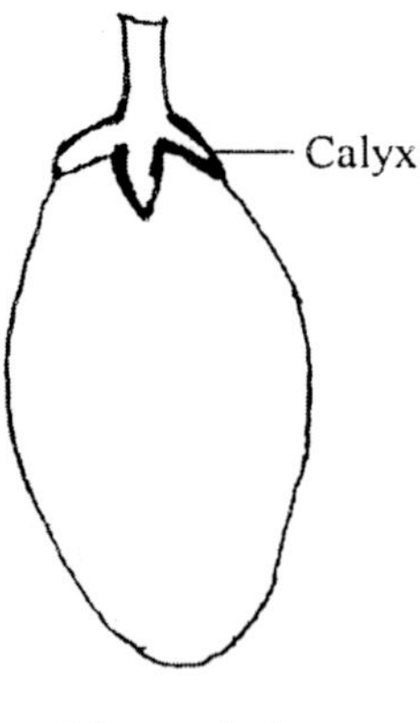

Mature fruit

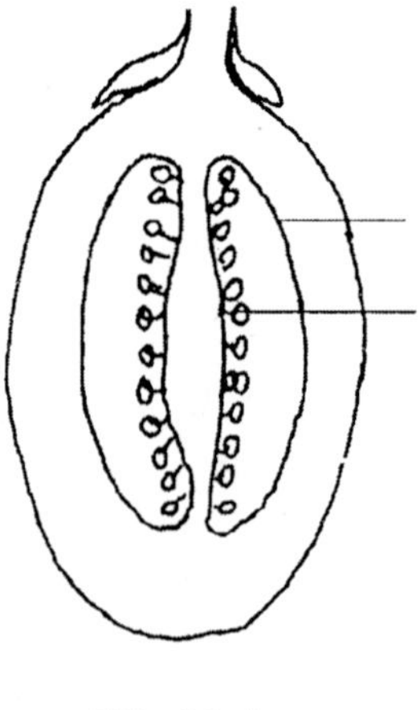

VS of fruit

succulent fruits are agreeable and eaten as raw. In central America it is called tomate and cultivated at an elevation of 4000 to 8000 feet. The crop loves rich loamy soil and grows best under irrigated conditions. In its native place Peru, of the several food plants cultivated by the local agriculturists, tree tomato or a spineless herb a bush fruit was planted in their gardens high on the mountain slopes, is also grown in the hilly region of India and Srilanka and number of other countries is regarded to be a plant of minor value. In most of the characters it resembles tomato to which it is closely related. The plant grows rapidly under irrigated condition and starts bearing is two to three years of planting, but does not relish high degree of atmospheric humidity.

Origin and Distribution

Tree tomato has its origin in south America specifically in the Andes ranges located in Peru. From its habitat, the crop was later introduced in most of the tropical, subtropical and in some marginal temperate regions all over the world.

Use

As tomato (vegetable) is subjected to number of uses, same is true for tree tomato. A variety of products can be prepared i.e. sweet and savoury dishes, curry, chutney from unripe fruits, salad, soups and stuffing from mature fruits. High quality jams, jellies, desserts etc from tree ripened fruits which have high aroma and flavour.

Description

Plant is a shrub or tree, generally short statured, but sometimes may attain a height of 8m. Tree trunk generally short with thick branches. Leaf shape typically cordate, ovate, simple alternate, margin entire with soft dense pubescence, veins prominent on the under surface, petiole long. Flowers are borne in small, axillary loose clusters, mainly on branch tips. Flowers are small (1cm diameter), fragrant, pentamerous pink to light blue in colour. Calyx persistent 4-5 segmented/ lobed, pale yellow, corolla 5 segmented generally bell shaped thick, succulent, pink tinged oblong lanceolate, inside also pink coloured (companulte), stamens five placed on corolla throat, anthers clustered around the pistil, stigma small, ovary 2 celled and each with many ovules. Fruit is a typical berry, ovoid or obovoid in shape, usually tapering at both the ends, stalk long. Fruit skin thin, smooth, shiny, initially white, turns to yellowish-orange red and finally is overlaid as purple reddish, flesh juicy and yellowish to black, sub acid to sweet in taste. Seeds many, thin, usually flat round and some what hard in texture.

At present no named cultivar is enlisted for the crop. Cultivars / varieties are classified on the basis of fruit colour, i.e. yellow or red. Fruits of red types are very attractive in appearance, preferred more for fresh fruit industry, but are more acidic in taste. Yellow types are employed for canning purposes.

❑❑❑

35
Rasbhari

Genus : *Physalis*
Species : *peruviana L.*
Family : *Solanaceae*
Chromosome No. : 2n = 28, 48

Physalis (Greek for Bladder, as the thin calyx enlarges and encloses the fruit). Fruit is called as husk tomato or ground cherry. Is a plant of warm and temperate regions and is mostly cultivated for its edible fruit which is sub acidic in taste and the flavour resembles that of tomato hence the name tree tomato or husk tomato. The genus comprises nearly 75 species mostly of American origin, however few species are also found in Asia and Europe. The species p*eruviana* and *pubescens* are cultivated for their edible fruit whereas *P. alkekengi* Linn and *P Franchetii* Hort., are highly valued for

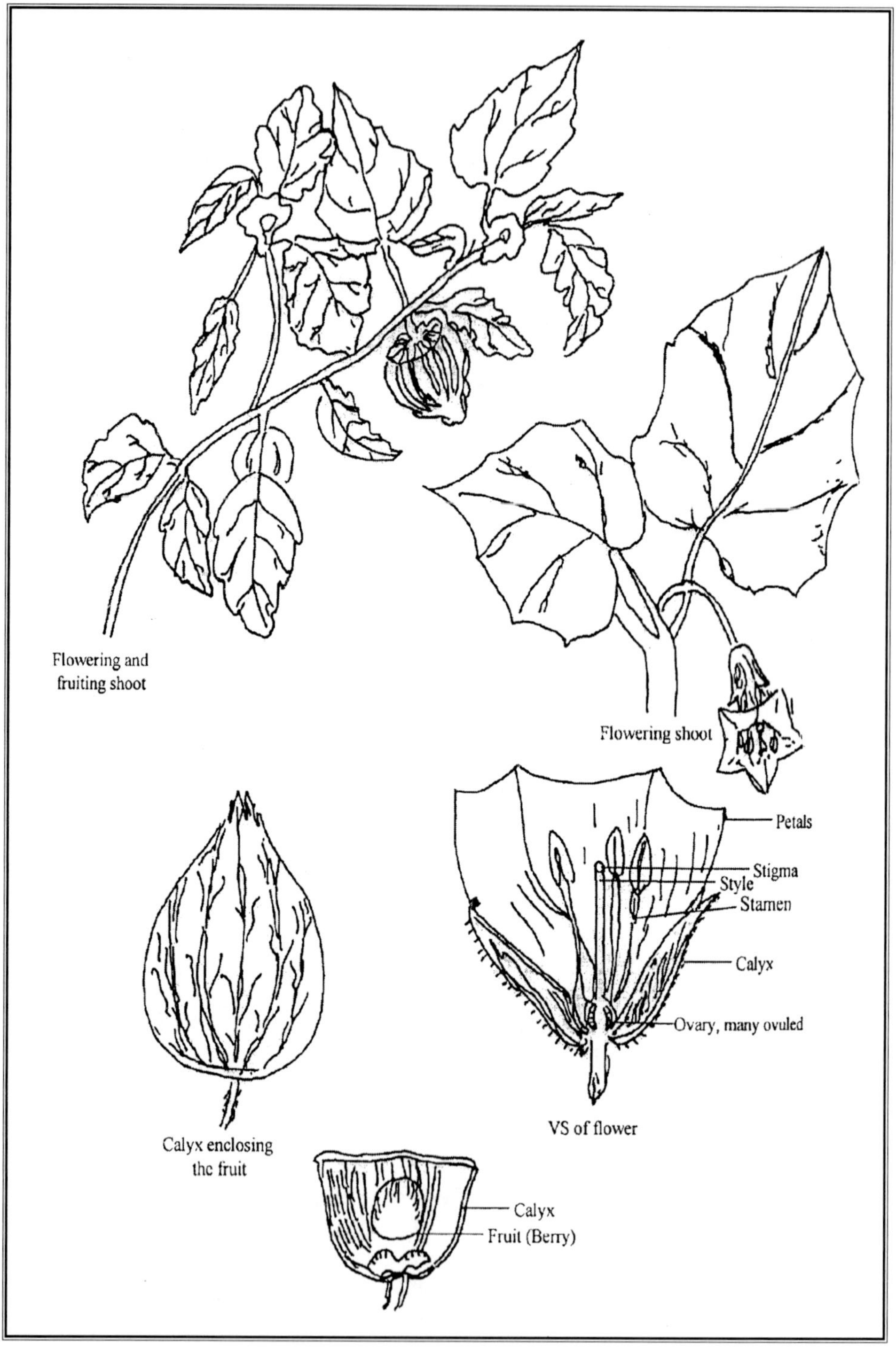

Flowering and fruiting shoot

Flowering shoot

Calyx enclosing the fruit

VS of flower

their shiny large red calices. Plant is a very old garden plant grown mainly for its brilliantly coloured calices, is perennial, grows zigzag with soft pubescence on stems. Leaves ovate, base broad and angular, petiole spreading at top. Flowers whitish with yellow anthers. Fruit red, the enclosing calices brilliant dark blood red in colour, very showy, only sometimes eaten.

Origin and Distribution

This tropical plant is commonly known by its vernacular name as Cape gooseberry, and occurs in the Chilean and Peruvian highlands. The plant has wide adaptability and now is widely spread in the tropical and subtropical and to less extent in the milder temperate areas in the south east Asia. The crop however is still not grown in regular orcharding fashion.

Uses

The plant has several uses. The berries (yellow orange to sometimes light brown) are generally consumed fresh, or can be used in salads and in preparation of different fruit products like ice- cream, pies, chutneys or puddings and can be processed or canned to make jam or jelly. At the places of its origin other plant parts especially the leaves are used as medicine to cure number of ailments.

Description

The genus *Physalis* L. comprises 100 herbaceous species which have both annual and perennial types and both have husk covered fruit. Beside the common fruit yielding types, some are also weeds i.e. *P. minima* and *P. angulata* but with edible fruits. These species as weed are common in many of the south east Asian countries. The plant behaves as perennial when grown in the tropics.

Cape gooseberry plant is a perennial herb, measuring about 0.5 to 2 meter tall. The branches are spreading and also with creeping tendency, purplish in colour, usually ribbed and densely pubescent.

Leaves are ovate sub-opposite, margin entire or very shallowly toothed, base cordate, apex acuminate, the petioles are about the same length as that of blades. Flowers usually solitary on short pedicel (10 mm long) in leaf axils, flower parts pentamerous, calyx (sepals) 5 toothed, petals 5, lobed, (measures about 2 cm in diameter), yellow in colour often with very dark brown - purplish spots inside, stigma subcapitate with purple anthers.

Fruit orange yellow, glabrous, globose berry (1-2 cm in diameter), which is enclosed in highly pubescent husk (calyx), flesh yellow smooth tender soft, pulp juicy. Seeds very small, numerous, yellowish in colour.

Brief description of some other species is given below.

Beside the two species which yield fruit of commerce and are edible, there are other species which are valued as ornamental for their attractive foliage. Brief description of some species are given below.

P. nivosus Bull. Snow Bush. Plant is shrub, branches zigzag terete or angled, leaves well developed 1-2 inch long, broad ovate-elliptical obtuse, variegated with white and green, flowers are small greenish with long pedicels that hang from leaf axils. Specie has number of botanical forms.

P. nivosus var. roseopictus Hort., Leaves are markedly mottled with different colours i.e. white, green.

P. nivosus var. atropurpureus Hort., grown mostly in tropics, leaves are dark purple coloured, with pink or red flowers.

P. grandifolius Linn. Indigenous to W. Indies and S. America. Tall tree, growth symmetrical foliage has metallic shine. The vegetative or foliage branch long and stout, when falls a big scar is seen of the branches. Leaves long 3-5 inch, oblong lanceolate, thick round or chordate at base. Fruit large 3 celled.

P. Ferdinandi Muell. A shrub of very attractive foliage, fruit thick and about ½ inch across, leaves oblong ovate, apex acuminate, thick 2-3 inch long.

P. pubescens Linn. (Dwarf cape gooseberry, Husk tomato ground cherry or strawberry tomato of vegetable gardens).

Plants are low trailing flat annual plant, or may be climbing to a height of about 1 foot. Leaves significantly notched, teeth blunt, smooth thin. Flowers small campanulate border white yellow upright, throat with 5 distinct large brown marks, anthers usually yellow. Calyx or husk almost smooth thin paper like, five angled and enclose small sweet yellow fruit.

P. ixocarpa Brot. Tomatillo. Native to Mexico. Grown for edible berry. Plants are erect 3-4 feet, leaves and branches smooth, leaves thin ovate or lance-ovate notched or toothed. Flower large broad and open, petals yellow, throat with 5 black brown spots, anthers purple, calyx purple veined, enlarges to enclose round large purplish sticky berry.

P. Franchetii Mast (P. Alkekengi var. Franchetii. Hort.). Chinese Lantern plant. Probably native of Japan. Very showy plant. Plant about 2 feet tall, shiny with short petioles. Calices are about 2 inch in diameter. Behaves as annual and perennial. Fruits are very brilliantly orange red coloured which ripens in autumn.

P. alkekengii Linn. Alkekengi. Bladder cherry, Strawberry tomato or Winter cherry. Indigenous to south east Europe, Japan, presently naturalized in many parts of the world.

Cultivars

No standard recommended cultivar are presently known and grown in India.

❑❑❑

36
Aonla

Genus : *Phyllanthus*
Species : *emblica* L.
Family : *Euphorbiaceae*
Chromosome No. : 2n = 28

The aonla (*Emblica officinalis* syn. *Phyllanthus emblica*) is native of tropical south-eastern Asia, and to central and southern India in particular. Aonla although considered as a minor fruit, but is of commercial significance primarily due to early high returns without much care. The small lateral branches resembles pinnate leaves of walnut or sumac and when fall off leaves behind a big scar. Flowers are axillary, apetalous mono or dioecious in small clusters or singly. Plants show great adaptability, are hardy and high bearer. The genus Phyllanthus (Greek-for leaf flower) has some 500

Flowering shoot

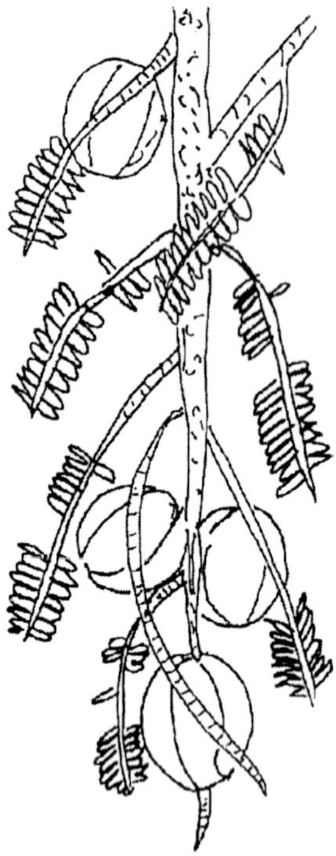

Fruiting shoot

Staminate/male flower

Pistillate /female flower

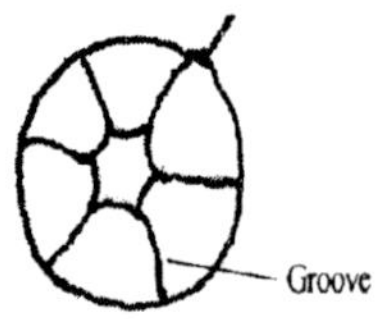

TS of fruit

Seed

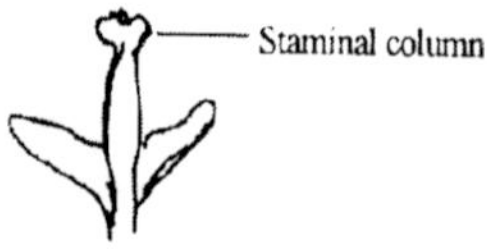

VS male flower

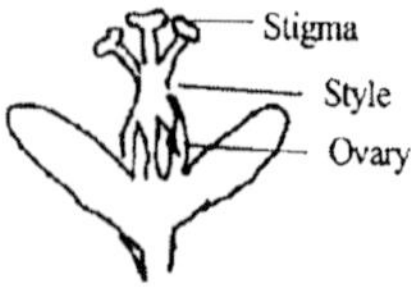

VS female flower

species or even more and two have been identified as important from fruit point of view i.e. *P. emblica* and *P. acidus.*

Aonla fruit is the richest source of vitamin C, except Barbados cherry, besides also has minerals in sufficient quantities namely iron, phosphorus, calcium, and Vit. B, nicotinic acid and fibre.

Origion and Distribution

Phyllanthus is regarded to be indigenous to the tropical parts of south eastern Asian countries and in particular to the central and southern Indian. Besides India it is also said to be native of Srilanka, Malaysia and southern China. In many parts it spreads from Himalayan base to Srilanka and from Malacca to south India the crop is found growing both in cultivated and wild forms.

Uses

Amla is a rich source of vitamin C, very nutritive, extensively used for pickling. Remarkable feature of the crop is that ascorbic acid and other constituents are well retained in the dry form hence is made into preserves, sauce, jellies, toffees, candy, dried chips, tablets etc. Various plant parts used to treat diabetes, dysentery, bronchitis, jaundice, cough, diarrhea, in dyeing and tanning industry, is also valued as an antiscordutic, laxative and antibiotic.

Description

Phyllanthus (Greek-for leaf flower, the flower of some species are borne apparently on leaves). This large genus comprises of about 650 species of evergreen and deciduous, herbs, shrubs or trees cultivated for their graceful foliage. Plant quite large tree, trunk usually crooked, branches thin spreading and scattered in all directions, bark ash coloured, scabrous. Leaves very small entire, alternate placed closely (2 ranked) on the small lateral branches make it to resemble like pinnate leaves or foliage which later shed or fall and leaves behind a prominent scar, petioles round and striated. Some species have cladophylls rather than real leaves and they become fringed with some flowers which give an impression that leaves are flowering. Flowers, appear in leaf axils, apetalous, monoecious or dioecious, appear singly or in small cluster, disc present, calyx 4-6 imbricate, without any pistil (even rudimentary) in the staminate flower. Stamens usually 3 but may range from 4-6. Styles slender, ovary superior 3 celled with 2 ovules in each cell. Fruit is a fleshy drupe, smooth, globular, six ribbed, striated. *P. emblica* and *P. acidus* are the two species which are important horticulturally, their import differiating characters are detailed as.

P. emblica. Myrobalan Emblica plants are highly branched shrub or small tree. Flowering appears at the basal region 3-12 inches of the branches each of these small flowering branches bear 100 or more stalkless to linear elliptical obtuse leaves, red tinged when young. Leaves are closely placed together and measure about ½ inch, stipule small and withering. Female flowers are small/ minute, short pedicelled, greenish yellow or red tinged, petalless, flowers appear during spring and summer. Nectary cupshaped, ovary superior, styles united, stigma 3. Male flowers, numerous in the axils of lower leaflets. Sepals of staminate flower 5-6 number, stamens 2-4 anther open vertically and filaments are connate (united) appear single. Fruit fleshy drupe smooth, six ribbed, ovary three celled with two ovules in each cell.

P. acidus. Otaheite or gooseberry/star goosberry. Is a handsome shrub or tree of about 10 meter high and usually grown for its edible fruits. Branches are stout with conspicuous scars of fallen flowering branches. Flowers appear on separate branches. Leaves are ovate acute measuring 1-2 inches long, pale green almost stalkless. In spring it bears dense clusters of small red flowers. Sepals of staminate flowers 4 anthers do not split vertically. Other characters of male and female flower are same as that of emblica. *Acidus* fruits are small 2 cm across flesh thick, fruit star shaped in cross sections.

Cultivars

The fruit is usually raised from seed, the resultant seedlings fail to be true to type. On the basis of genetic variability exhibited by the seedlings some useful types have been selected which are multiplied on seedling stock to maintain the genotype. Some important cultivars of aonla are characterized on the basis of names of places, fruit size and colour. Brief description is given below.

Banarsi : Plant has upright growth habit and there are usually three branchlets emerging from single node. The fruits are large in size, skin smooth yellowish, segments distinct and raised in three parts, apices flat, oblong. Flesh light yellowish green, fibres moderate, soft can resist transportation. Female flowers are less in number, self incompatible, shy bearer, contains high levels of ascorbic acid.

Chakaiya : Plants are spreading, bears profusely productive, has more than 4 female flowers per branchlet. Fruits are small-medium skin smooth, russetted, greenish both the apices flattened, flesh hard crisp and fibrous.

Francis : Plant not very tall growing, branchlets drooping, medium cropper. Fruits are large, both ends flattened, skin smooth, greenish yellow flesh soft, nearly fibreless, does not keep for very long, very susceptible to fruit necrosis.

Kanchan : Said to be chance seedling of Chakaiya, spreading tree bears heavily, fruits are small-medium, surfaces flattened, oblong in shape. Skin smooth, yellowish. Recommended for making pickles.

Krishna : Chance seedling of Banarasi moderate in production. Fruit medium to large, conical somewhat angular, apices flattened, skin yellowish very smooth, shiny with red blush on sun facing sides, flesh fibreless hard can withstand transportation.

37
Barbados Cherry

Genus : *Malpighia*
Species : *punicifolia L., glabra Millsp.*
Family : *Malpighiaceae*
Chromosome No. : 2n = 28

The barbados cherry, a member of the Malpighiaceae, is an interesting example of a fruit that rose like Cinderella, from relative obscurity about 40 years ago. It was at that time the subject of much taxonomic confusion, having been described and discussed previously under the binomical *Malpighia glabra* L. which properly belongs to a wild relative inhibiting the west Indies, tropical America and the lowlands of Mexico to southern Texas, and having smaller pointed leaves, smaller flowers in peduncled umbels, styles nearly equal, and smaller fruits. *M. punicifolia* L.

(*M. glabra Millsp.*) has been generally approved as the correct botanical name for the Barbados cherry, which is also called west Indian cherry, native cherry, garden cherry, French cherry, in Spanish, acerola, cereza, cereza colorada, *cereza de la sabana, or grosella* in French, *cerisier, cerise de St. Domingue* in *portuguese, cerejeira*. The name in Venezuela is semeruco, or cemeruco in the Netherlands Antillers, *shimarucu* in the Philippines, malpi (an abbreviation of generic name).

Origin and Distribution

The barbados cherry is native to the Lesser Antilles from St. Croix to Trinidad, also Curacao and Margarita and neighbouring northern south America and as far south as Brazil. It has become naturalized in Cuba, Jamaica and Puerto Rico after cultivation, and is commonly grown in dooryards in the Bahamas and Bermuda and to some extent in central and south America.

The plant is thought to have been first brought to Florida from Cuba by Pliny Reasoner because it appeared in the catalog of the Royal Palm Nursery for 1887-1888. It was carried abroad rather early for it is known to have borne fruit for the first time in the Philippines in 1916. The plant was casually grown in southern and central Florida until after Word War II.

Uses

Barbados cherries are eaten out of hand, mainly by children. For dessert use, they are delicious merely stewed with whatever amount of sugar is desired to modify the acidity of the particularly type available. The seeds must be separated from the pulp in the mouth and returned by spoon to the dish. The cooked fruits must be strained to remove the seeds and the resulting sauce or puree can be utilized as topping on cake, pudding, ice cream or sliced bananas, or used in other culinary products. Commercially prepared puree may be dried or frozen for further use. The fresh juice will prevent darkening off bananas sliced for fruit cups of salads. It can be used for gelatin desserts, punch or sherbet, and has been added as an ascorbic acid supplement to other fruit juices. The juice was dried and powdered commercially in Puerto Rico. The fruits may be made to syrup or, with added pectin, excellent jelly, jam, and other preserves. Cooking causes the bright red colour to change to brownish red. The pasteurization process in the canning of the juice changes the colour to orange red or yellow, and packing in tin cans brings on further colour deterioration. Enamel lined cans preserve the color better.

The bark of the tree contains 20-25% tannin and has been utilized in the leather industry. The wood is surprisingly hard and heavy. Trials have

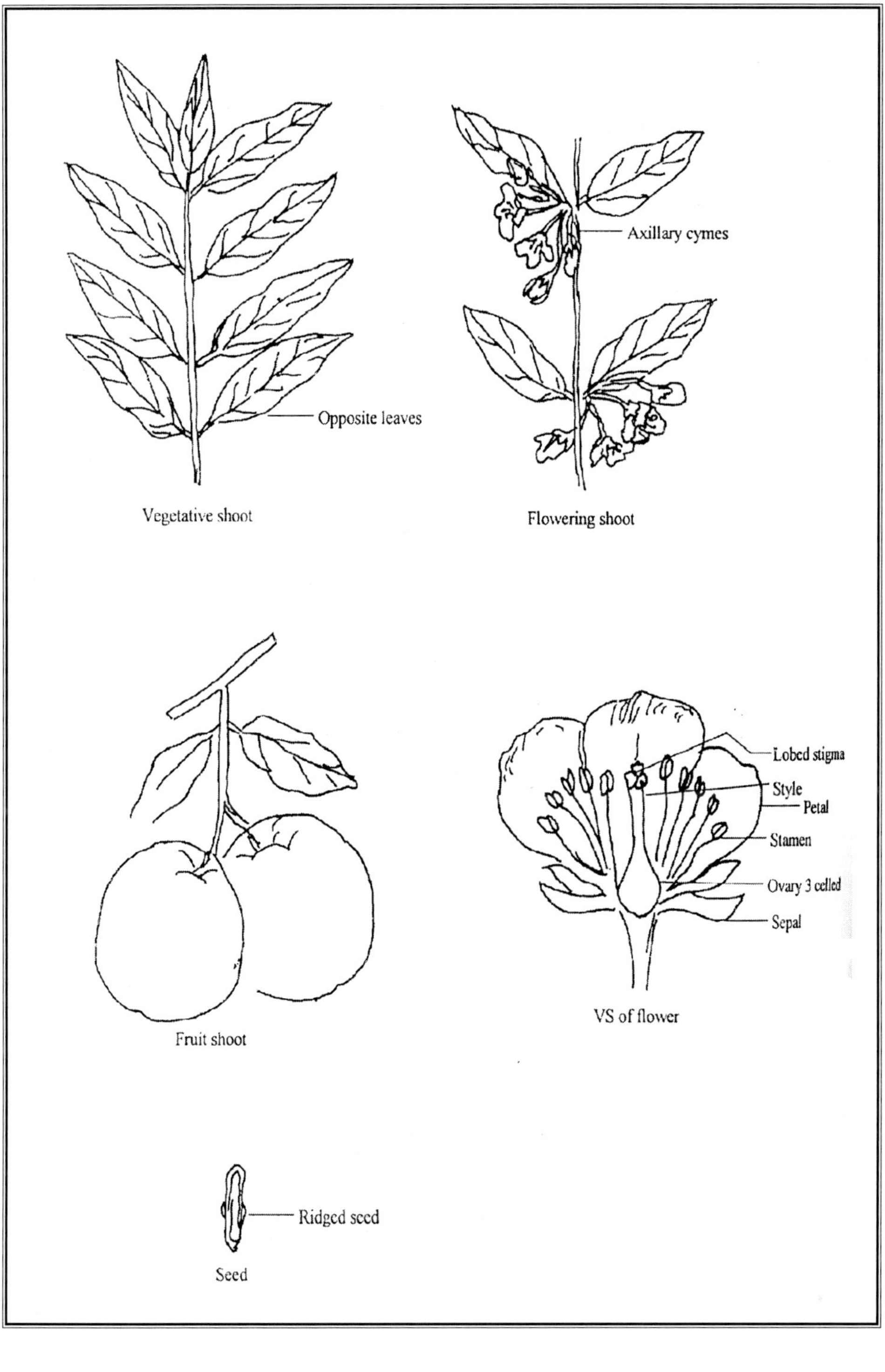

Vegetative shoot

Flowering shoot

Fruit shoot

VS of flower

Seed

demonstrated that it refuses to ignite even when treated with flammable fluid unless perfectly dry.

The fruits are considered beneficial to patients with liver ailments, diarrhea and dysentery, as well as those with coughs or colds. The juice may be gargled to relieve sore throat.

Description

The barbados cherry is large, bushy shrub or small tree attaining upto 20 ft (6cm) in height and an equal breadth, with more or less erect or spreading and drooping minutely hairy branches, and a short trunk to 4 inch (10 cm) in diameter. Its evergreen leaves are elliptic, oblong, obovate, or narrowly oblanceolate, somewhat wavy, ¾ to 2 ¾ in (2-7 cm) long, 3/8 to 1 5/8 inch. (9.5-40 mm) wide, obtuse or rounded at the apex, acute or cuneate at the base, bearing white, silky, irritating hairs when very young, hairless, dark green, and glossy when mature. The flower occur in sessile or short peduncled cymes, have 5 pink or lavender, spoon shaped, fringed petals. The fruits, borne singly or in 2's or 3's in the leaf axils, are oblate to round, cherry like but more or less obviously 3 lobed ½ to 1 inch (1.25-2.5 cm) wide, bright red, with thin, glossy skin and orange colored, very juicy acid to subacid pulp. The 3 small, rounded seeds each have 2 large and 1 small fluted wings, thus forming what are generally conceived to be 3 triangular, yellowish, leathery coated, corrugated inedible stones.

Varities

Not too many established cultivar are known of this crop. After preliminary evaluation a superior type has been selected and named as Florida sweet, a clone that was observed to have an upright habit of growth, large fruits, thick skin, apple like, semi sweet flavors and high yield.

On the basis of fruit size, yield and vitamin content A-1 and B-17 were identified but these were later found to be inferior to B-15 in ascorbic acid level and productivity. Yields of 10 clones (A-1, A-2, A-4, A-10, A-21, B-2, B-9, B-15, B-17 and K-7) were compared over two year period (1955-56) in Puerto Rico and B-15 exceeded the others in both years.

❑❑❑

38
Paw Paw

Genus : *Asimina*
Species : *triloba* (L.) Dunal
Family : *Annonaceae*
Chromosome No. : 2n = 18

Uses

The delicious pawpaw fruits are used as dessert. The red or brown streak wood is light in weight hence not used for furniture or allied work. On account of its handsome foliage, large leaves and attractive form, it provides additional attraction when planted in gardens.

Description

Pawpaw plants are large shrubs or small trees, crown open, broad topped, trunk smooth straight, brown to dark brown usually with prominent grey blotches and with small hairy projections which become prominent as the tree matures. Branches, narrow slender, spreading, pubescence is in the form of hairs, fine and rust coloured, branchlets same as branches except that young growth is red tinged. During dormancy buds are covered with hair like pubescence, rusty brown in colour. Leaves simple, deciduous oblong lanceolate broadest near the tip and tapering towards the base pale beneath light green above usually covered with rusty brown pubescence which is in the form of fine hairs, with age become smooth. Flowering takes place in leaf axils on last year's growth, flower usually stalked, companulate (bell shaped) green turns deep purple later, calyx 3, broad at base, apex pointed, pubescent, corolla 6, rounded thick, outer 3 spreading, stamens numerous, carpels variable in number 3-12 sometimes even more. Fruit are aromatic green, yellow later turns to black, flesh soft pear to somewhat cylinder shaped, seeds long and dark brown in colour.

❑❑❑

Section - II
SMALL FRUITS

39 Grapes

Genus : *Vitis*
Species : *vinifera L.*
Family : *Vitaceae*
Chromosome No. : 2n=38, 40

Grapes are considered to be the oldest plants on earth which dates back to 90-95 million years, a time when dinosaurs ruled this planet, hence grape is as old as mankind. Grape is one of the most ancient crops. The Aryans knew grape cultivation as well as to prepare beverages from it. Grape cultivation in India dates back to 11th century BC. Later Muslim invaders from Iran and Afghanistan in the 12th century also introduced grapes to India, further in the 16th century Emperor Akbar the Great introduced and encouraged grape cultivation in various parts of India.

Documents regarding general viticulture and wine making in Egypt dates back to 5000 to 6000 years. Prior to the start of Christian era grapes and wine were thought to be very significant in the day to day life of Middle East and Mediterranean people. Grapes have high caloric value being nearly 20-25 per cent sugar, which in dried grapes may be as high as 80 per cent and no fruit amongst the present ones constitute such a strong source of sweet material that can be easily transported and stored over a prolonged time period. In the region between Black and Caspian sea this prehistoric fruit still grows wild and in all possibilities is the native home of the old world grape *Vitis vinifera* L. From its native place in the Near East, grapes first spread and established around the Mediterranean region and later extended to all parts of the world. Today it is widely distributed in the tropics, subtropics and some forms even extending into the temperate region.

The family Vitaceae comprises of 11 genera and about 600 species. From this vast taxonomic material, the genus of greatest economic importance and the only one having food plant is *Vitis*. This important fruit genus has two subgenera Muscadine (2n=40) and Euvitis (2n=38), further the species of Euvitis are interfertile and are separated by geographic, ecologic and phenologic barriers. The wild specie *vinifera* is further divided into two subspecies, spp. caucasia (S. Russia, Armenia, Iran, Turkistan and Kashmir) and spp. sylvestris (south and central Europe, west Turkey, Palestine, N.W. Africa). These are the store house of wide range of characters i.e. dessert, wine and raisin the characters which are also the part of pure *vinifera. V. vinifera* has resulted in large number of cultivars that have been the back bone of viticulture in most grape growing areas of the world. Grape in general exhibits vide variation as species have originated in different parts of the world under varied climatic conditions.

Origin and Distribution

The geographical area extending from north eastern Afghanistan to the southern borders of the Black and Caspian sea is regarded to be the place of origin of grape (*V. vinifera*).The species in turn has resulted in large number of outstanding world famous cultivars. From here the crop migrated to Mediterranean region, where it further developed . Later grape was introduced in western Europe, China ,India and Japan. With the organised voyages of Spanish and Portuguese the crop was introduced to the new World, where probably it hybridized with the native species, resulted in the evolution of new types which were highly adapted to the local conditions, and the crop got established in new areas of China, Japan and America. On account of its wide adaptability it is grown all over the world.

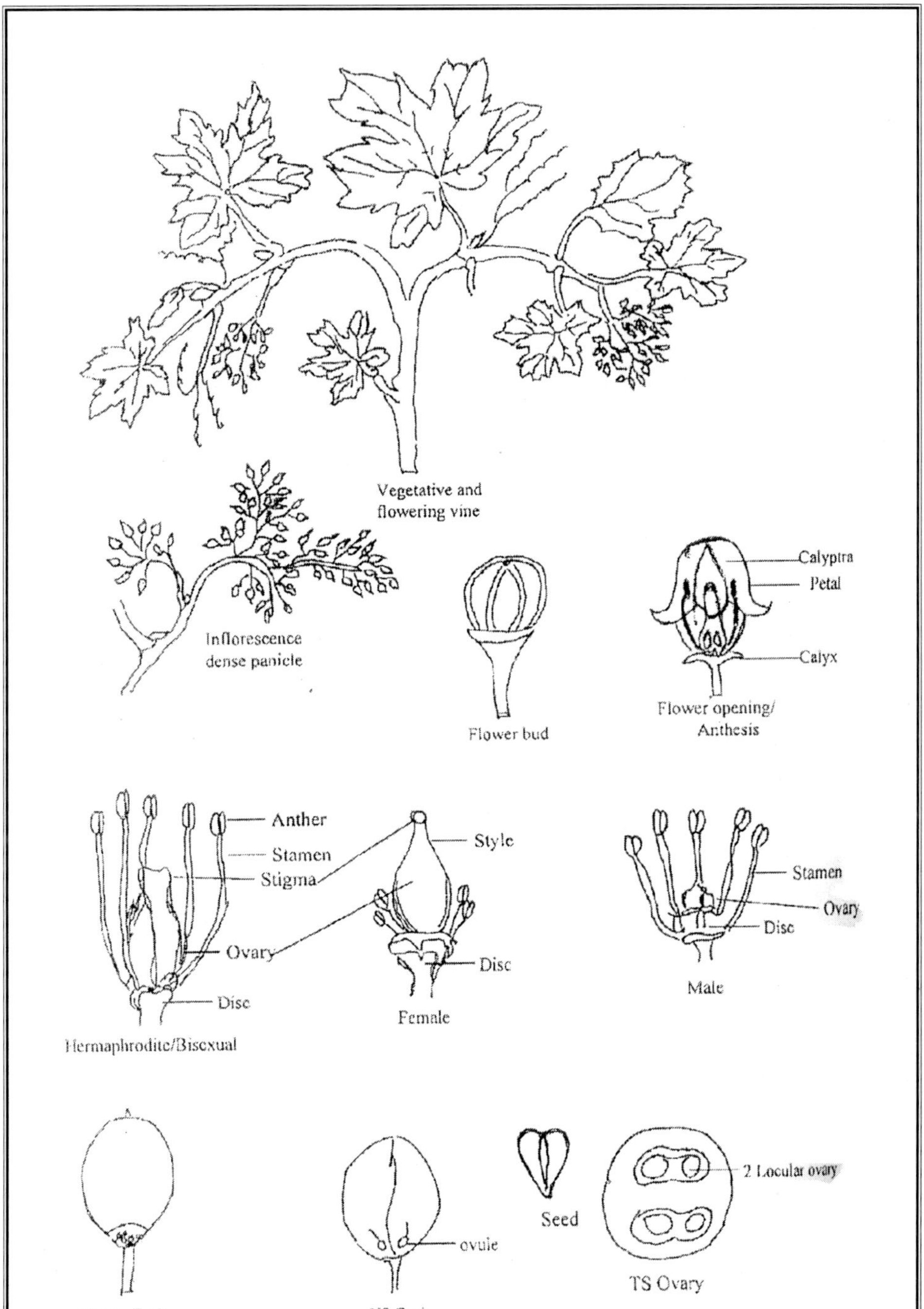
Vegetative and flowering vine
Inflorescence dense panicle
Flower bud
Calyptra
Petal
Calyx
Flower opening/ Anthesis
Anther
Stamen
Stigma
Ovary
Disc
Hermaphrodite/Bisexual
Style
Disc
Female
Stamen
Ovary
Disc
Male
Mature Fruit
ovule
VS Fruit
Seed
2 Locular ovary
TS Ovary

Uses

The fruit can be used in a number of ways , is most widely used to make wine (well developed wine industries), fresh and as dried fruit (raisins) are ranked second and third as use product. In the south east Asia, majority of the fruit is consumed fresh and wine industry is likely to flourish.

Description

Plants are woody, perennial, climbing, vigorous vine, may reach a height of 15-20m or more if unpruned, young growth is pithy, crooked/zigzag growth with swollen node. Vine has number of distinct parts i.e. growing tip, nodes, internodes, buds, tendrils and laterals. Leaves simple, light green, stalk thin, long, round or round ovate, palmately lobed (10-25cm wide),coarsely toothed or dentate, light to dense pubescence, some times in the form of long hair. Leaf arrangement is distichous and alternate. Petiole usually cylindrical, flattened or grooved on the upper side. The petiole divides into 5 large veins each extending into the 5 leaf lobes. Veins divide to form network of veins. Opposite each leaf, tendrils are present with apex bifid, trifid, intermittent (not at each node) to coil around support. Inflorescence is a panicle that appear in leaf axils of current growth that take the place of tendrils. Flower, pentamerous borne in clusters, small, mildly fragrant, usually unisexual functionally, (perfect in varieties of *V. vinifera*) light yellow to green in colour, calyx 5 lobed (light green), corolla 5 small (5mm long) united at the top to form a cap like structure (calyptra).At anthesis calyptra falls off from the disc, disc small prominent and glandular. Stamens 5 opposite to each petal, anthers are bilobed and each has 2 pollen sacs. Pistil single, ovary enlarged and contain four ovules. Pollen fine mass yellow in colour, very sticky when fresh. Ovary 2 celled, 2 ovuled. Fruit is a berry globose, round or ellipsoid of various colours, green yellow, red, dark blue-purple, skin smooth thin to thick, bloom thin to tomentose, flesh white/creamish, red or purple, juicy soft, sweet, sub acid or acidic in taste, seed 3-4,round oblong or pyriform.

Species which are of commercial significance and of horticulture importance are pomologically described below.

V. cardifolia. Michx (True Frost Grape, Raccoon, Chichen or Winter Grape). Amongst the American species it is the most fast growing type and the trunk may be 1-2 feet across, diaphragms strong and thick. Leaves long or triangular cordate, base rounded, may be lobed or unlobed, if lobed then 3 distinct angled lobes are seen, sinus narrow and deep, teeth on margin sharp and vary in size. Ventral surface shiny, lower bright green with slight greyish pubescence along the veins. Petiole quite long. Fertile stamens reflexed and sterile are erect. Flower cluster with numerous flower. Bunch loose, berries black and many, small, skin thick slightly shiny pulp scanty, seed broad.

V. aestivalis Michx. (*V. labrusca*, var. aestivalis, Regel. *V. bracteate* Le Conte). Summer, Bunch or Pigeon Grape. Vines are vigorous growing and stout, diaphragms thick, petioles sometimes pubescent. Leaves thin later thick and rough, ovate-cordate to round-cordate, lobes/sinus deep broad and open, lobes 3-5, apex broadly triangular, dull on upper surface under surface distinct with red brown pubescence. In fertile flower stamens are laterally reflexed. Flower cluster with long pedicels and long in size. Berries are glaucous (shiny), black, small with thick skin, pulp may be dry or juicy, seed 2-4 medium in size.

V. labrusca Linn. (*V. blandii.* Prince). Fox Grape, Sunk grape. Vines are strong climbing, young shoots tawny and pubescent, leaves thick and large, veins very conspicuous beneath, cordate-ovate in form, 3 lobed towards tip, petiolar sinus usually open and shallow, margins with teeth and end with short pointed teeth. Upper surface dull green and shiny, under surface covered with tawny pubescence(red-brown). In sterile flowers stamens are erect and long, short and recurved in fertile flowers, raceme short. Berries are usually large spherical, of varying colour red-brown, purple, purple-black to amber green, drop when ripe. Skin thick seed thick and large. Mostly sweet with musky flavour, may be astringent too.

V. reticulata Pampinini. (*V. Wilsonae*, Veitch). Native to China. Climbing shrub, branches shiny, leaves ovate-cordate, base cordate-truncate, apex short acuminate, margin 3 lobed or entire shiny above, pubescent below. Inflorescence racemose, flowers are small, fruit black shiny above ½ inch across.

V. cinerea Engelm. Sweet Winter Grape. Vines high climbing, internodes long, diaphragms strong and thick, leaves broad cordate ovate, sinus round and wide, margin entire or with small teeth or 3 lobed at the apex, upper surface pubescent later turns glossy and dull green in colour, under surface ashy grey. In sterile flower stamens are thin and long, in fertile they are short and recurved. Cluster or bunch loose, has many small berries, slightly shiny. Ripens very late, later develops pleasant sweet taste.

V. munsoniana Simpson. Mustang Grape of Florida, Bird, Ever bearing or Everlasting Grape. Vine very thin loves to run on ground, low bush, mostly evergreen. Leaves thin small, shiny almost rounded, apex less pointed teeth broad and more spreading. Cluster large and more thyrse like, berries usually small, skin thin black shiny, pulp soft, juice more acidic, free of musky odour.

V. rupestris Scheele. Sugar, Bush, Sand, Currant or Mountain Grape. Grown widely in N. America. Small shrub, about 6 feet tall, slight to high climbing with none to few tendrils. Leaves reniform or ovate reniform, smooth, shiny on both the surfaces, thick base rarely with distinct sinus, margins coarsely toothed apex short pointed often with 2 lateral teeth indicating lobes. In fertile flowers stamens are recurved laterally. Cluster generally small, thin open branched. Berries very shiny, purple black in colour with pleasant flavour and taste, seed usually small and broad.

V. vulpina Linn. (Syn. *V. riparia* Michx, *V. odorate*, Hort., V. *cordifolia* var. riparia Gray). River bank or frost grape.

The vines are high climbing vigorous in growth, foliage bright green young shoots shiny, diaphragm very thin. Leaves cordate-ovate, thin broad and with distinct sinus, inclination towards 3 lobes, pubescent along mid and lateral veins, margin irregular and deep. Stamens in fertile flower curved while in sterile they are usually erect, flowers with sweet flavour, cluster somewhat large on short stalks prominently branched. Fruit berries are small with dense blue bloom, purple black in colour, sour in taste, ripens late, seed small somewhat pyriform in shape.

V. rubra Michx. (*V. monosperma*, Michx) Red or Cat Grape. The specie is distinct for its red colouration on various plant parts. Vine thin and vigorous, shoots and petioles red. Leaves somewhat small to medium ovate, apex pointed shiny, dark green above, pubescent along the veins below, leaves either entire or may be lobed, if lobed the sinus is broad (obtuse) may be even coarsely notched. In sterile flowers, stamens are long and erect. Flower cluster/panicle is with long peducles, branched and loose. Berries are late in ripening small without bloom, black, juice scant, seed usually solitary broad and large.

Some Asian grape species and their detail is given below.

V. Davidii Foex. (*V. Davidiana*, Dipp. *V. Vinfera* var. *Davidii*, Hort.) Native of China. On the stems and petioles straight or somewhat recurved spiny prickles are found, vines shiny, tendrils intermittent or interrupted. Leaves large thin broad ovate, base cordate with long petioles, shiny, later under surface becomes glabrous, margin serrate apex acuminate. Peduncles surpass the leaves. Flowering occurs very early, fruit/berry black globose and large.

V. amurensis Ruper. (*V. vulpina* var. amurensis. Regel). Indigenous to Manchuria Amoor region of China. Is considered to be a very hardy specie planted on limited scale. Vines angled and striate, young shoots loosely floccose or webby, but becomes shiny with age. Leaves round cordate may have 3-5 lobes, both surfaces green and somewhat shiny, flower cluster with thin stalks or peduncle. Fruit small with 2-3 seeds.

V. flexuosa Thumb. Native of Japan, China and Korea. Stems or vines slender flexuose at maturity, become glabrous, new growth densely pubescent-tomentose. Leaves cordate-ovate with petioles, margin entire or 3 lobed unequally dentate, glabrous above and hairy beneath at maturity. Flowers in cluster which are peduncled and branched. Fruit small with 2-3 seeds. Has one botanical form *V. flexuose*, var. parviflora, Gagnep, cultivated mainly for its coloured leaves which is bronze coloured above (metallic shine) and purple below.

V. lanata Roxb. Indigenous to Himalayan region and China. A tropical or warm temperate species. Stems/vines and inflorescence are densely pubescent/tomentose. Leaves cordate-ovate, long almost glabrous but some soft pubescence is present which is more below. Flowers green and small in a paniculate cyme, petal do not separate at top. Fruit small purple in colour with 4 seeds.

V. coignetiae Pulliat. Native to N. Japan. Vines are stout and strong growing, foliage attractive and showy, tendrils intermittent, branches with floccose- tomentose when young. Leaves cordate ovate, 3-5 lobed, margin shallow toothed, dull green above, under surface with thick grey pubescence, stalk small. Fruit small, globular and inedible, but after frozen it in eaten.

Cultivars

As grapes are used for various purposes like table grapes, raisin juice, wine, canning, hence specific cultivars have been selected for different uses. More than 10,000 grape cultivars are known all over the world which are well documented. Cultivars are broadly classed as seedless and seeded types. Brief pomological description of some very important cultivars of both types are described below.

Seedless cultivars

Thompson Seedless: originated in Asia Minor, cultivated first by William Thompson. Also called as Sultaniana (Australia, S. Africa) and Oval Kishmish (Mediterranean region). Table and raisin cultivar, also used to make white wine. Vines medium in vigour, bunch medium-large, compact markedly shouldered, conical-cylindrical in shape, berries well filled, small ellipsoidal elongate, yellowish green to golden yellow. Skin smooth thick adhere to pulp, eating and keeping quality excellent, mid season cultivar and ripens uniformly.

Perlette: Is of hybrid origin (Scolokerlekhiralynoje 26 x Sultaniana Marble), name signifies small pearl because of translucence of mature berry. Vines are vigorous, tendrils 2-3 fid thin and short. Bunch medium large, conical compact to very compact, shouldered. Berries medium sized yellowish green spherical glossy, sometimes small berries are also seen on the bunches (defect) ellipsoidal in shape, flesh soft with mild muscat flavour.

Beauty Seedless : Vines are medium in vigour, tendrils 2-3 fid and sometimes upto 8 fids, early ripening but not uniformly, prolific bearer. Bunch medium-large conical-cylindrical well filled to compact and shouldered. Berries bluish black variable in size, spherical, bloom heavy on berries lessen with ripening. Skin thin adhering, flesh fairly soft, sweet in taste, flavour mild.

Pusa Seedless : Is a selection from unknown origin made at IARI, New Delhi India and resembles closely to Thompson Seedless in number of characters. Vines are vigorous, fair in production, bunch medium sized, conical to cylindrical shouldered, berries are more elongated than Thompson Seedless, yellow green to golden yellow. Eating and raisin making qualities are good, keeps well, overall a good cultivar.

Delight : Is a seedling of Perlette, which further is of hybrid origin. Selection was mainly done for its very early ripening in the season and for its strong muscat flavour. Vines are medium in vigour, bunch also medium sized, conical shouldered compact attractive in appearance, berries are small, round green to light yellow, skin adhering thin, table quality good and transports well.

Kishmish Charni : Cultivar is known for its sweet berries and very good keeping qualities, though only fair in production. Vines medium in growth, bunch medium to large, well filled to compact, conical and well shouldered. Berries medium large, round to slightly elongated, bright red in colour.

Himrod : Is of hybrid origin, mid season cultivar, vines are vigorous, bears heavily. Bunch medium large with well filled berries, shouldered and compact. Berries yellow-green, skin thick resistant to major grape diseases and pests.

Seeded Cultivars

Bangalore Blue : Evolved as a result of cross between *Vinifera x labrusca,* hardy and resistant to disease and is frequently used in crop improvement. Vines are vigorous tendrils continuous, short, leaves with thick pubescence below, bunches are small and compact with few berries. Berries small-medium, bluish black, skin thick tough, free from pulp, seeded, used as dessert and extensively for juice and wine making.

Black Champa : Selected at Bangalore from introduced material. Vines are vigorous with medium yields, tendrils short thick 2-3 fid, bunches are small medium, short conical branched (2 bunches from same rachis), well filled. Berries are bluish black or purple, seeded spherical to oblate, small, bloom thick skin adhering, pulp soft tender melting with musky flavour, suitable for table, juice and for red wine preparation, susceptible to cracking and rot during rains.

Anab-e-Shahi : Originated as a bud sport. Vines are very vigorous, tendrils 2-3 fid, bunches are medium-large, cylindrical to long conical well filled. Berries amber/straw yellow, ovoid-elongate, large with thick bloom, sticky. Skin adhering to pulp, medium thick pulp, firm crisp, flavour not very strong slightly insipid.

Pearl of Casaba : This cultivar has been used extensively in breeding programme for the fact that it ripens very early under north Indian condition-mid May onwards. Highly seeded. The vines are poor in growth and low in yield, bunches are small and compact, light green to greenish yellow berries, well filled soft tender very sweet with watery juice and Muscat flavour.

Bhokri : Vines are vigorous, yield capacity is very high, prolific bearer, bunches are large, shouldered, compact. Berries are green oval, seeded, pulp soft, ripen uniformly and late, poor in keeping quality has tendency to crack with onset of rains.

Cardinal : Evolved from a cross between Tokay x Ribier, American grape. Vines medium in vigor, ripens very early red coloured, seeded table grape. The bunches are loose, medium-large in size, berries conical spherical bright red in colour, attractive on ripening colour changes to reddish black.

❑❑❑

40
Kiwifruit

Genus : *Actinidia*
Species : *deliciosa* (Plonch)
Family : *Actinidiaceae*
Chromosome No. : 2n = 58, 170, 174

The remarkable fruit of the previous century Kiwifruit is known by various names like China's miracle fruit, the wonder fruit of New Zealand, Kiwi berry, Chinese gooseberry, Yangtao, Grosella and Mihoutao. In a short period of about two decades the crop has attracted huge inclination towards its cultivation and has become an established fruit crop of different world markets, which is unparallel in history of commercial fruit culture. In China the fruit has been growing over centuries in its wild form. The name Mihoutao was used for this crop by the Tang dynasty which indicated peach of the Macaque monkey indicating both the presence of fruit in wild stands and the love of the local monkey for this fruit.

By the start of the twentieth century, small seed consignment reached New Zealand from its native place China. Through these seeds vines were raised by the New Zealand nurseryman, who in fact were first to realize its potential as a new crop in a short span of 10-15 years time. In the next 25-30 years through systematic evaluation of the seedling population several cultivars were developed, which was followed by development of standard cultural practices that helped in the expression of its genetic variability. The improved types were selected and planted in number of vineyards. When these vineyards came into fruiting, serious thoughts were implied towards its usefulness and importance as a commercial fruit. As of today both kiwifruit culture and industry is most developed in New Zealand. The fruit has been aptly named Kiwifruit or Kiwi berry because of hairy appearance and brownish colour that resembles flightless bird 'Kiwi' which is the national symbol of New Zealand. The fruit defies the conventional attractive bright appearance, as it is brown in colour with stout hairy growth. However when the fruit is mature and flesh observed in a cross section provides very attractive sight of creamish rays radiating from the centre against the light green background of flesh, with numerous soft small black seeds embedded in the flesh and along the rays that finally disappear towards the fruit periphery. The distinct and delicate flavour appear to be a mixture of melon, gooseberry and strawberry. In India, Kiwi culture was first started in Bangalore, where it failed to fruit. Thereafter it was introduced at Shimla where it was an instant success. The crop has established itself in number of places in Himachal Pradesh.

Origin and Distribution

Kiwifruit is native of south-western China where it is present abundantly in wild forms in temperate forest, hills and mountains at its native place. Besides other species are also found growing over wide geographical and climate range spreading from Siberia to down in south east Asia (Indonesia). From its native place, *Actinidia* species (seeds & plants) were introduced to Europe, USA and New Zealand in last quarter of nineteenth and beginning of twentieth century. In New Zealand the crop gained popularity and became domesticated and later spread to different parts of the world.

Uses

Fruits of Kiwi are ready and rich source of vitamin-C, is valued highly for its distinctive fresh flavour which becomes aromatic as fruit ripens. Used as fresh/dessert, rich source of minerals like calcium, magnesium, potassium, crude and dietary fibre, juice, canning etc.

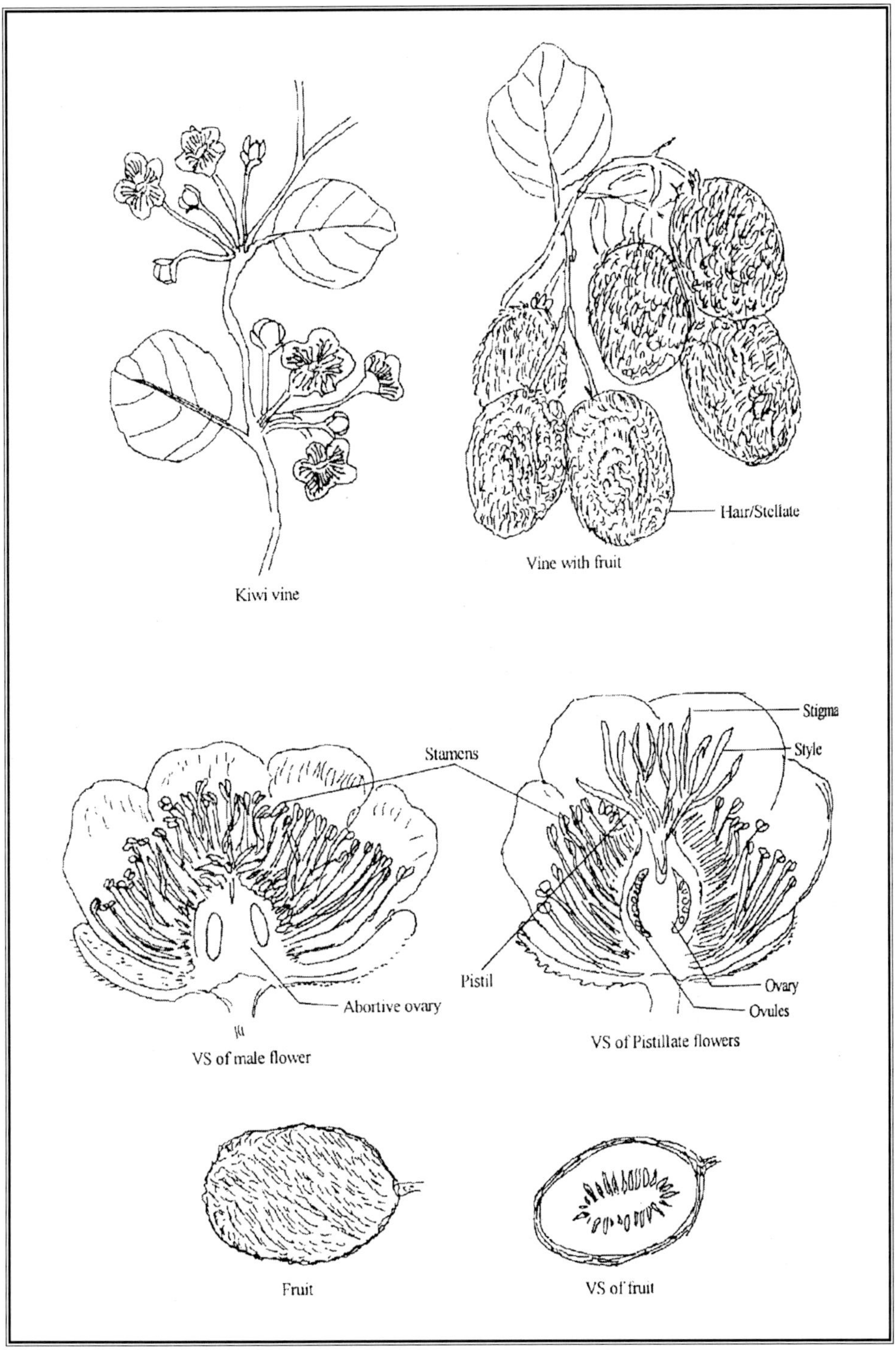

Kiwi vine

Vine with fruit

VS of male flower

VS of Pistillate flowers

Fruit

VS of fruit

Description

Kiwi plants are straggling/dangling plants or perennial climbers but lack organs for climbing. The long and weak shoots help to form tangles which helps in climbing. New growth appears from axillary buds formed on last years growth. The young growth may be glabrous or pubescent, pubescence of varying extent. Old shoots usually greyish brown to chestnut brown, lenticels white, prominent. Leaves usually simple, alternate deciduous petiole long with stiff hairs, usually entire, sometimes also toothed, green above hairy and lighter beneath. Leaves exhibit variation in shape, size and pubescence (hairs). Flowers are dioecious, male and female flowers appear on different vines. Dioeciousness is not complete as self pollinating and self setting are available. Flowers appear axillary (previous seasons growth), singly or in few clusters. Calyx usually 5, brown in colour, hairy (pubescent) forms a cup-like structure. Corolla, usually 5 sometimes more, thin usually white, may be yellow or pink, stamens numerous, with long slender filaments, which arc attached to the centre of anthers. Anthers variable in colour usually yellow may be brown, purple or black. Carpels-many (15-30) united or fused below, ovary superior styles as many as carpels, each carpel with two rows of ovules attached along the central axis. Fruit is a many seeded berry, flesh soft and juicy. Skin brown covered with hairy growth, flesh at skin is green due to presence of chlorophyll, but may be complete yellow, or yellow green to complete green. Various members of kiwi exhibit variation in number of attributes like shape, size, hairy skin, taste etc.

On the basis of several radiating divisions of the style, the genus has been assigned the name *Actinidia* which further is derived from Greek 'aktis' meaning rays. The genus belongs to the family Dilleniacea (Actinidiaceae) has over 50 species and more than 100 different taxas are known.

The kiwifruit of commercial importance are the large fruited selection of *A. deliciosa*, and all *Actinidia* species are usually straggling or climbing perennials. Fruit and vegetative characters show variation both within and between different taxas. However the flowers are alike although there is inherent variation in colour, size, morphology and inflorescence size etc. The genus has been divided into four distinct sections primarily on the amount and type of pubescence, fruits with or without lenticels. Section Leiocarpae is further parted into two – the fruit pith being solid or Lamellate and Stellatae on the basis of persistent and type of hairs on the dorsal surface of leaf. Brief description of various section and species in each and their natural distribution is given below.

Section 1.

Leiocarpae (Dunn) Li. Leaves generally glabrous or sparingly pubescent, fruit not spotted. Ser. Lamellatae (Pith lamellate) C.F. Liang. This section has nine important species and some with botanical forms are of significance pomologically .All are native of China and few are native to Japan, Korea and eastern Siberia. Various species are described briefly.

A. arguta (Sieb. et Zucc) Planch. ex Miq. var. arguta.

------------------- var. nervosa C.F. Liang.

------------------- var. purpurea (Rehd.).

------------------- var. cordifolia (Miq.) Bean, Japan, Korea

A. rufa (Sieb et. Zucc.) Planch ex Miq. Japan, Korea

A. hypoleuca Nakai, Japan.

A. Kolomika (Maxim et. Rupr.) Maxim.easten Siberia, Japan,Korea.

A. tetramera Maxim.

Ser. Solidae C.F. Liang. (Pith not lamellate)

This sub group has three species along with some varietal forms. All are indigenous to China.

A. polygama (Sieb et. Zucc.) Maxim.

A. valvata Dunn var. valvata

var. longipedicellata

var. boehmerifolio.

A. macrosperma C.F. Liang. var. mumoides and var. macrosperma.

Section 2

Maculatae Dunn. Leaves usually shiny or slightly pubescent fruit with spots.

This section is very wide and diverse and consists of 16 species and some even have botanical forms or varieties which are native to India (North India), Tibet, Indo China and Malaysia. Most of the species are however indigenous to China. Some important species are listed below.

A. fasciculoides C.F. Liang.

– – – – – – –var. cuneata.

– – – – – – –var. orbiculata.

A. ulmifolia C.F. Liang.

A. callosa Lindl. var. callosa. North India, Indo China, Indonesia and Malaysia

– – – – – – – –var. acuminate.

– – – – – – – –var. discolor.

– – – – – – – –var.strigillosa.

– – – – – – – –var. formosana Fin. et Gagn. Taiwan

A .venosa Rehd. f. venosa. Tibet, India.

A. gracilis, C.F.Liang.

A. indo chinensis, Merr.

A. sabifolia Dunn., *A chrysantha* C.F. Liang., *A. gracilis, A rubricaulis* Dunn. *A. leptophyllum* C.Y. Wu.

Section 3

Strigosae. Li. Leaves and branches woolly or with dense hairs, hard coarse, simple hairs.

This section has some thirteen distinct species along with some botanical forms. Important ones are listed below.

A. petelotii Diels (Indo-China), *A. strigosa* Hook f.ex. Thoms (Sikkim), *A. arisanensis* Hayata (Taiwan). Other species are distributed in various provinances of China, namely *A.rubus* Levl , *A. rudis* Dunn., *A. henryi* Dunn., *A. hemsleyana* Dunn., *A. melliana* Hand- Mazz.

Section 4

Stellatae Li. Leaves and branches woolly or hairy, fine soft hairs, stellate on under surface of leaves.

Ser. Perfectae C.F. Liang. Under surface of leaf covered with thickly persistent stellate hairs.

This section comprises of thirteen distinct and well described species, some have botanical forms also.

Species *A. latifolia* is found in Taiwan, Indonesia, Indo-China and Malaysia. *A.setosa* (Li) Taiwan. All other species are however indigenous to China, *A. lanceolata* Dunn, *A. suberifolia* C.Y.Wu *A. fulvicoma* Hance. *A. rufotricha* C.Y.Wu. *A. cinerascens* var. tenuifolia, C.F. Liang.

– – – – – – – – – – – –var. longipetiolata C.F. Liang.

A. Chinensis var. Planch chinensis.

--------- var. rufopulpa (C.F. Liang et. R.H. Huang.)

A. liangguangensis CF.Liang.

A. deliciosa (*A.chev.*) C.F. Liang et A.R. Ferguson var. deliciosa.

---------var.longipila (C.F. Liang et R.Z. Wang) C.F.Liang et . A.R.Ferguson.

---------var. chlorocarpa (C.F.Liang) C.F.Liang et A.R. Ferguson.

Ser. Imperfect C.F. Liang (under surface of leaf has imperfectly stellate hairs or scattered deciduous hairs).

Imperfectae consists of about eight species all being native to China and one *A.pilosula* Stapf. Ex Hand . Mazz is native to Burma .Important species part of this group are *A. guilinensis* C.F. Liang *A. sorbifolia* C.F.Liang and *A.zheziangensis* C. F.Liang., *A. obovata* Chun ex C.F. Liang., *A. grandiflora* C.F. Liang.

At present three intraspecific Kiwi varieties are well established / recognized and widely cultivated are described as.

1. *A. chinensis* Soft haired kiwifruit (name derived from Chinese name Zhonghua,mihoutao ruanmao). This is the original kiwi variety which has three distinct forms.

   A.chinensis Planch .var. chinensis. Fruits nearly spherical covered with soft hairs which are lost when fruit is ripe.

   A. chinensis var. jinggangshanensis. Branches and petioles are covered with rough dense pubescence and is persistent.

   ------------------------ var. rufopulpa. Fruits are small, stalk long reddish from inside.

2. *A. deliciosa* is called Zhonghua mihoutao yingmao (stiff-haired chinesemihoutao) and different varieties of *A.deliciosa* are briefly described below

   ------------------ var. deliciosa. Kiwifruit of cultivation.

   ------------------ var. chlorocarpa. Shoots coarse with thick pubescence, stiff hairs from fruit fall off on ripening and pericarp is dark green.

   ------------------ var. longipila. Bristles/stiff hairs on branches are longer and numerous stellate hairs on under surface of leaves are separate and not interlocked.

------------------ var. hispida. Fruits oblong covered with stiff hairs even when fruit was ripe. Number of cultivated varieties belong to this group.

3. *A. chinensis* var. setosa. Spiny kiwifruit (Chinese name mihoutao cimao). Botanical form setosa has now been raised to specie status ie. *A. setosa*.

Detail of some distinguishing characters of Kiwi species, *Actinidia chinensis, A. deliciosa* and A. *setosa* are summarized below:

Character	A. chinensis	A. deliciosa	A. setosa
Bud	Base small, spherical covered with scales, hairs are very short yellowish brown.	Base large, projected out bud embedded in the bark, hairs are long greyish white.	-----
Stem	Young growing points yellowish green, ashy white short hairs shed and makes bark smooth. Old growth more straight and reddish brown in color.	Young growing points with attractive red hairs. On mature growth hairs are stiff long yellow - brown, on shedding a scar or stub remains. The thick dark brown branches tend to twist.	Covered with rust coloured stout spine like hairs.
Leaves	Obovate cordate, small thick and leathery, margin short serrate and brittle under surface greyish white with stellate hairs, petiole short yellow green on exposure reddish with white short hairs, shed rapidly.	Obovate to cordate apex pointed to cuspidate thin and larger leaves. Young growth with bright red hairs, margin uneven serration long under surface of leaf has greyish brown stellate hairs, petiole long purple red tomentose with yellow-brown hairs that do not fall readily.	Long, more thin and narrower than *A.chinensis,* apex long pointed to short acuminate. Petioles rust coloured has stiff hairs which are spine like.
Flower	Flowering shoot 4 - 5 cm long with soft white hairs	Flowering shoots long 15-20 cm with persistent	Flowers are smaller in size 1.8cm across, stamens

Contd...

Table Contd...

	shed easily. Flower smaller than *A. deliciosa* (2-5 cm). Female flower larger than male, stamens 50, less than *A. deliciosa*, pollen round and smooth.	stiff long yellow brown hairs. Flower bigger (3.5 cm), female (5.5 cm) larger than male, stamens more 130, pollen round flat finely reticulate.	more sharp pointed like arrow, sagittate.
Fruits	4-5 cm across spherical covered with greyish-white hairs that shed early, surface than becomes smooth, scar yellow-brown, pulp variable green, yellow or greenish yellow.	Ovoid elongated or cylindrical 5-6 cm long, hairs are persistent, hard yellow brown, if shed number of brown spots are visible, pulp light or dark green.	Elliptical or spherical in form 3.5 cm long, hairs long and stiff.

Cultivars

Hayward : Most popular cultivar grown all over kiwi-growing areas of the world. Fruits are large in size, broad and flat, fruit weight may range from 80-120 g/fruit, keeping quality very good, and has good storage (4-6 months), superior flavour, high sugars and fairly high ascorbic acid contents, core is very large compared to any other cultivar. Flowers late, bears solitary flower on fruiting canes, shy bearer, has tendency of alternate bearing, requires pollinizer for good yields. Has high chilling requirements.

Allison : Resembles very closely to cv. Abbott, except that both the apices are more flat, fruit weight varies from 50-75 g. The corolla of this cultivar are overlapping and crimped over the margins. Bears very heavily and ripens early, has performed well under mid hills of Himachal Pradesh, sweet in taste, acidity and ascorbic acid contents are somewhat less.

Bruno : The cultivar was selected at much early time. In contrast to Allison, it has very high titratable acidity and ascorbic acid contents. Requires less chilling hours, and is a good cropper. Fruits are smaller to Hayward, but larger than Abbott and Allison. Fruits taper towards the stem end and in length are the longest compared to any kiwi cultivar, usually dark brown in colour with short, bright very dense hairs, suitable for low hill areas.

Monty : The fruit apices are flat as in Allison and Abbott. Is a late flowering type but fruit ripens in mid season, individual fruit weight varies from 60-70 g., very prolific and productive, fruit small to medium, to get good sized fruits thinning is required, has high acidity, whereas it is medium in sugar and ascorbic acid contents. The fruit is distinct as it is broader at the blossom end.

Staminate Cultivars

Tomuri : Tomuri means late in Maori language. In this cultivar the flowers appear in clusters of five (1-7), flowers earlier and is considered to be a very effective pollinizar for Hayward, however in New Zealand it flowers late. The hairs on the peduncle are short. Over all is a good pollinizer for early and mid-season pistillate cultivars.

Matua : The cultivar is a good pollinizer for early and mid-season pistillate cultivars as it flowers over a long period and also flowers profusely. The pubescence or hairs on the peduncles are usually short and are borne in clusters of usually three (1-5).

Qinmei : New cultivar selected from wild and released in China, has high adaptability to both low and high temperatures, very productive and starts bearing early. Fruits are large (110g.), oblong in shape. For most of the characters, it is regarded to be superior to Hayward.

□□□

41
Strawberries

Genus : *Fragaria*
Species : *ananassa* Duch.
Family : *Rosaceae*
Chromosome No. : 2n=14, 25, 42, 56

Strawberry is relatively small genus comprising of usually low perennial creeping herbs, wild forms found wide spread in temperate and subtropical parts of the world. Cultivated strawberry is highly polymorphic and wide range of diversity present in the wild germplasm has resulted in the development of present day strawberry. Today it is grown throughout the world and is rated high in the list of liked foods. The breeders have contributed enormously towards the development of cultivars with wide adaptability and through the incorporation of desired traits helped to improve its quality dramatically over its wild forms. Today strawberry is known for its distinct delicate flavour, attractive fruit shape and colour and has made a special niche in the soft fruits culture. The cultivated strawberry *Fragaria x ananassa* Duch has resulted from the crossing of two local America species *F. chiloensis* (L) Duch. and *F.virgianana*, Duch. The cultivated strawberry originated in the area around Brest a place in France around 1750. Later in the early nineteenth century Thomas A. Knight from controlled hybridization probably between *F. virgianana* and *F. chiloensis* obtained two outstanding cultivars namely 'Downton' and 'Elton'. This provided a suitable platform for the other breeders and extensive strawberry improvement programmes were initiated throughout the world. Consequently today there are types that grow successfully in wide range of conditions varying from very cold to very hot, types which are not influenced by photoperiod, types that flower and fruit throughout the year, and for other plant characters including resistance to insect pest and diseases, adaptation to different soil conditions, yield and fruit quality.

The back bone of strawberry improvement, the wild species are said to have 17 chromosome number and are classed in four chromosome groups with basic chromosome number 'x' as seven. The ploidy break up of these wild species is nine are diploid, three tetraploids one hexaploid and four are octaploids.

Origin and Distribution

The cultivated strawberry is widely distributed throughout north temperate Europe, Asia and America and also occurs in north Africa and south America. The species present polyploid series from diploid to octaploid. In Asia diploid types have limited distribution and some occur wild in Europe and Asia. *F. ananassa* (octaploid) occurs continuously along the Pacific coast from Aleutian Islands to California and in Chile, where it is cultivated upto an altitude of 1600m. Romans cultivated it (*F.vesca*) in gardens as early as 200 B.C. and was later planted in England and France during (*F.vesca*) fourteenth, fifteenth and sixteenth centuries. European species *F. viridis* and *F. moschata* were also used. In America, American octaploids *F. virginiana* was used before Europeans settled in America. In Chile the same spp. was cultivated before the Spanish arrival.

Uses

Strawberry fruit is a rich source of vitamins and minerals. Fruits are highly flavoured and is due to volatile esters. Fruit is eaten fresh or processed includes jams, conserves, canning, drinks and freezing.

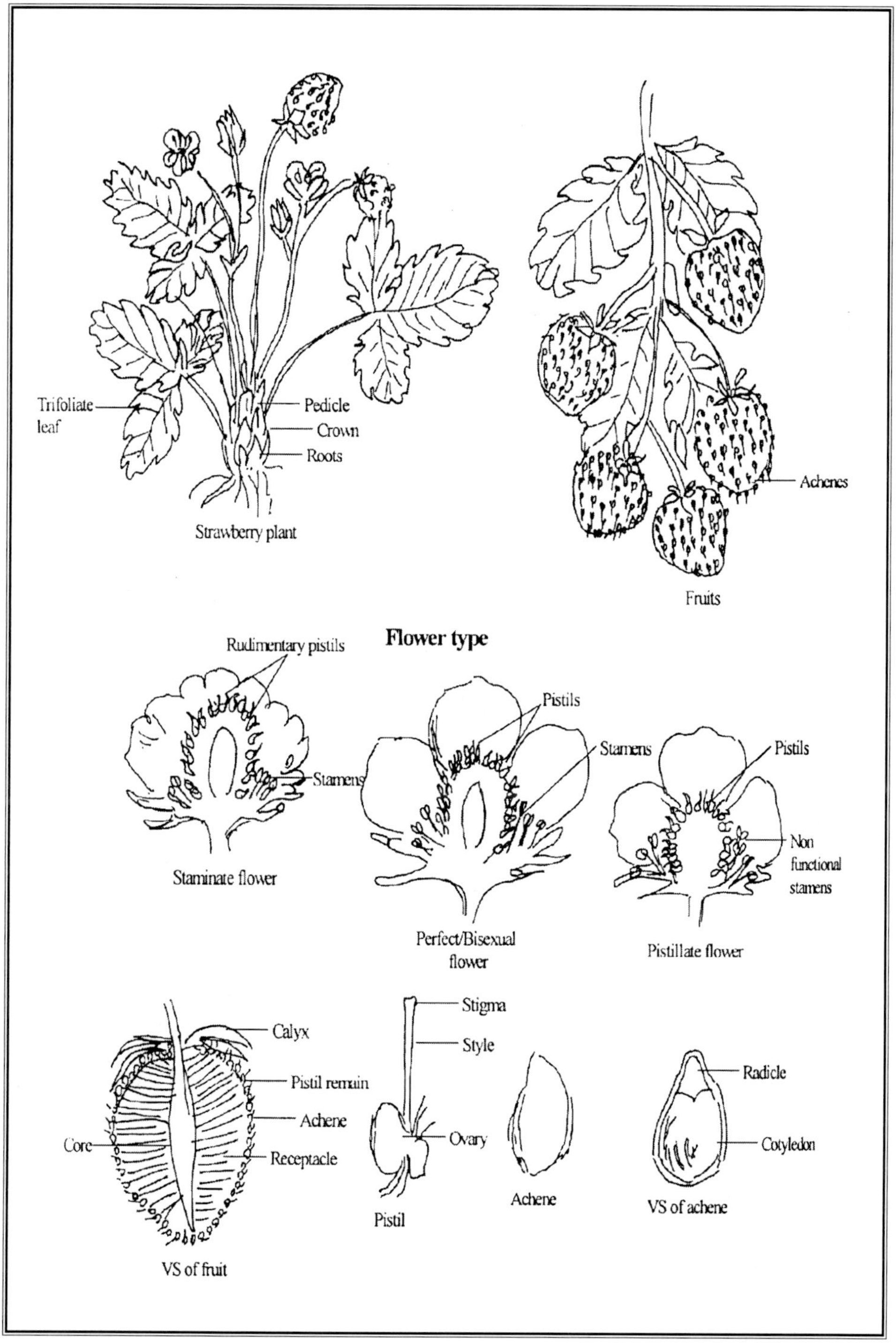
Trifoliate leaf
Pedicle
Crown
Roots
Strawberry plant
Achenes
Fruits
Flower type
Rudimentary pistils
Stamens
Staminate flower
Pistils
Stamens
Perfect/Bisexual flower
Pistils
Non functional stamens
Pistillate flower
Calyx
Pistil remain
Achene
Core
Receptacle
VS of fruit
Stigma
Style
Ovary
Pistil
Achene
Radicle
Cotyledon
VS of achene

Description

The strawberry (*Fragaria*) is a low growing almost evergreen perennial herb distributed in the wild state in the temperate and sub-tropical regions of the world, producing profuse runners or stolons from a scaly rootstock, which readily roots after coming in contact with the soil. The dark green pubescent leaf has three leaflets (trifoliate) markedly toothed, the pedicels are also pubescent. The flowers appear on crown buds, bisexual usually perfect, flowers are generally white in colour very rarely pink or pinkish tinged, borne usually on cymes on leafless stems, calyx and corolla pentamerous, calyx 5 lobed with five branchlets alternately inserted in between calyx, green pubescent alternate to petals and are closely attached to the fruit, corolla five white, orbicular or elliptic, free with twenty or more stamens (though some species and varieties are pistillate also) apocarpous placed at the foot of the dome or receptacle, pistils many (40-60 sometimes even more) and are borne on the surface of the enlarged fleshy receptacle. Every carpel is depressed in the receptacle, has a style which ends with a capitate stigma. The strawberry fruit is an aggregate fruit resulting from fleshy receptacle with many seed-like achenes present on the fruit surface. These achenes may be prominently projecting from the surface (superficial) or sometimes may be indented into the surface. The individual achenes are with persistent styles and aggregate fruit is with persistent calyx. In literature about 150 names have been applied to the *fragarias* but of these only four species are of prime interest to the horticulturists, namely *F. chiloensis*, progenitor of ordinary cultivated strawberry of America

F. virgianiana (early domesticated)

F. moschata

F. vesca (the alpine strawberry.)

Brief description of species is as follows.

Diploid (2n = 2 x = 14)

1. *Fragaria vesca* L. Wood strawberry. Native of north America, northern Asia, Africa and Europe. Most widely distributed specie. Plants are erect, leaves thin, light green, petioles slender. Flowers are mostly bisexual and self fruitful. Fruit very aromatic and delicious, bright red and long ovate in shape. The specie has four subspecies and are self fertile.
2. *Fragaria viridis* Duch. Is native of Europe and eastern and central Asia. Plants are erect, slender, leaves dark green. Inflorescence has small and bisexual flowers, fruits are small light red to red in colour, flesh firm and aromatic, produces few runners.

3. *F. nilgerrensis* Schlecht. Native of southeast Asia and China. Plants are stout and spreading, leaves dark green, pubescent, veins very prominent. Inflorescence small with bisexual flowers. Fruit small round light pink in colour, with many seeds, inferior in quality. Produces good strong runners.
4. *F. daltoniana* J. Gay. Occurs in Sikkim and foothills of the Himalayas at an elevation of 10,000 feet and more. Plants are vigorous, leaves are shiny and coriaceous, flower single. Fruits somewhat elongated bright red, but inferior in taste.
5. *F. nubicola* Lindl. Plants resemble *F. vesca*, are slender and dioecious and not hermaphrodite. Found in Sikkim, at foot hills of Himalayas, Bokhara in USSR at 1500-4000 m. height.
6. *F. imumae* Makino. Important specie found in Japan, plants are erect, slender, flowers in cluster of 6-8, white petals, leaves shiny margin coarsely dentatoserrate. Fruits very small 1.5 cm. long, ovoid, achenes imbedded in fruit flesh.

F. mandschurica Staudt. Is native to Manchuria. Flowers are hermaphrodite, but self infertile, fruit very acidic, achenes yellow placed in shallow pits.

F. nipponica Mikano. and *F. yesoensis* Hara. Both are native to Japan but of no horticultural significance. In the former, fruit is about 1 cm across, taste unpleasant.

Tetraploids (2n=4x=28)

Three species *F. orientalis, F. moupinensis* and *F. corymbosa* represent tetraploid strawberry species. Indigenous to eastern Tibet, Yunnan and western China. Plants resemble *F. nilgerrensis,* leaves trifoliate has small leaflet on the petiole, flower pedicle longer than leaf pedicle. Flowers 2-4 per inflorescence, fruit small orange red in colour.

F. orientalis Losinsk. Wide spread in Manchuria, Mongolia western Siberia and Korea. Plants of all sex, staminate, pistillate and hermaphrodite are found in wild, plants upright small, leaves almost sessile ovate in form light green, margin deeply serrated, flowers are large but few per inflorescence, fruit small soft and conical.

F. corymbosa. Less known species has only male plants.

Hexaploid, 2n = 6x = 42

F. moschata, Duch. Native of central and northern Europe, Siberia and Russia. Plants are dioecious, tall, vigor in growth, leaves and flowers are large with thick conspicuous veins, dull green in color, flower pedicle longer than leaf petiole, fruits are dark red in color globose or ovoid, flavour aromatic. Runners produced freely.

Octoploids (2n = 8x = 56)

In *F. chiloensis* and *F. virgianana* leave usually are over top the flowers and achenes generally sunken in berry flesh.

F. chiloensis, (L) Duch. Is indigenous to Pacific Coast, central California, Chile, Peru and Ecuador. Plants are vigorous low spreading dioecious and hermaphrodite are also found, leaves are dark green, shiny and thick. Fruits are large, light to bright red, flesh white flavour unpleasant. The species has number of ecotypes. Prolific runner producer. *F. chiloensis* var. ananassa Hort. Pine strawberry or common Garden strawberry (syn. *F. ananassa, F. anaanssa, F. tincta, F. calyculata, F.grandiflora).* Plants are usually vigorous tall growing, leaves larger, usually light green on both the leaf surfaces, thin. Fruit also large but with variable shapes and forms.

F. virginiana, Duch. Scarlet or Virginian strawberry found in abundance in eastern and central north America. Like *chiloensis* this species also produces dioecious as well as hermaphrodite plant. Plants are slender, fruits are round small 1.5 cm across soft light red in colour, flesh white at core, pinkish at periphery. Seeds or achenes embedded in the flesh.

The varietal wealth and variation of strawberry cultivars are well documented. Brief description of some important cultivars is described below as.

Tioga : Has been evolved from a complex cross involving number of cultivars. Originated in California. Fruits are long conic to wedge shaped, weighs 12-15 gm, skin thick, tough, shiny, red, flesh medium red, firm with pleasant flavour, soluble solids vary 7-10%, acidity medium, produces heavily, ships well, tolerates salinity.

Torrey : Originated from a cross between Larsen and California selection and tested as a sibling of Fresno. Fruit large uniform, wedge shaped, skin darker than Fresno, calyx cap loose, flesh very firm quality good. Plant are vigorous leaflets deeply incised, lobes irregular leaf blade wrinkled, short rest period, tolerant to virus and salinity.

Senga Sengana : Markee x Sieger, developed in Germany. Plant erect type and high cropper, starts runner production in late August. Fruit large which remain till later picking, skin red, bright glossy attractive, flesh medium firm, calyx cap loose and usually remains on the plant.

Gorella : Originated in Holland from a cross between Juspa x US selection. Plants are erect type leaves thick tough sometimes 4-5 foliate, trusses very firm, fruits are large conical, maintains the size till end of the season, skin shiny bright red attractive firm with good flavour, ripens early.

Catskill : Originated in New York from a cross between Marshall x Howard17, grows under varied soil conditions. Fruits are large long conic somewhat irregular skin bright red, shiny attractive flesh light red fairly firm not good for shipping, eating quality is very good, good for freezing . Plant productive and vigorous, runners produced freely.

Selva : Evolved from complex cross. Fruits are large medium to long conic in some fruits the apex may be wedged or flat, skin dark red glossy attractive, flesh light red in colour, core somewhat hollow, firmer than Tioga, Tufts, Pajaro. Plants are large and semi erect.

Chandler : Evolved from a cross, Douglas x California selection. Self fertile excellent runner producer, suitable for summer and winter plantation. Fruits large long conic, apex flat or wedged, skin glossy red coloured, flesh light red, firm, flavour excellent, ripens early, used both as dessert and processing.

❑❑❑

42 Currants & Gooseberries

Genus : *Ribes*
Species : *nigrum L. americana Mill. bracteosum Dougl. niveum Lindl. divaricatum Dougl.*
Family : *Saxifragaceae / Grossulariaceae*
Chromosome No. : 2n = 16

The genus Ribes belongs to the family Saxifragaceae mainly comprises of woody shrubs distinctly marked with spines and prickles and those which have attractive handsome foliage, flowers and fruits. The genus name Ribes is likely to be derived from ribas, the Arabic name for Rheum Ribes. Some are of the view that it is the latinized form of riebs, an old name of currants in Germany.

Nearly 150 species are found in cold and temperate regions of north and south America, Europe, north Africa and north and central Asia. The genus

Ribes is sometimes divided into two distinct groups, the true *Ribes* and the *Grossularia*. *Ribes* in general are characterized with unarmed stems, flowers occur in racemose with joint pedicels, whereas in *Grossularia* the stems are with spines and prickles, flowers appear solitary or in 2-4 floral raceme and pedicels are not jointed.

Both currants and gooseberries in general are low growing shrubs medium sized usually with lobed leaves, flower small insignificant greenish or reddish, appear in spring with leaves, but in some species they are brightly coloured with varying shades of yellow, red, orange or scarlet. Fruits also exhibit variation in colour and attraction in varying shades of greenish or yellowish, black, purple or red in colour.

Although currants exhibit variation in fruit colour but black and red currants by far are more important. Black currant is a rich source of vitamins and minerals, rich in catechin content, the vitamin C varies between 180 to 240 mg. per 100 g of fruit while in red currants it is much lower (30-40 mg). Likewise total amino acids are higher in white currants while red currants are said to be richest source of glutamine acid. Despite having abundance of nutrition ingredients the black currants are not preferred as dessert for its strong flavour. The fruit is mainly processed into various products like jam, jellies, pie, liquor besides preparation of juice.

The genus *Ribes* on the basis of morphological feature, has been segmented into subgenera, section or series. The black currant has been evolved from *Ribes nigrum* a specie widespread in Europe and northern and central Asia. On account of strong smell / flavour of foliage and flavour is considered as low food, but way back during medieval times wine colouring was done through black currant.

The black currant (European black currant along with the black or brown type with resinous glands) have further been placed under Eucorcosma which comprise number of species like *R. nigrum* (European black currant), *R. americana* (American Black currant), *R.petiolare* (Western Black Currant), *R. hudsonianum* (northern black currant), *R.bracteosum*, *R. missouriense R. dikuscha*, *R.viburnifolium*.

Red currants on the basis of inherent variability are placed under subgenus Ribesia, and have been derived from three species namely *R. rubrum* (central and northern Europe and northern Asia), *R. sativum (R. vulgare*) western Europe and *R. petraeum* (western and central Europe). Further white currants are said to be some sort of botanical forms of red currants. Species of this group are *R. multiflorum* (Native of Eastern Europe), *R. triste* (N. Asia and America) and *R. longeracemosum*.

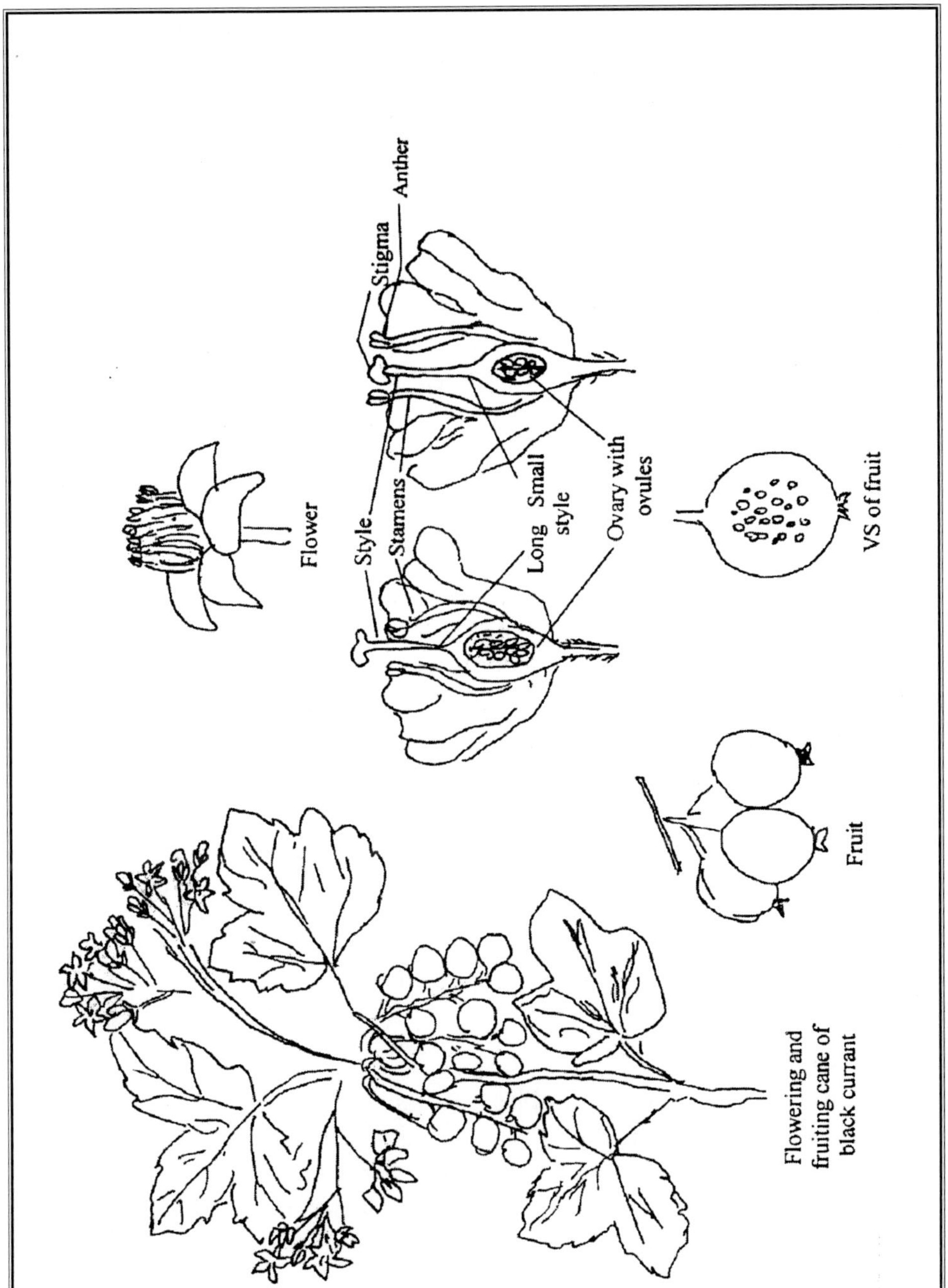
Flower
Style
Stamens
Stigma
Anther
Long
Small style
Ovary with ovules
VS of fruit
Fruit
Flowering and fruiting cane of black currant

Gooseberry baring few differences resemble much with the currants. The American gooseberry are forms derived from *Ribes hirtellum* Michx and are characterized by medium sized fruits and more resistant to powdery mildew. European gooseberry have been obtained from *R. grossularia* and are cultivated considerably in Europe, fruits are larger and are susceptible to powdery mildew. Popularly gooseberries are called as British fruit and rightly so as the most development of this fruit took place in Britain. *Ribes grossularia* is widely distributed in the Caucasus and north Africa, while American gooseberries *R. hirtellum* are found growing in abundance in America. American gooseberries are usually grouped with other species like the western Chinese (*R. stenocarpum*), the Japanese *(R. grossularioides*) and N. American (*R. divaricatum, R. niveum R. cynosbati*) and their cultivars are of hybrid origin involving *R. grossularia* and *R. hirtellum*.

Origin and Distribution

Important species of currants are *R. nigrum* (European Black currant L.), R. rubrum (Red currant L.), *R. americanam* Mill. (American Black currant), *R. aureum* Pursh. (golden black currant), *R. sativum,* Common or garden (red/ white) (Reichenb.) Symecurrant. The genus Ribes, comprises of about 150 species, distributed mainly in northern temperate regions of N. America and Europe and few in Asia, northwest Africa and in the Andes ranges.

Uses

Currants on the bases of different fruit colour have been named differently like red, black, golden or white,. The fruit is a berry is eaten fresh or used primarily in canning, or making of juice, jam, jellies also valued for high level of vitamin C and anthocyanins etc. and fruit currant should not be mistaken with the dried grapes which are commercially referred to as currants.

Description

The currant plant unlike brambles is usually spineless, stout woody shrub about 5-7 feet in height but should be trained to lower heights for quick and easy fruit harvest. Being native to cooler parts of the world, hence are considered more resistant to cold thereby fails to establish under dry and hot climates. Currants exhibit wide variation i.e. some species are fragrant and some exudes undesirable flavour.

Leaves simple, long pedicelled leaves which are broad ovate, slightly lobed, margins serrated (sharply) to dented, green, rough to feel, pubescent below, pedicel long and medium thick with few short stiff spines, veins very prominent, pinnate, impressed above and raised below. In black and red currant the flowers are generally small and forms a cup like/saucer like (open campanulate) shape. The whitish or greenish pentamerous flowers appear in few to many flowered racemes. Calyx 5 usually coloured, corolla 5, generally smaller than calyx and may sometimes be absent. Stamens are very short and arise from the base of petals and curve towards the style. In most of the species the stamen length is about the length of style but in some cases stigma extends beyond the anthers length. Ovary inferior, one celled multi ovuled. Fruit is a berry with many seeds, 1-2cm in diameter, oval, round, juicy, texture soft, colour variable, red yellow, black, purple.

Important species of genus Ribes (currants) with distinguishing characters are detailed below as,

R. odoratum Wendl. Missouri Currant, Golden Currant. Small shrub all young growth pubescent, leave ovate to orbicular narrow at base 3-5 lobed margin dentate, racemes 5-8 flowered, flower yellow fragrant, calyx oblong half the length of calyx tube, corolla half the length of calyx. Fruit large ¾ inch across ovoid black in colour.

R. aureum Pursh. The specie resembles *R. odoratum*, but is small and slender for all plant characters, young growth may be shiny or pubescent, leaves 13 lobed, flowers occur in raceme 5-15 flowered clusters, flowers are yellow showy and fragrant, fruit red or black, globose.

R. cereum Douglas. Flower white or greenish, petals very small round, style hairy very smooth or with glands. Fruit bright red ½ inch or more across. Densely branched small upright shrub, leaves 3-5 lobed with soft pubescence on undersurface.

R. sanguineum Pursh. Bush tall growing about 10 feet, young growth glandular and pubescent, leaves like other specie, dark green, slightly pubescent above densely tomentose below, racemes with many flowers, flowers are red and pubescent. Fruits slightly glandular, blue-black in colour and with bloom on surface.

R. nigrum Linn. European Black Currant. Found in abundance in Europe, north and central Asia, Himalayan region. Medium growing upright shrub 6 feet, branches thick and stout with unpleasant flavour, leaves 3-5 lobed, slightly pubescent.

R. americanum Mill. American Black Currant. Foliage distinct with strong odour, bush upright, 5 feet, young growth slightly pubescent and glandular, leaves 3-5 lobed, pubescent on and along the veins. Flower variable in colour greenish or yellowish, calyx tube bell shaped. Fruits black in colour and smooth.

R. triste Pall. Swamp Red Currant. Short shrub, young growth slightly pubescent and glandular, leaves round thin 3-5 lobed, dark green shiny above, white tomentose below, flower purplish, calyx tube saucer shaped, petals reddish smooth and red.

R. bracteosum Douglas. California Black Currant. Upright growing shrub, young growth slightly pubescent and with resinous dots, leaves thin cordate 3-7 lobed, raceme long and narrow, flowers greenish or purplish, calyx tube cup-shaped, fruits edible, black with white bloom and resinous dots.

R. petraeum Wulf. Native of central and south Europe, N. Asia and Caucasus. Shrub about 8 feet tall, upright, young growth and branches shiny, leaves rounded 3 lobed, pubescent below, raceme 4 inch long and dense, flower red or pink, calyx tube bell shaped, fruit dark red with acidic taste.

R. rubrum Linn. Northern Red Currant. Native to central and north Europe, and N. Asia. Small shrub, 6 feet, young growth generally shiny, leaves truncate 3-5 lobed almost shiny, racemose spreading shiny, calyx tube saucer-shaped, flower greenish or brownish, fruit juicy and red coloured with flower reminents persistent at basal end.

R. vulgare Lam. Red or Garden Currant. Native of western Europe. Fruits distinctly five angled at base with flower reminents at base, fruit red with stripes of white, shrub upright 5 feet, young growth pubescent and glandular, leaves thin and cordate 3-5 lobed, racemes pendulous, with many flowers, calyx tube saucer shaped, flower greenish with purple tinge inside.

Grossularia - Gooseberries

The important goosebery species with short pomological description is summarized below.

R. niveum Lindl. (*Grossularia nivea*, Spach). Native of N. America. Very attractive with its white and numerous flowers, relatively tall 8 feet, upright shrub branches reddish brown in colour armed with stout brown prickles/ spines about ¾ inch long, bristles absent, leaves suborbicular 3-5 lobed each with few teeth, may be shiny or pubescent. Flowers in clusters of 1-4, pedicels thin, calyx tube campanulate, calyx narrow lanceolate, stamens longer than calyx. Fruit shiny bluish black, glabrous subacidic in taste.

R. hirtellum Michx. Is the most important American gooseberry with edible fruits and several hybrids of it are in cultivation. Small shrub 4 feet, branches without prickles, slender but small spines present at base of rapid growing shoots. Leaves sub orbicular, 3-5 lobed, shiny or slightly pubescent. Flower in 1-3 clusters greenish, calyx tube narrow companulate. Fruits globose, black or purple, smooth, sometimes with glands.

R. grossularia Linn. (*Grossularia reclinata*, Mill.) Indigenous to Europe, N. Africa and Caucasus region. Very small shrub, 3 feet, spines ½ inch long usually in group of threes. Leaves suborbicular 3-5 lobed. Flowers in 1-2 cluster greenish, calyx tube short campanulate. Fruit globose-ovoid, pubescent and with glands.

R. cynosbati Linn. Native to N. America. A shrub medium in height, 5 feet tall, basal branches spreading , spines in cluster of 1-3, ½ inch or more in length bristles few weak or none, leaves orbicular 3-5 lobed, pubescent below, margin dentate, flower in cluster of 1-3 green, petiole thin, calyx tube campanulate broadly, fruit ovoid-globose, with prickles, edible, red coloured.

R. divaricatum Douglas. Wide spread from British Columbia to California. Tall shrub, 10 feet or more branches grey to brown armed with stout spines about ¾ inch long, bristles present sometimes. Leaves suborbicular with 5 lobes, margin crenate dentate, pubescent more along the veins, flowers in 2-4 clusters, green or purple, calyx tube campanulate. Fruit globose, purple or black, smooth.

Salient features of some gooseberry cultivars are described briefly.

Red Gooseberry

Abundance : Originated in North Dakota, Parentage is *Ribes missouriensis* x Oregon Champion. Bush strong growing, very productive and hardy. Fruit medium in size, round occur on long stems, purplish in colour, skin thick and tough helps in long storage and transportation.

Clark : Originated in Burlington, Canada. Chance seedling, a natural hybrid of European and American species. Bush moderate in vigour, medium in production, hardy. Fruits are very large, skin thick and quite tough, initially greenish yellow turns red at maturity, easy to pick, mid-late in ripening, quality good.

Jahns Prairie : A new disease resistant high quality dessert gooseberry developed at Corvallis, Oregon. Plants are upright with some drooping branches, high in production spineless or with few spines less than 4mm, resistant to American powdery mildew leaf spot, aphid and sawflies. Fruit-large globose in shape, red in colour with slight pink tinge, ripen in mid to July end.

Rokula : Originated in Germany from a cross Mauk's Early Red x mildew resistant R. UVA-Crispa. Bush medium in vigour, slightly drooping, medium in production, ripens very early in season. Resistant to American powdery mildew and other leaf spots, fruits are red, round large, flavour and quality both outstanding, cracks in rain.

Green Gooseberry

Pembina Pride : Originated in Manitoba, Canada. Bush very winter hardy like Thoreson, and productive. Fruit green in colour at maturity, medium in size skin thick and tough, quality wise fair.

Mucurines : A new green cultivar selected from R. uvr-crispa. Bush high growing vigorous, upright, suckers freely from base, suited for commercial production and home gardens for its regular production size and yield traits. Fruit, large green broad elliptic berries in form, with good flavour easy to pick, resists cracking, hangs well on bush.

Malling Greenfinch : Developed at East Malling Research Station, England. Evolved from a double cross, Careless x (Whinman's Industry x Resistent). Plant very productive compact and with upright growth, spineless, resistant to American powdery mildew, moderately to leaf spot. Fruit medium in size, green in colour, attractive, flesh somewhat soft, flavour good, good for fresh and home garden. Canning and freezing quality moderate.

Malling Invicta : Plant vigorous spreading spines many. High yielding powdery mildew resistant type. Resistenta x Whinman's Industry. Fruit large light green in colour, skin somewhat bristly, irregular in shape, flavour good, suited for processing.

White Gooseberry

Keepsake : Also known as Berry's early Kent. An excellent white gooseberry. Plants are usually spreading and very tender to frost damage. The fruit ripens early in the season, and very good cultivar for fresh consumption, very productive heavy and regular cropper. Berries are usually white in colour and tinged with green.

Whitesmith : Cultivar is very prolific bearer, bears regularly and consistently, fruit ripens early, berries are quite large in size, oval in shape pale green tinged with light yellow, taste and flavour good.

Careless : Cultivar is extensively used for making jam, canning and ripe green fruits in market have added attraction for fresh consumption. Widely cultivated white gooseberry ripens in mid season, very heavy cropper producing greenish white fruits.

Yellow Gooseberry

Pilot : A very useful cultivar, does well where Leveller fails to do well, highly splendid cultivar, but not grown extensively, may be due to the fact that it is not heavy cropper. A dessert gooseberry, produces fruits which are big in size, berries are usually oblong-oval in form, attractive yellow, with good flavour.

Golda : A new yellow gooseberry cultivar, developed at Netherlands from a cross between Whitesmith x May Duke. Plant is a medium bush, high yielding but susceptible to American powdery mildew. Fruits are yellow in colour, large, ripen mid-late season.

Kristin : A newly developed black currant cultivar at Norway from a cross, Ben Tron x (Hedda x E.M. 1428/70). Plant quite upright, vigorous resistant to powdery mildew and leaf spot susceptible to gall mite and pine blister rust. Fruit large, black in colour occur on long strings, firmness medium, with low acid and high sweet contents, easy to hand picking and less for mechanical harvesting, quality maintained for long time. Flowers early and ripens in mid season.

Currant

Ogden : Originated in Ottawa, Canada. Bush vigorous medium in size and very productive. Fruits are black in colour ripens in mid season, bears in small to medium clusters, skin quite tender/soft, small to medium in size, flesh sub acid, quality fair, ripens irregularly.

Fleming : Originated in north Dakota. Is regarded to be one of the better fruiting types obtained from *Ribes americanum*. Fruits are black in colour, better in quality and production than most other black currants cultivated earlier.

Tor Cross : Originated at Bristol, England from a cross between Baldwin x Rogue. Bush medium in vigour high yielding. Fruit black in colour, ripens early in season resembles Mandip Cross, has more resistance to splitting and dropping.

Albatross : Originated in Netherlands, Red Lake x Fay's Pride. Plants produce moderate to high yields, flowers early, moderate in growth. A white currant with excellent flavour, fruits medium sized, attractive pale yellow, easy to harvest sweet in taste, flavour excellent ripens early in season.

Detvan : Originated at Bojnice, Slovakia from a cross between Joukheer van Tets x Heinemann's Rote Spatlese. Bush tall growing and produces regularly, vigorous and adapts to warmer places. Fruits are large attractive and produced on long shoots, ripens early, a high yielding large fruiting red currant.

Malling Red Start : Developed at East Malling Research Station, England from a cross between Red Lake x B 156/34 [(Ribes *sativum* x R. *multiflorum*) x Red Lake]. Plants are resistant to wind, high yielder, erect with moderate growth vigor, fruit occur on long strings high in production, ripens late, red in colour, acidic flavour, good for processing.

Red wing : Evolved from a cross between Red Lake x a selection from *Ribes longeracemosum* and *R. multiflorum* at East Malling Research Station, England . Plants with erect growth, vigorous, crop well distributed on the bush, resistant to leaf spot susceptible to American powdery mildew. Fruit small-medium in size, bright red in colour, easy to pick and harvest, ripens late thereby fresh market season is prolonged, makes very pleasant jelly.

Rondom : Selected from R. *multiflorum*, in Netherlands. Plant upright growth vigorous, flowers in mid season with very good production, harvesting easy, somewhat susceptible to leaf spot, American powdery mildew. Fruit bright red, medium in size.

❑❑❑

43
Brambles

Genus : *Rubus*
Species : *idaelus L. (European Red Raspberry) strigorus* Michx. (Red Raspberry *occidentalis* L. *(Black Raspberry)*
Family : *Rosaceae*
Chromosome No. : 2n = 14 (variable 28, 42, 56, 84)

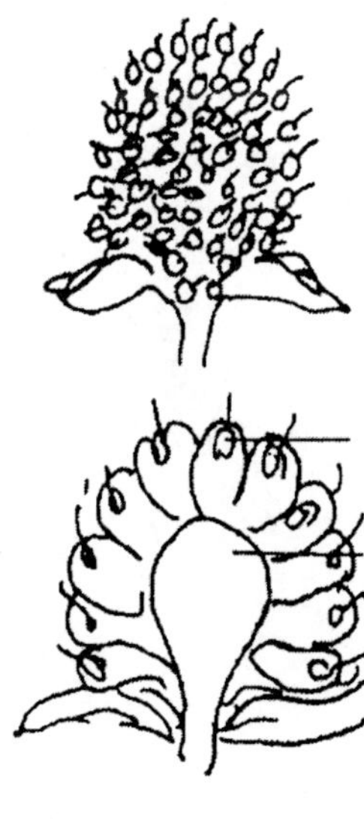

The blackberries, raspberries, dewberries, boysenberry are collectively referred to as bramble fruits. Mostly found in the north hemisphere. They belong to the family Rosaceae and genus *Rubus*. The family has nearly 250 species besides also has large number of apomictic segregates. Usually shrub, the woody species has biennial canes i.e. in the first year remains vegetative and in the second year it fruits and

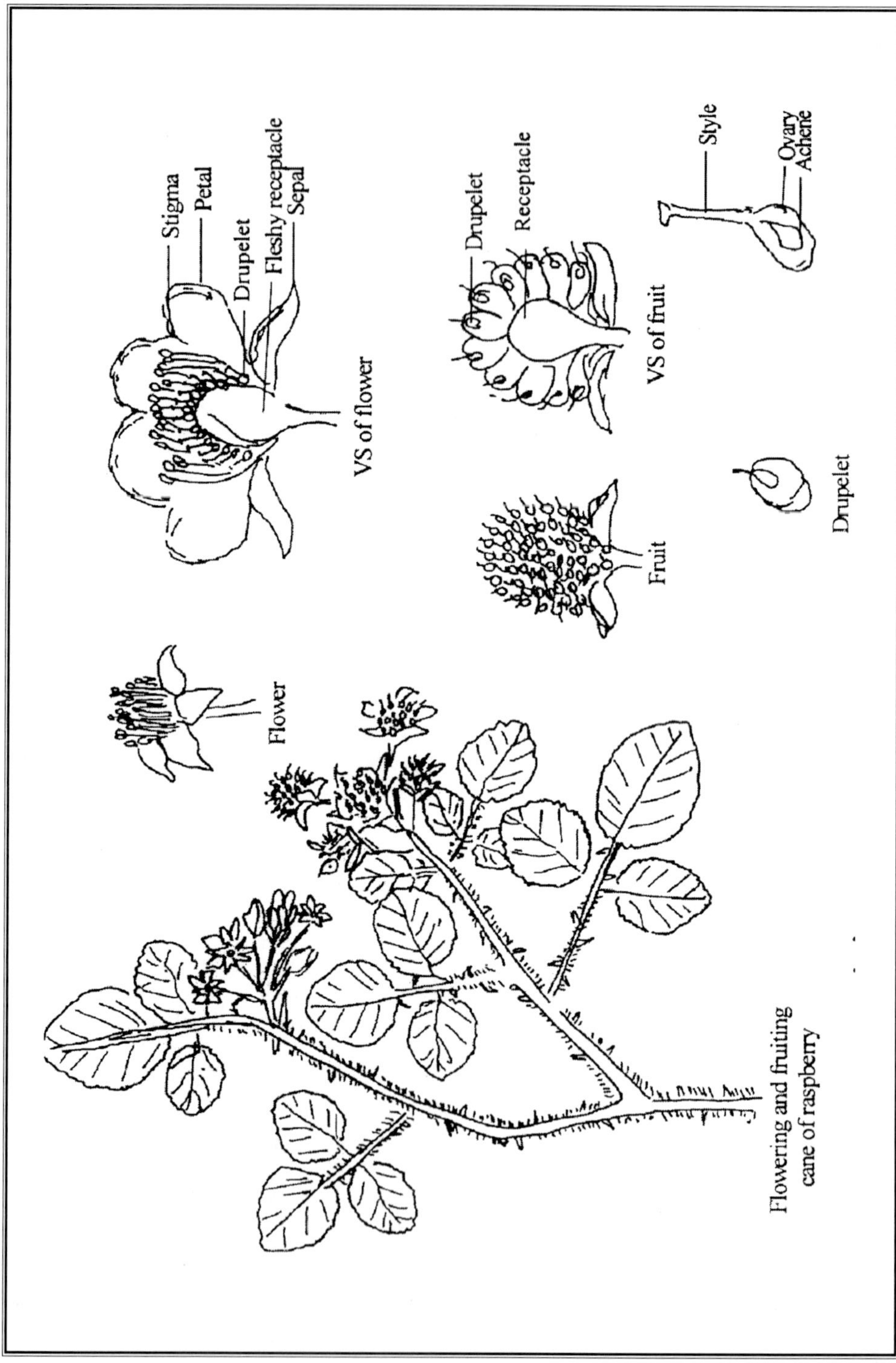
Stigma
Petal
Drupelet
Fleshy receptacle
Sepal
VS of flower
Flower
Drupelet
Receptacle
VS of fruit
Fruit
Style
Ovary
Achene
Drupelet
Flowering and fruiting cane of raspberry

dies. Leaves variable in form from simple leaf to pinnately or palmately compound. Flowers usually occur in clusters and are variable in colour, white pink to red. Fruit is an aggregate of number of small drupes. The basic difference between raspberries and blackberries or dewberries is that in the latter types the drupelets remain attached to the receptacle and is usually eaten as fruit part, but in the former-raspberries, the receptacle readily separate from the drupelets as there is no fusion between the receptacle and the drupelets. Dewberries are vines that run over the ground but require support when planted in orchards. The canes of these bramble fruits must be removed to the ground level soon after fruiting so as to encourage fresh vegetative growth.

The bramble fruits contain almost all the important nutrients in substantial quantities and not being very rich in any individual nutrient component. Fruits are usually eaten fresh, or processed in products like jam, jelly, yoghurt, syrup, canned or frozen, for flavouring and also in wine making.

The cultivated raspberries have been originated from three diploid species or subspecies *R. idaeus* var. vulgatus, *R. idaeus* var. strigosus and *R. occidentalis*. Rubus is a highly variable genus, and few species are also found growing in abundance in India. From horticulture significance point of view the genus has been divided into four sections namely Idaeobatus - Raspberries, Eubatus - Blackberries, Cylactis - comprises of alpine species and Anoplobatus - flowering raspberries.

Origin and Distribution

Uses

Is one of the most tastiest wild fruits, used as dessert and in early summer the fruit is liked by all. Fruits have pleasant blend of sweetness and acidity and has excellent eating quality, due to rich flavour and attractive colour it can also be used in making of jam, juice, preserves and for squash making.

Description

Brambles comprises of more than 250 species excluding a large number of apomictic segregates including boysenberry youngberry, dewberry, and longanberry (red rasp berry) are grouped as black berries. Plants are evergreen shrubs, nearly herbaceous, the woody species with biennial canes which arise from a perennial root system. The cane usually produces growth in the first year, bears fruits in the second year and then dies back to the crown, canes usually prickly with stiff thorns, about one quarter inch long along the branches and smaller ones on the stems and leaves. The canes non flowering and fruiting in the first year are called as primocane, and second year when it flowers and fruits the cane is called floricane. Leaves are compound, trifoliate with reticulate pinnate veination, usually deciduous, in some cultivars they are persistent, leaflets irregularly toothed, terminal leaflet larger and stalked and lateral leaflets nearly sessile and smaller in size. Flowers generally white, pedicellate, complete, hermaphrodite, regular, 1-1.4 cm across in diameter, inflorescence a corymbose panicle, number of flowers variable more in erect blackberry (10-20) flowers) and lesser in trailing type (1-10 flowers). Calyx 5 tomentose, obovate, apex acute, persistent green, gamosepalous, free at terminal end. Petal usually 5 polypetalous usually white, actinomorphic. Stamens numerous (50-100), closely clustered and overshadows about equal number of pistils. Gynoecium polycarpellary apocarpous. Fruits is an aggregate of eterio of drupes, usually borne on a thalamus which is nipple shaped. (In *R. ellipticus* the yellow coloured fruits (drupelets) are easily detachable from the thalamus). In blackberry the thalamus remains attached with the drupelets, whereas in raspberry drupelets separates readily from the thalamus.

Pomological detail of some species of each of the four sections are given below.

Section Idaeobatus

Rubus idaeus L. Red raspberry, European raspberry, Framboise. Red raspberry belongs to this species. An erect shrub 3-5 feet tall, prickles variable non to many, sometimes with bristles, leaflets 3-5 broad ovate margin dentate, upper surface glabrous, with greyish or white pubescence below, Flowers occur terminally or laterally, individual flower small white, few in number. Fruit, conical or oblong usually dark red but white and yellow are also found. Has some botanical forms like *R. idaeus* var. canadensis Richardson. Young canes are with ashy dense pubescence - tomentose var. strigosus (Michx) Maxim. American red raspberry. Hardy, new canes and inflorescence with bristles and glands, conical or hemispherical light red in colour Another form purple cane raspberry, the fruits are purple and probably is of hybrid origin between *R. idaeus* and *R. occidentalis.*

R. pubescens Raf. Dwarf raspberry. Native to N. America. Herbaceous, small soft, without prickles, leaflet 3-5 lobed ,thin, margin serrate shinning slightly, flowers 1-3, white ½ inch across. Fruit red-purple in colour.

R. phoenicolasius Maxim. Wine berry. Native of China and Japan. Fruits are small bright red in colour edible, flowers occur in loose clusters, small, white or pink coloured. Canes usually long, re-curving and roots at tip, prickles straight and with red-brown glandular hairs, leaflets usually 3, cordate-ovate margin dentate.

R. niveus Thumb. (*R. albescens* Roxb.). Hill raspberry, Mysore raspberry. Native to west India and China. Flowers are small attractive red-purple in panicle-like clusters, fruit red. Shrub 6 feet tall, young growth pubescent later becomes glabrous, prickles hook like, leaflets 5-7 in number, elliptic ovate, margins coarsely serrate.

Section Eubatus

R. occidentalis L. Black cap, Thimbleberry, Black raspberry. Indigenous to N. America. Has been the progenitor of number of black raspberry cultivars. Canes shiny, erect, small 3-5 feet, canes bend and roots at tip. Leaflets usually 3, ovate, apex pointed, margin double dentate, white densely pubescent below. Flowers are small occur on dense short spiny clusters, white, fruit firm, black hemispherical very shiny.

R. ulmifolius Schott. Native of Europe. Canes are thick and stout, arching with stout prickles base broad hairy, leaflet 3-5 densely pubescent, white below, inflorescence elongated with white or pink flowers. A botanical form of the species is var. inermis (Willd). Evergreen thornless blackberry. Canes very long without prickles, leaflets 3-5 ovate to elliptic ovate thick, margin serrate, fruit black, globose.

R. macropetalus Dougl. ex Hook. Blackberry, Dewberry. Native to N. America. Canes erect later becomes vine-like, shiny with weak prickles, leaflets 3, thin apex sharp pointed margin sharply toothed, green on both surfaces. Flowers are white, unisexual occur on short panicles, pedicelled. Fruit black and shiny.

R. flagellaris Willd. American Dewberry. Native to Canada and Gulf States. Trailing type rooting at cane tips, without glands, prickles recurved. Leaflets 3-5, ovate, margin serrate dentate, apex acute. Flowers few in branched inflorescence. Fruit edible black, round to oblong.

Section Cylactis

R. arcticus L. (Syn. *R. stellatus*). Crimson Arctic Blackberry found in Arctic and sub-arctic Eurasia region. Herbaceous in nature 3-15 inch long, rhizomes slender, lacks aerial stolons leaves with 3 leaflets, central one longer and broader than the lateral ones, margins coarsely toothed, flower pink, single, fruit red in colour.

R. chamaemorus L. Cloudberry, Salmon or Yellow berry. Native to sub-arctic, circumpolar region of north hemisphere. Herbaceous, mono or dioecious, 3-10 inch high. Leaves 2-3 almost round, broadly lobed. Flower white, single 1 inch across. Fruit red or yellowish.

Section Anoplobatus

R. coronarius Sweet. Brier Rose. Most probably has originated in Asia widely grown in warmer region. Young growth and leaves are shiny and distinctly marked with resin dots, prickles numerous and stout, leaflets small, narrow at base. Flowers white, 1 inch across, double flowered type.

R. deliciosus Torr. Rocky Mountain flowering raspberry. Grown mainly for its attractive rose coloured flowers. An upright shrub 6 feet, usually without prickles, spreading, leaves orbicular-ovate with 3-5 shallow lobes. Flowers solitary white, large in size 2 inch across fruit dark purple or red in colour.

R. odoratus L. Flowering, Purple-flowering raspberry or Thimbleberry. Flowers are large attractive whitish to rose-purple 2 inch across, occur in loose clusters. Relatively tall shrub 6 feet without spines, bark shreds, leaves simple cordate 3-5 lobed margin finely serrated, pubescent below. Flower 2 inch across white to rose purple in loose clusters. Fruit dry flat red, inedible.

R. parviflorus Nutt. Salmonberry or Thimbleberry. Resembles *R. odoratus*, but flowers are white and few in number.

Few Rubus species indigenous to India, their description in detailed below as.

R. ellipticus Sm. In India the specie is found growing in plenty in warm climates. Canes are strong, 10-15 feet tall, evergreen, prickles thick and stout, stems with reddish-brown hairs, leaflets 3 ovate to elliptic with dense greyish pubescence below, margin serrate. Flowers occur in panicles, white, fruit yellow.

R. moluccanus L. Indigenous to India and Malay peninsula. Flowers are white in terminal clusters, fruit edible, canes strong stout densely pubescent with red hairs, prickles curved. Leaves simple, broad ovate tomentose below 3-5 lobed.

Cultivars

The Raspberry cultivars presently being cultivated are divided into many groups on the basis of the ripening, early mid or late, their flowering time, fruit colour yellow, red, purple or black, processing juice making, canning, freezing etc. Salient features of some cultivars are given below.

September : Bush vigorous, reliable bearer, crops well. Fruit medium sized, skin bright red, attractive, flesh firm, fruitlets do not crumble, quality of summer crop only fair, but autumn berries are good in quality.

Glen Clova : Originated from a complex cross at Scotland. Bush erect, dense, hardy productive, tolerant to cane spot, susceptible to mildew. Fruit medium in size, short conical, skin red, slightly downy, flesh firm quality very good for jam making, canning and freezing.

Willamette : Developed from a cross Newburgh x Lloyd George. Canes medium-large laterals strong, straight, very productive, susceptible to root rot. Fruit very large almost round, dark red, flesh firm somewhat acidic, flavour mild compared to characteristic raspberry flavour, good for canning, excellent shipping only fair for freezing.

Bramdywine : Is of hybrid origin between New York 631 x Hilton. Fruit large 5-6 gm in weight, reddish purple, round-conic, pubescence moderate, very acidic, quality very good, ripens late. Bush vigorous, very productive, strong and upright and erect susceptible to verticillium wilt and raspberry aphid.

Comache : Canes are vigorous, erect, productive, suckers freely, suited to mechanical harvest, fruit large, black, attractive, flesh firm, quality good and ripens about 4 days earlier than Cherokee. Obtained from Darrow x Brazos at Arkansas.

Cherokee : Evolved from a cross between Darrow and Brazos, at Arkansas. Bush fast growing, erect suckers produced readily, productive, very suitable for mechanical harvest. Fruit large, skin black, attractive, quality good of both when frozen or canned, ripen early.

Comax : (Creston x Willamette) x Skeena is a double hybrid. Winter hardy bush, very productive, vigorous and upright in growth, with drooping laterals - Primocanes numerous shiny with purple prickles, floricanes yellowish brown, firmly erect, cracking at base. Fruit very large, skin red, conical firm, drupelets many and separate easily from receptacle, suited for mechanical harvest, ripens late than Willamette and Skeena.

❑❑❑

Section - III
NUT FRUITS

44
Almond

Genus : *Prunus / Amygdalus*
Species : *dulcis* L. *communis, Fritsch.*
Family : *Rosaceae*
Chromosome No. : 2n = 28

Almond tree amongst the deciduous temperate fruits is the first crop to bloom in spring as it responds readily to warm temperature and also for its relatively low chilling requirements. The almond is thought to be one of the oldest nut crop used by man since very long. However, its specific climatic requirements have greatly restricted its commercial production to limited regions of the world. The almond blossoms are large attractive pink coloured very sensitive to spring frost, but at the same time the nut require high heat quantum at maturity, cool rainy weather during flowering restricts

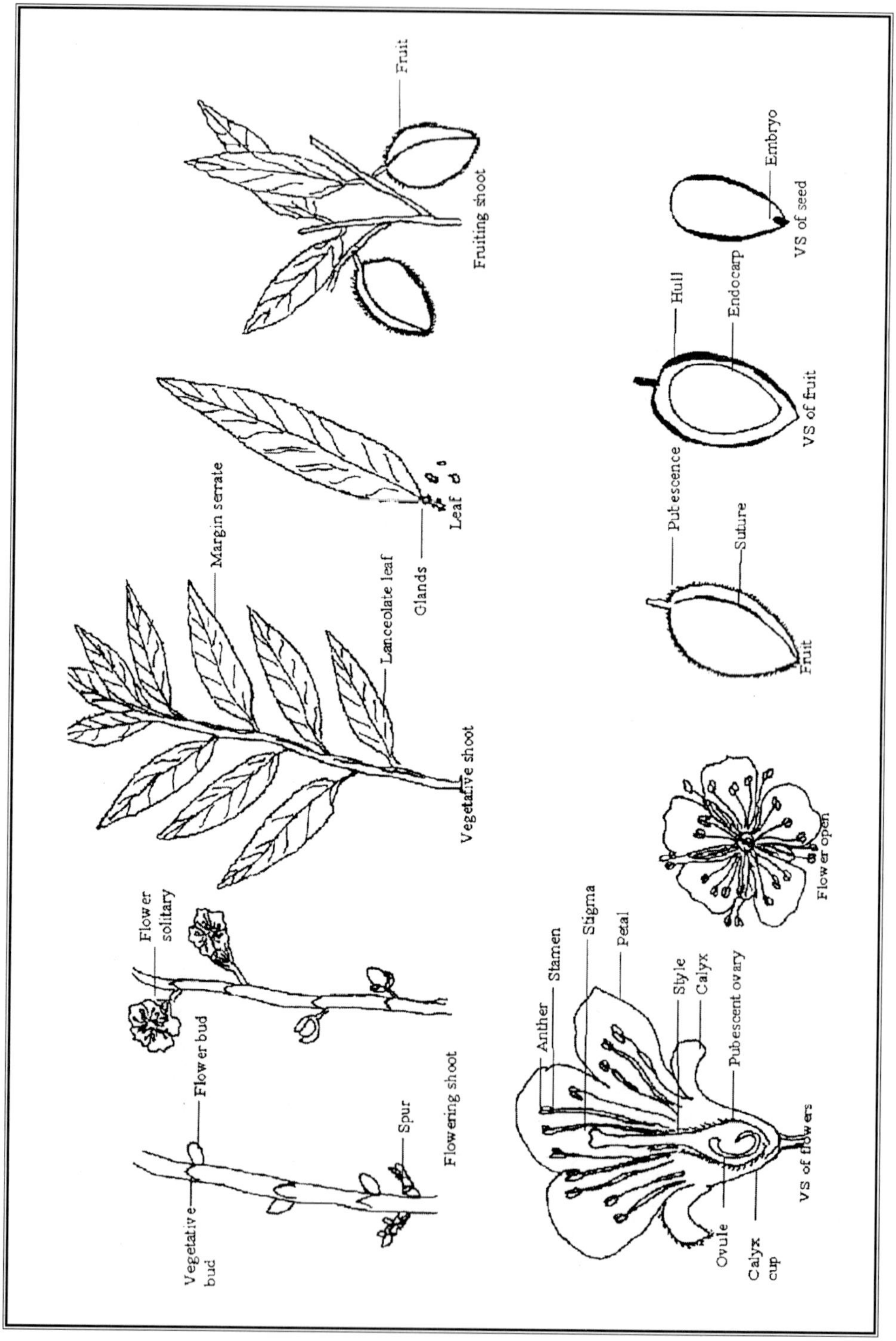
Vegetative bud
Flower bud
Spur
Flowering shoot
Flower solitary
Vegetative shoot
Lanceolate leaf
Margin serrate
Glands
Leaf
Fruiting shoot
Fruit
Anther
Stamen
Stigma
Petal
Style
Calyx
Pubescent ovary
Ovule
Calyx cup
VS of flowers
Flower open
Fruit
Pubescence
Suture
Hull
Endocarp
VS of fruit
Embryo
VS of seed

cross pollination and induces fungal infections. In the mild winter regions almond trees are limited by the fact that distinct chilling requirement for growth and flowering are not met with. Therefore most ideal condition for successful almond culture is the Mediterranean type of climate which has mild winter combined with frost free and rainless spring and warm summer for nut ripening.

The almond kernels for their high oil content are rich source of concentrated energy. The oil is mostly unsaturated and comprises of oleic and linoleic fatty acids, also has considerable contents of protein, some vitamins like riboflavin, thiamin and niacin and minerals mainly calcium, phosphorus, potassium and magnesium.

The cultivated almond *Prunus amygdalus* is an important member of family Rosaceae sub family prunoideae. The genus Prunus has more than 30 species which show wide range of morphological and geographical forms.

Origin and Distribution

The almond shows wide range of morphological and geographical forms which are found growing in most of the parts of southwest and central Asia, Turkey, Cyria, Caucasus, mountains of southern Russia, from Iran to the deserts of Tian-Shan, Afghanistan and Hindukush mountains. Has nearly 30 species but on account of polymorphism and the presence of number of geographical basis, variation within species is not very distinct.

Uses

The fruit is important for its highly esteemed nuts which has high food value with distinct flavour. Excellent in eating quality both as green and in ripe form. The sweet nuts are eaten fresh or cooked either blanched or unblanched, salted, roasted, used extensively in confectionary or processed products like ice cream, chocolates, candies, in baked goods, biscuits and toping of sweets etc. The oil is very valuable and used in perfume products, baby oil, good for human skin and massaging. The nut husk is used as animal feed.

Description

The presently cultivated almonds are generally of two distinct groups, one grown for its nut or kernel and the other one as the ornamentals or the flowering type. The former belongs to *P. dulcis* and the later to *P. triloba, P. glandulosa* and *P. japonica* and are valued for their white, pink or rose, single or double flowers. Almond fruit is a dryish drupe with two distinct varieties recognized, the bitter and sweet almonds.

Almond plants are woody trees about 10-25 feet tall, semispreading, much like peach, bark usually greyish, early growt smooth but with age becomes rough and somewhat scaly. Leaves alternate, deciduous, simple, firm, glossy, narrow lanceolate, margin sharply serrated, petiole long thin green in colour, with glands but number is lesser than peach. Flowering occurs laterally on last season's growth and on spurs, before the leaves open. Flower about 1 inch across, sometimes even more, single per bud, receptacle cup-shaped, sepal 5, pubescent, purple-red in colour, petal 5, variable in colour white, pink or rose coloured, round at tip. Fruit is a compressed drupe with hard hull or husk (exocarp and mesocarp are fused). When green, the exocarp is velvetty. At maturity the husk splits and nut containing the kernel is exposed, endocarp distinctly pitted, shell may be thin to very thick, apex pointed or with short apex. Kernel may be sweet or bitter in taste.

The named species have been divided into four groups or sections namely Euamygdalus, Chamaeamygdalus, Spartiodes and Lycioides. Species with their distinct characters are described briefly.

Section Euamygdalus

Trees are medium sized 3-4 m. with spines.

1. *Prunus fenzliana* Fritch. Tree small very hardy, leaves elliptical nuts are flat, hard shelled. Grows in Turkey and north west Iran at a height of 1500 meters.
2. *Prunus argentea* (Lam.) Rehd. Leaves are silvery white, shell netted, not pitted. Plants are shrubs, grows in Syria and north west Iran.
3. *P. bucharia*. Fedtchenko. Trees are small and nut small flat pointed, smooth, polished kernel usually bitter, but some sweet forms also found. Dominant specie in Russian, Turkistan, grows at 800-2500 m. elevation.
4. *P. ulmifolia* Franchet. Small bushy trees, shoots and young leaves are pubescent. Old leaves pubescent on upper side, strongly toothed. Indehiscent fruits are small.
5. *P. communis* Fritsch. Tree and nut size is larger than any other specie, includes bitter and sweet, soft and hard shelled types. Native of south eastern Russia and Afghanistan. Natural hybrid of *P. fenzliana, bucharia* and *ulmifolia.*

Section Chamaeamygdalus

1. *P. nana* Stokes. A small dense bushy shrub 1-1.5 m. tall. Leaves lanceolate shiny, dentate. Fruit small, nuts flat, very bitter, very polymorphic, widespread in steppes of southern and central Russia.

2. *P. ledebouriana* (Schlechtendal). Though close to *P. nana,* but tree, leaves and nuts are larger.
3. *P. georgia* (Desfontaine). Small shrub 1 m. tall, leaves lens like, large flowers, nuts large and flat. Native of Georgia and west Trancaucasia.
4. *P. petunnikowkii* (Litwin). Rehd. Small shrub, 1 m. tall, nuts symmetrical. Native of Tian-Shan and steppes of Turkistan at 1400-1800 m. elevation.

Section Spartiodes

1. *P. spartiodes* (Spach) Schneid. Small shrub, 1.5 m. tall, branches pliant/ flexible long and green, leaves small. Indigenous to arid mountains of central Iran at 1500 m. height.
2. *P. scoparia* (Spach) Schneid. Similar to *P. spartiodes,* leaves narrower, tree small, native of Iran.
3. *P. arabica* (Olivier). Indigenous to the deserts of Arabia.
4. *P. agretis* (Boissier). Similar to *P. spartiodes,* leaves are narrower native of Syria.

Section Lycioides

1. *P. spinosissima* (Bunge). Tree small, 2-2.5 m. xerophytic and trees very spiny, fruits very large, kernels with 60 percent oil. Native of Iran, grows at 300-1500 m. height.
2. *P. brachuica* (Boissier). Shrub or small tree 1-2 m. tall, xerophytic very shiny leaves are small, nut with small apex.
3. *P. nairica* (Fedtchenko). Small bush, 1 m, very spiny, leaves very small, fruits very large. Grows in Armenia at 1500 m. elevation.
4. *P. webbi* (Spach). Small shrub, leaves pubescent, nut smooth.

Two fruits almond and peach of tribe pruneae, genus Prunus are close relative for similarity of various characters which are summarized as, leaves are generally folded along the mid rib lengthwise in the bud, fruit surface usually soft velvetty or with hairs, stone usually pitted or furrowed, receptacle cup-shaped and wide spread. Flowers occur laterally on last year's growth and before the leaves, fruits are usually sessile.

Some recognized species of almond other than cultivated almond (*P. amygdalus*) are described briefly:

Prunus triloba Lindl. (*Amygdalus pedunculata,* Burge *P. ulmifolia,* Franch) Flowering Almond. Native of China. A desirable bush, leaves are broad ovate or obovate, pubescence soft, apex pointed, margin doubly serrate sometimes may be 3 lobed. Flowering takes place before leaf emergance, pedicle short, and flowers are solitary per bud, variable in colour white to pink. Calyx tube pubescent from inside. Calyx shiny outside but slightly pubescent inside. Fruit small hairy when young, turns glabrous later.

P. orientalis Koehne (*Amygdalus orientalis*, Mill). Oriental Almond. The specie is found growing in plenty in Asia Minor and Syria. Plants are small shrubs upto 10 feet tall, young shoots densely pubescent. Leaves oval oblong or narrow-obovate, base round or short acute, margin may be entire or finely serrate (serrulate). Flower single per bud, and open before the leaves, pink or rose coloured. Fruit oblong/ovate, apex pointed, pubescent becomes glabrous later.

P. fenzliana Fritsch. Indigenous to Caucasus. Early flowering and very showy type. Compared to cultivated almond this specie is low growing more thorny and bushy. Leaves are smaller, bluish green or grey-green in colour. Flower usually white in colour. In form, the fruit resembles peach, hardly any flesh, stone small, short and round.

P. nana Stokes (*Amygdalus nana*, Linn.). Russian Almond. Very hardy specie. Found in abundance in Russia and west Asia. Distinctly a low bush, 3-5 feet tall, leaves thick stiff, narrowly elliptic or elliptic lanceolate, apex not pointed, veins conspicuous below, light green in colour, margin with saw-like teeth. Flower single per bud, white or pink about 1 inch across, open before leaves. Fruit hairy, hard and small, bitter in taste, seed relatively large wringled, apex pointed. Has 2-3 botanical types.

P. nana georgica DC., leaves are narrower, flower small dark rose in colour.

------- var. campestris Hort., flowers are white in colour and larger in size

----------------- var. rubra Hort., flowers are red coloured, small in size.

----------- var. cochin-chinensis Hort., flowers are white and plant is large.

P. sweginzowii Koehne. Found growing in Turkestan and adjoining areas. Resembles *P. nana*, but differs from it being shiny small shrub, stipules are relatively large and leaf-like, margins doubly serrated with sharp teeth. Flowers are deep rose coloured, calyx-tubular in form, lobes grandular and oblong. Corolla oblong-obovate.

Almond is cultivated particularly for its delicious and nutritive nuts and also grown as an ornamental tree. Has number of botanical forms for example.

P. amygdalus var. alba plena. Flowers may be in double whorls and white in colour.

----------------- var. rosa plena. Flowers are red in colour.

----------------- var. pendula. Branches are drooping type.

----------------- var. compacta. The tree is markedly dwarf.

----------------- var. variegata. The leaves are distinctly with white and yellow marks.

Apart from the botanical forms these forms have further been grouped into two following classes. *P. amygdalus* var. typica, Schneid. The hard-shelled almonds are usually grown as an ornamental although sweet (dulcis) and bitter (amara) kernel type are found. P. amygdalus var. fragilis, Schneid which comprises of soft or brittle shelled almonds.

Almond has a large number of ecotypes which exhibit extreme variation in shell thickness, softness, kernel quality, flower colour, self compatibility and number of other characters. The most cultivated and potentially suitable cultivars are briefly described below.

Ne Plus Ultra : Tree vigorous spreading, bark rough branches long medium wide, leaves large, flat, bears alternately and irregularly. Hull light green with some dark green spots, nut irregular ovate, apex pointed, shell very soft, kernel medium to large, not well filled, double kernel ranges 14-17 %.

Non Pareil : Tree medium in size vigorous spreading, flowers in midseason, self incompatible, compatible with Drake, Ne Plus Ultra, Marcona, ripens early, medium in production. Nut cordate-oblong shell very thin, shelling 65-70% very few double kernel, light coloured shrivels slightly.

Texas : Tree upright vigorous bears both on spurs and laterally on 1 year old shoot, self incompatible, pollen compatible with Cristomorto, Ferragnes, medium to high cropper, matures very late. Nut elliptic oblong, shell semi hard. Kernel dark with about 40% double kernel.

Merced : Is of hybrid origin Texas x Non Pareil. Tree medium, vigorous, upright, productivity very high, cross compatible with Non Pareil. Nut well sealed, shell thin light in colour than Non Pareil, kernel averages 58-63 % of nut.

Peerless : Plants are fast growing with spreading branches moderate in yield, flowers appear mostly on spurs and on 1 year old shoots also. Self incompatible, pollen compatible with Non Pariel, Ne Plus Ultra, bloom time mid to late. Nut elliptic shell semi hard, shelling about 40%, kernel colour intermediate with nearly 10% double kernel.

Tuono : A new bred self compatible cultivar also pollen compatible with Ferranges, Filippo Ceo. Tree vigorous, branches spreading bears only on spurs, flowers late in season, escapes spring frost injury, short maturity period as nuts are picked early, produces regularly and heavily. Nuts oblong elliptic shell well sealed medium thick and hard, kernel nearly 40% of nut, light coloured nearly 30% nuts with double kernel.

Filippo Ceo : Self compatible mid season with high productivity bears only on spurs. Tree growth moderate upright branches semi spreading, compatible with Genco and Tuono. Flowers very late to escape spring frost damage. Nuts elliptic in shape, shell hard well sealed, kernel light brown, constitute 35-37% of nut, slightly wrinkled with 40-45% double kernel.

❑❑❑

45 Walnuts

Genus : *Juglans*
Species : *regia L.*
Family : *Juglandaceae*
Chromosome No. : 2n=2x=32

Walnut (*Juglans regia* L.) among the nut crops is regarded to be a rich source of proteins, fats and minerals and is said to be a concentrated source of energy. It is on this account that wild walnuts have been utilized as food for human and animals since pre-historic times. It contains good amount of different vitamins of B group and is richest in vitamin B6 compared to almond, pecan, filbert, chestnut and Brazil nut. After the last ice age, records indicate that walnut spread from Asia Minor to other parts of the world, while some opinion it to be even older than man as revealed by fossil studies.

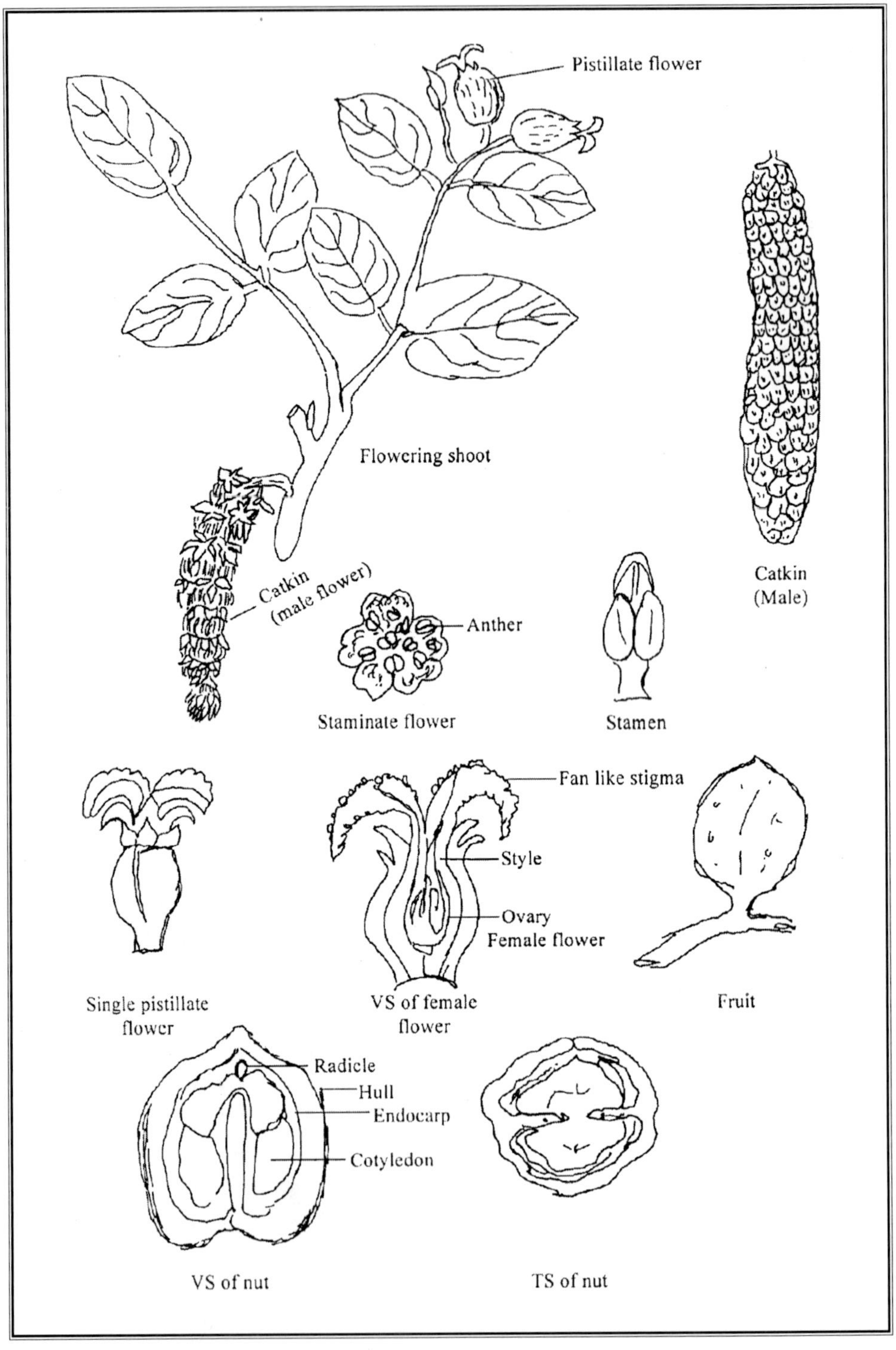
Pistillate flower
Flowering shoot
Catkin (male flower)
Anther
Staminate flower
Stamen
Catkin (Male)
Fan like stigma
Style
Ovary
Female flower
Single pistillate flower
VS of female flower
Fruit
Radicle
Hull
Endocarp
Cotyledon
VS of nut
TS of nut

The wild walnut seedlings are found growing in abundance from the Carpathian mountain in the eastern Europe, in Turkey, Iran, Iraq, southern USSR, Afghanistan and India in the north western Himalayas. The common name Persian walnut rightly suggests that walnut originated in Iran and surrounding areas. The armies of Alexander the Great on their return introduced the walnuts carried by them to Europe. Walnut is known by several other names like 'Carpathian Walnut', 'Persian Walnut' and 'English Walnut'. The last name being the most misleading as walnut is in no way is related with England. Another belief is that the name Walnut has been derived from 'Gaulnut' and in early time France was called 'Gaul', which was finally altered to walnut in English. The name is also thought to be derived from 'Welsh-nut' an old English word, and also on account of drying the nuts on the top of English garden walls.

Walnut belongs to the family Juglandaceae. The genus Juglans has twenty one species of which *J. regia* is commercially grown around the world, and eighteen more species are native to north and south America, Manchuria, central China and Japan.

Origin and Distribution

Walnut is considered probably native to wide range of areas spreading from Carpathian mountain across Turkey, Iran, Iraq, Afghanistan, south Russia to hilly regions of northern India. From Persia, walnut was introduced to the Greeks, which further moved to Rome. Later from Italy it spread to France, Spain, Portugal and southern Germany. The crop was present in England by 1952 and was brought to America by early European settlers.

Uses

Nuts are highly valued for its very high food value and also for its distinct flavour. The endocarp or shell (s) is employed for metal cleaning and polishing, filler in polishing and as texturizer in paints. Kernels extensively used in baked goods, ice cream, candies and in a whole lot of processed products. Produces world's finest cabinet wood, has very high properties for working and finishing.

Description

Walnuts are usually medium to large sized trees, with deep taproots, wood dark coloured, hard, compact and durable, trunk short, branches appear low on the trunk, crown broadly round, open at top. Bark smooth greyish white when young and turn brown to dark brown with black scales, which later develop deep round long ridges. Branches thick stout, spreading young branchlets reddish brown with dense rust colour pubescence/hairs later become smooth, lenticles pale or white, conspicuous, pith chambered. Leaves are deciduous alternate, scar rounded to triangular, winter buds almost rounded or with shoot tip usually two in number one is bigger and other smaller pubescent measures 5-7 mm long. Leaflets 7,9 rarely 15 or 17, broad oblong to lance-shape somewhat curved, pointed at top, usually rounded at base margins with fine serrations. Leaf colour light or yellowish green above and pale beneath, mid rib along with veins are more pubescent. Male flowers/catkin green yellow in colour appears in spring usually solitary or in pairs, 5-8 cm long. Individual staminate flower comprises of irregular perianth segments, which encloses variable number of stamens, usually 4-6 in number.

Female flower small appears at the tip of current season growth, 1-4 per terminal. Single pistillate flower has 4 perianth segment that surround 2 lobed style, stigma fan shaped, ovary inferior, unilocular formed from 2 fused carpels. Fruit variable, large oval egg shaped, globe or pear shaped, covered with glabrous dark green dotted husk (epicarp and mesocarp fused) that splits at maturity. Nuts almost round with pointed tip, are flattened at base and slightly compressed along the sides, indented with shallow to deep irregular grooves, shell thin to thick, light to dark brown in colour, kernel sweet, loosely or tightly adhered to the deeply contoured inner surface. Endocarp encloses seed with deeply lobed cotyledons, which is covered by a thin brown seed coat.

Important old and new world species and their brief description is given below.

Old world species

a) *Juglans regia* Linn. English or Persian Walnut.

b) *J. sieboldiana* Maxim. Japanese butternut.

c) *J. cathayensis* Dode. Chinese Walnut.

d) *J. mandschurica* Maxim. Manchurian Walnut.

New World Species

a) *J. californica* Wats. Southern California Blacknut.

b) *J. cinerea* Linn. American butternut.

c) *J. hindsii* Japan. northern California Blacknut.

d) *J. nigra* Linn. eastern Black walnut.

Brief description of important species is dealt below

1) *Juglans sieboldiana* Maxim. Japanese butternut is native to Japan and central Asia. Trees are tall, 60 feet, round topped, leaves are very pubescent on the under surface and glandular. Fruit ovoid to globose, nut thick shelled very rough, suture thick, ovoid to subglobose. Has one botanical variety *J. sieboldiana* var. cordifornis.

2) *J. cathayensis* Dode. Chinese Walnut. Native of central China. Trees are tall growing 75 feet or more. Both leaves and branches are pubescent and glandular, leaflets 9-17. Fruit and nut ovoid pointed, shell thick and kernel is small.

3) *J. mandschurica* Maxim. Manchurian walnut. Distributed widely in Manchuria. Tree tall growing, 60 feet, round topped head, leaves, petioles heavily pubescent and glandular beneath and shiny above. Fruit ovoid or subglobose. Nut ovoid apex pointed.

4) *J. californica* Wats. southern California black walnut is native to coastal southern California. Plants are shrub-like or small tree, crown wide spreading, suckers heavily with several branches at or near the base. Fruit small, globose, smooth, thin to thick shelled, hull adhering. Nut globose.

5) *J. hindsii* Jeps. northern California black nut is native to central and northern California. A tall erect tree 90 feet, bark smooth leaves somewhat smaller. Fruits subglobose with soft pubescence. Nuts nearly globose thick shelled with adhering hull.

6) *J. nigra* Linn. eastern Black Walnut is indigenous from Atlantic Ocean to Texas, Kansas, Nebraska, Great Lakes region to Gulf of Mexico. Trees are tall growing140 feet or more, trunk straight and large, remains dormant for long and is first to shed the foliage. Fruit large globose, husk pubescent. Nuts are thick, hard brown or nearly black, rough shell ovoid and pointed.

7) *J. cinerea* Linn. Butternut is native from Georgia, Arkansas to New Brunswick, is most cold hardy specie. Plant slow growing wide spreading may attain a height of 100 feet or more, bark deeply fissured. Fruit ellipsoid bears in clusters of 2-6. Nuts are elongated, ovoid - oblong, has very hard rough shell with adhering hull.

Walnut trees grown in India are mostly of seedling origin and are named locally. In Jammu and Kashmir, walnut exhibits tremendous variation for the characteristics tolerance to pest and diseases, bearing potential, nut and kernel quality. These collections need to be evaluated systematically for selection of superior types.

Description of some cultivars grown world wide is given below.

Hartley : Trees are moderate to large in size and leaf opening takes place 12-15 days after in 'Payne'. Is a commercial cultivar of California and is of seedling selection, performs well under deep drained and irrigated conditions. The nuts are large, base broad apex pointed, endocarp or shell light coloured thin well sealed, kernel light in colour. Starts bearing late and production increases with age of plants. Tolerant to codling moth and blight.

Payne : Popular Californian cultivar originated as seedling selection. Tree has moderate vigour, round topped, very productive because the lateral buds are fruitful, pollen and stigma matures together hence can be grown in a single block. Nuts are small to medium, well sealed. Susceptible to codling moth and blight.

Franquette : An old and commercially grown cultivar of France. Leafing out is late, escapes spring frost, late in bearing, laterals unfruitful. Trees are upright and large. Nuts are small, shell seal good, kernel light in colour quality good, maturity is late.

Ashley : Seedling selection, parentage unknown, Persian walnut. Trees are smaller, starts bearing early, medium in chilling, pollen shedding coincoides stigma receptivity of pistillate flowers. Nuts are large ovoid, shell seal fair, ripens early kernel light tan in colour plumb about 54% of nut, quality and flavour good. High yielding, about 90% of lateral buds are fruitful.

Chico : Originated as a cross between Sharkey x Marchetti. Persian nut with moderate plant growth about 80% lateral buds produce pistillate flowers, pollen shedding is late and does not synchronize with its pistillate flowers. Nuts are round, small-medium, shell seal fair to good, early to mid season in maturity, kernel about 49% of nut, light in colour quality excellent.

Serr : Originated in California, Persian nut type Trees are very vigorous and grows very fast when young, leafing out about one week after Payne, adapts well to warmer areas. Nuts are large, matures early to mid season, shell thin, seal fair to good, cracking quality good. Kernel quality good, light coloured, resistant to sunburn. About 50% lateral buds produce pistillate flowers.

❑❑❑

46
Pecan

Genus : *Carya*
Species : *illinoensis (Wangenh) K. Koch.*
Family : *Juglandaceae*
Chromosome No. : 2n = 32

Unlike other temperate nut fruits which have been in cultivation for long history of pecan cultivation is just about one hundred years, and in this short period pecan nut has gained huge popularity on account of its sweet nutty flavour and superb tree characteristics that is symmetrical form, luxuriant dark green shinny foliage with added ornamental value. Pecan nut is a remarkable highly valuable gift of North America to the horticulture of the world, and presently occupies fifth position among the leading nut fruit crops. In United States of America, pecan is recognized as 'Queen of Nuts'.

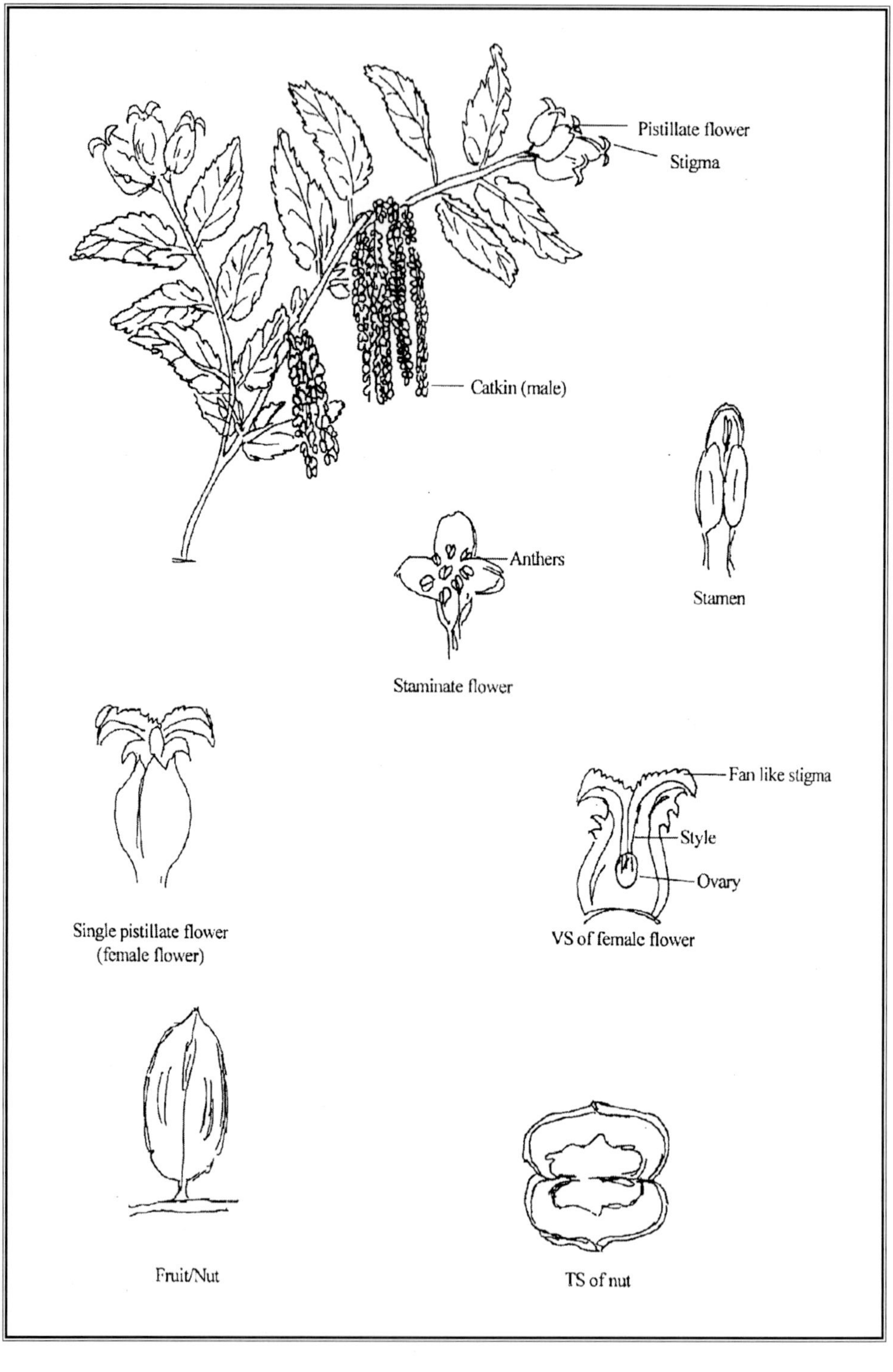
Pistillate flower
Stigma
Catkin (male)
Anthers
Stamen
Staminate flower
Fan like stigma
Style
Ovary
Single pistillate flower
(female flower)
VS of female flower
Fruit/Nut
TS of nut

Like other nut crops pecan is also a rich source of fat, protein, carbohydrate and minerals. Pecan is regarded as a high source of nutrition and energy giving nut fruit which generally is either roasted or salted and is taken in normal diet.

Although pecan is native of N. America, but some do not agree with this contention and believe that it must have been originated further north and the seed was probably water borne to lower region down south where it spread over a wide region. The Red Indians living there and the north American crows were instrumental in further spread of pecan and as of today besides the USA, pecan cultivation is mainly confined to Canada, Australia, Egypt, India, Israel, Mexico, Peru, Morocco, Turkey and S. Africa because initial pecan growth is very slow.

Pecan was known by several old names like *Juglans pecan, J. allionensis,* Carya pecan, Hickory pecan etc., but today these names are no more accepted and botanically pecan is *Carya illinoensis* (Wang.) K. Koch., member of genus *Carya* and belongs to the family Juglandaceae. The genus has twenty species and except four all are native to the Americas. Of these only pecan is important commercially while the other species like *C. ovata* (Shagbark), *C. laciniosa* (shellbark) and *C. tomentosa* (mokernut) also produce kernel which are sweet in taste, but being thick shelled are poor in cracking quality.

Origin and Distribution

Economically pecan is regarded as the most significant member of *Carya* and is considered as the most valuable nut crop of North America. Genetic base being narrow improved trees were selected which formed the basis of future tree selection. From the native pecan ranges extending from Canada Ontario, Atlantic coast of Virginia, California, Mexico (south of Oaxaca) its cultivation has spread all around the world i.e. Argentina, Brazil, Peru, Egypt, Australia, S. Africa, Israel and number of other countries.

Uses

Important for its highly valued edible, sweet, tasting flavoured edible nuts, seed/kernel eaten raw or consumed after being cooked or roasted, mostly used in a large number of processed products i.e. candies, chocolates, ice-cream, baked goods etc., oil used in soap and perfume products.

Description

Pecan trees are moderate to tall growing (60m) monoecious with dark coloured hard wood and deep tap roots, crown open, spreading at the top, trunk large straight round with longitudinal furrows, bark-greyish brown to light brown when young, later turns dark-reddish brown, develops irregular ridges, bark thick with prominent lenticels. Branches erect to slightly spreading stout, branchlets usually stout, light brown and thickly pubescent or hairy when young and become reddish brown and smooth with many elongated orange brown lenticels. Leaf scar 3 lobed large and elongated, winter buds flattened at base pointed at top, usually two at each node yellow orange-brown pubescence or hairy. Leaves compound alternate on branchlets with 9-17 leaflets each measuring 4-8 inches in length and 1-3 inches in width, usually lance shaped dark green above, margin crenated curved, apex long pointed, leaf lamina base unequally rounded, leaf curved, margin coarsely toothed, dark green above, smooth sometimes with slight pubescence, lower surface paler, more pubescent, pubescence more along mid rib and veins. Flower male (catkins) slender appears on previous season growth laterally. Usually in cluster of three, light green measures about 4-6 inches in length each perianth segment/cluster with 5-6 stamens. Female appear in few to several flowered cluster terminally on current season growth, individual flower small yellow pubescent, 4 angled or lobed involucre enclosing 1 celled ovary, stigma finely divided. Fruit oblong to globular broad at base pointed at apex, clustered in groups of 3-6, husk thin narrow 4 winged and 4 angled dark brown. Husk splits along wing or suture to expose the nut. Nuts oblong to cylinder shaped, broader at base apex short pointed brown to reddish brown with few black spots on the thin soft smooth shell, kernel is sweet and with more oil content.

Other important species of pecan with short pomological description is detailed below.

C. ovata (Mill) K. Koch. (Shagbark). Plant 40 m. tall, trunk shaggy when old, with grey bark that splits into long plates. Leaves large shiny has 5-7 leaflets 10-15 cm long, dormant buds are large and overlapping, late to break dormancy, growth ceases 7-8 weeks thereafter. Flavour and quality highest among native hicory nuts. Nuts are small, whitish distinctly 4 angular, thin shelled hull thick, opens to the base, shell has longitudinal ridges.

C. laciniosa (Michx.f.) Loud. Shellbark hickory, resembles Shagbark hickory, that distinction between two is sometimes very difficult. Leaves are longer (30 cm) with 7 pairs of leaflets, hull large (7.5 x 3 cm). Nut yellowish white, four angled does not open along suture lines, kernel not well filled, sweet in taste.

C. cordiformis (Wang.) K. Koch. Swamp hickory. Tree about 30 m. tall, foliage fine, leaf has 5-9 leaflets, fruits in clusters of two or three rounded 2 to

4 cm in length. Suture flange towards the apical end and hull splits only about half the nuts length. Shell and hull thin, kernel poor in quality due to astringency and bitterness.

C. glabra (Mill) Sweet pignut, *C. ovalis* (Wang) Sarg. Sweet pignut. Red hickory. Both trees and nuts are similar in both the species, kernels are sweet in latter and rarely so in former specie. Trunk shaggy, 5-7 leaflets (3-7 in. long). Nut small, shell and hull thin. The hull splits to half way in *C. glabra* and to nut base in *C. ovalis*.

C. tomenstosa (Lam.) Nutt. Mocker nut. Trees are large well shaped, produces dense shade, bark grooved, without scales. Leaflets 7-9 pairs 8-18 cm long, dark green, pubescent. Hull thick splits to the base, nuts pear shaped thick shelled, sweet and edible, difficult to shell out as halves.

C. myristicaeformis (Michx. f) Nutt. Nutmeg hickory. Tree 30 m. tall, branches with yellow brown scales which turns to reddish brown. Leaflets 5-11, stalk solitary. Male catkin pedicelled, fruit solitary, reddish brown with irregular spots and stripes, shell thick, kernels are sweet in taste.

C. cathayensis Sarg. Cathay hickory. Native to eastern China. Trees are small (20 m), with 5-7 leaflets 10-14 cm long. Fruit ovoid small four angled, kernel sweet.

S. aquatica (Michx. f.) Nutt. Water or bitter hickory. Plant medium tall (30 m) bark light brown splits into long thin plates. Buds are dark reddish brown in winter, leaflets 7-13, sessile, 8-12 cm long. Nuts borne in clusters of 3-4, ovoid (2.5 to 3.5 cm) hull thin, splits completely, four angled dull reddish brown in colour, kernel bitter in taste.

Like walnut much of the pecan plantation comprises of seedling selection. However seedling plantation are now gradually decreasing with the advent of new improved cultivars. The need of cultivar for a particular site is based on market needs, chilling and maturity time and incidence of diseases etc.

Salient features of some cultivars are described below :

Cheyenne : Originated in Texas from a cross between Clark x Odom. Trees are tall, crown round, leaf oblique ovate dark green above pale below, highly pubescent beneath, very precocious and productive protandrous type. Nut medium in size with round ends, apex short pointed, brown in colour with some dark stripes or spots at apical end. Kernel, bright light brown in colour crisp in texture, somewhat wrinkled highly flavoured, lie loose in shell, shelling 57-61% of nut.

Mohawk : Is of hybrid origin, Success x Mahan. Tree vigorous, branches spreading, leaves large, catkin shed pollen when pistillate flowers are receptive. Nut large somewhat elongated, shell thin attractive with conspicuous markings, matures early. Kernel surface smooth, grooves shallow parallel, shelling 60% of nut, septum separates easily from kernel, quality very high.

Mahan : Parentage unknown. Plants have vigorous growth, starts bearing early and heavily. Nuts are long and large, shell thin filling poor on old trees, shelling nearly 60%.

Shawnee : Obtained from seedling of Schley x Barton. Tree medium tall very productive, vigorous, starts bearing early leaves medium dark green, flower protogynous. Nuts somewhat elongated, rounded at center and base, apex flat or short pointed, shell thin light brown with dark stripes, septum loose, kernel smooth bright in colour keeping and eating quality very good, flavour excellent.

Stuart : A very old and famous cultivar takes long to bear. Trees are moderately vigorous, leaves medium in size. The nuts are large in size light brown in colour, kernel light brown and constitutes 45-50% of nut, quality high very good for in -shell trade. Being protogynous it should be planted with protandrous cultivar to have good yields.

Desirable : This cultivar is of hybrid origin, Success x Jewett. Plants being small in size and compact are useful for high density plantation, starts bearing early, produces regularly and prolifically, scab resistant. Nuts are larger than Stuart, kernel quality very good and average shelling is 52 per cent of nuts, meaty.

❑❑❑

47
Chestnuts

Genus : *Castanea*
Species : *avellana*
Family : *Fagaceae*
Chromosome No. : 2n = 24

Before the history was started to be written chestnut nuts in Japan and China were harvested, collected and eaten and aptly has been an important food for man around the world especially in Asia, Asia Minor, Europe and North America for hundreds of years. As early as 50 B.C., chestnut was brought to southern Europe by Greeks from Asia Minor and red Indians in North America prior to its discovery. Used nuts of chestnut was a staple food especially during famine times.

Genus Castanea has 13 species and belongs to the family Fagaceae which also has important forest plants like Oak (Quercus), and beech (Fagus). Species

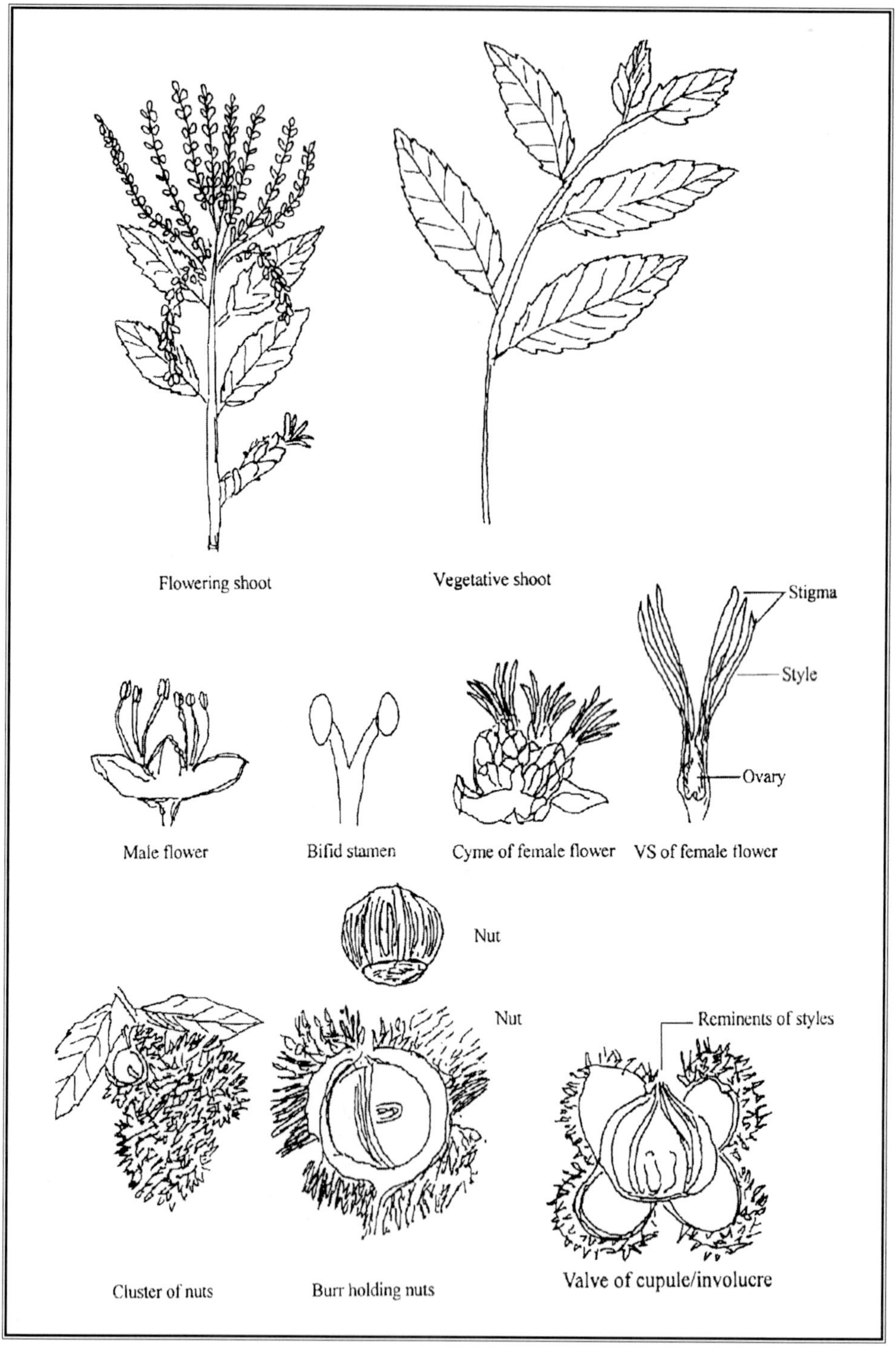
Flowering shoot
Vegetative shoot
Stigma
Style
Ovary
Male flower
Bifid stamen
Cyme of female flower
VS of female flower
Nut
Nut
Reminents of styles
Cluster of nuts
Burr holding nuts
Valve of cupule/involucre

of this genus are grown for ornamental and also for their edible nuts. Three species of chestnut are economically important as they produce edible nuts, are highly variable and selection has been practiced for hundreds of years to select high yielding, large fruited type. The species other than C. *dentata* (American chestnut) are called 'Chinkapins'. Chestnut is also unique among other nut trees, as it has very short juvenile period, takes as little as three years from seed to fruiting.

Origin and Distribution

The family comprises of 5 genera and about 350 species. Widely distributed in the temperate zone of the northern hemisphere, particularly in eastern region of north America, southern Europe, northern Africa and Asia.

Uses

Important for its edible nuts and also for durable wood. The American chestnuts produce excellent large sweet nuts used in confectionary etc. The nuts are palatable raw, usually consumed after cooking, often roasted in shell, boiled or steamed. Often used as chestnut flour and ground flour, dried nuts used in bread and pastries.

Description

Chestnut plants are large trees (30 or more m. tall), deciduous, trunk straight rough fissured, bark dark brown separating into broad flat ridges, broad rounded crown, branches stout, brown and rough, yellowish green pubescent when young later turns dark brown and smooth. The terminal winter buds are absent, lateral buds are small usually broader at base, slightly or short pointed at tip, usually covered with 2 or 3 overlapping dark brown scales (protects buds in winter). The leaves are simple, deciduous, alternate usually broadest near the base, pointed at the tip, with large coarse sharp pointed teeth on the margin, thin and papery, leafstalks stout, slightly angled (1-1.5 cm long). Flowers are unisexual in pendulous compound catkins may be solitary or clustered. In some species of *Castanea* male and female catkins occur at different region of the same plant, usually androgynous, male appear at apex and female at the base of the inflorescence. Male flowers are densly clustured on upright to spreading catkins, white to cream, strong scented, has 4-7 imbricate scaly perianth leaves with 4 to many free stamens, occur at the axil of a bract with bracteoles which forms partial inflorescence, later are disposed in pendulous catkins thereby inflorescence becomes a compound catkin of cymes, filament filiform, anthers bithecal split longitudinally, pistillode present in male flower. Female flowers are scattered at the base of the uppermost catkins on the branchlets, flowers usually 6, sometimes 4, perianth leaves small scaly, epigynous ovary inferior, 3 carpellate, trilocular, each loculus has 2 ovules (pendulous anatropous), style 3. Fruit mature in one growing season, therefore appear on current year branchlets. Fruit one seeded nut surrounded by enlarged persistent hard spiny bracteoles which forms the cupules 4 parted husk that surrounds 1-3 nuts. The nuts are globose shaped slightly compressed, pointed at the tip, shiny leathery, bright chestnut brown in colour.

Important species of chestnuts are

1. *Castanea dentata* Borkh. American chestnut.
2. *C. pumila* Mill. Eastern Chinkapin.
3. *C. crenata* Sieb and Zucc. Japanese chestnut
4. *C. sativa* Mill. European Chestnut and *C. mollisima* Bl. Chinese chestnut.
5. *C. mollissima* Bl. Tree about 60 feet tall, leaves 6 inch long coarsely toothed, pubescent below. Nut 2-3 inch long and1 inch across, blight resistant specie.
6. *C. sativa* Mill. Spanish, European or Eurasian chestnut. Trees are tall growing, 100 ft., leaves 8 inch long, round or broad at base, teeth coarse and spreading, pubescent below. Nuts 1-3 inch long and 1 inch across. Blight susceptible.
7. *C. crenata* (Sieb & Zucc.). Japanese chestnut. Plant medium tall, 30 feet, leaves 6 inch long, closely toothed, with rounded sinuses, tomentose below. Nuts 2-3 inch long and 1 inch across. Blight resistant specie.
8. *C. dentata* (Borkh.). American chestnut. Plant tall growing 100 feet or more, leaves 10 inch long, narrow at base, coarsely toothed, pubescent below. Nuts 2-3 inch long and 1 inch across. Hardiest of all chestnuts and produce high quality nuts. Susceptible to blight.
9. *C. pumila* Mill. Tree small, 45 feet tall. Leaves 5 inch long, coarsely toothed tomentose below. Nut generally solitary and small ½ inch across.

Pomological significant traits of some chestnut cultivars are detailed below.

Manoka : Originated in British Columbia Canada. Evolved from open-pollinated seedlings of the Chinese chestnut (*C. mollissima*). Tree very hardy, more like timber type. Nuts medium in size, shell dark brown, kernel yellow, well flavoured, quality good, husk small and thin.

Kelsey : Originated at Connectcut from the seedling of *Castanea mollissima.* Tree of Chinese type i.e. light coloured scaly bark, twigs silvery, bark on large branches rough free of blight. Nut small very dark brown in colour, squarish, pellicle separates readily, kernel yellow sweet firm, crisp, crunchy.

Orrin : Originated in Lockhaven Maryland. Tree precocious blight resistant has short growing season hence suitable for more northern areas. Nut medium large ,shell dark-mahogany sheen, shiny, attractive, but with some pubescence. 2-3 nuts per burr which is uniform in thickness, not wedge shaped, keeping quality very good compared to other cultivars.

Toumey : Evolved from a complex cross between different species *Castanea crenata, C dentate* and *C. mollissima.* Plants are fast growing upright Chinese type but not very fruitful. Nut medium in size, bay to chestnut in colour, darker on ridges, shape variable, plumb fair in flavour.

Crane : Originated in Maryland selection made from seedlings of *Castanea mollissima* nuts. Named in honour of Dr HL Crane. Trees are precocious, flowers in mid season cross compatible with 'Nanking', slightly protandrous. Nut small-medium in size, shell dark cherry red, almost completely shiny, seal good, kernel light in colour, firm crisp, eating quality excellent, keeps well compared to other cultivars.

Jersey Gem : Evolved from a cross Origin x Nanking. Tree moderate in vigour, very productive. Nuts are medium sized, spherical shell shiny red mahogany in colour, seal good, kernel sweet in taste, crisp crunchy yellow with very good flavour, pellicle removes easily, matures late and in short time all nuts drop.

Revival : Trees are upright, vigorous, upper branches spreading, crops regularly every year. Foliage dark green and shiny. Nuts are large, dark reddish brown in colour about 25 g in weight, 2-3 nuts per burr are seen, pellicle removes easily from kernel which is sweet in taste, responds to shake harvest.

Au-Cropper : A new chestnut cultivar, originated at Auburn University Alabama. Plant, Chinese type resistant to chestnut blight, produces consistently and high yields. Nuts are dark chocolate-brown in colour, with 2-3 nuts per burr, which opens well. Each nut weighs about 10-12 g, has midseason maturity.

48 Hazelnut/Filbert

Genus : *Corylus*
Species : *avellana L.*
Family : *Betulaceae*
Chromosome No. : 2n=2x=28

European filbert *Corylus avellana* L, is known by several names common being Cob, cobnut, hazel, Spanish or Lambert nut, which has been due to its origin, fruit shape, husk and the general appearance. 'Filbert' has also come from name 'full beard' as the nut of commerce is fully enclosed by the long husk. It is this husk or involucres which very often has been used to distinguish between various species and cultivars. *Corylus* the genus has been derived from *korys* which indicate bonnet, hood or helmet and there are several other views of its name being derived after the name of some renowned places. The word 'hazelnut' is from anglosaxon *haesel* meaning hood. Further the term hazelnut

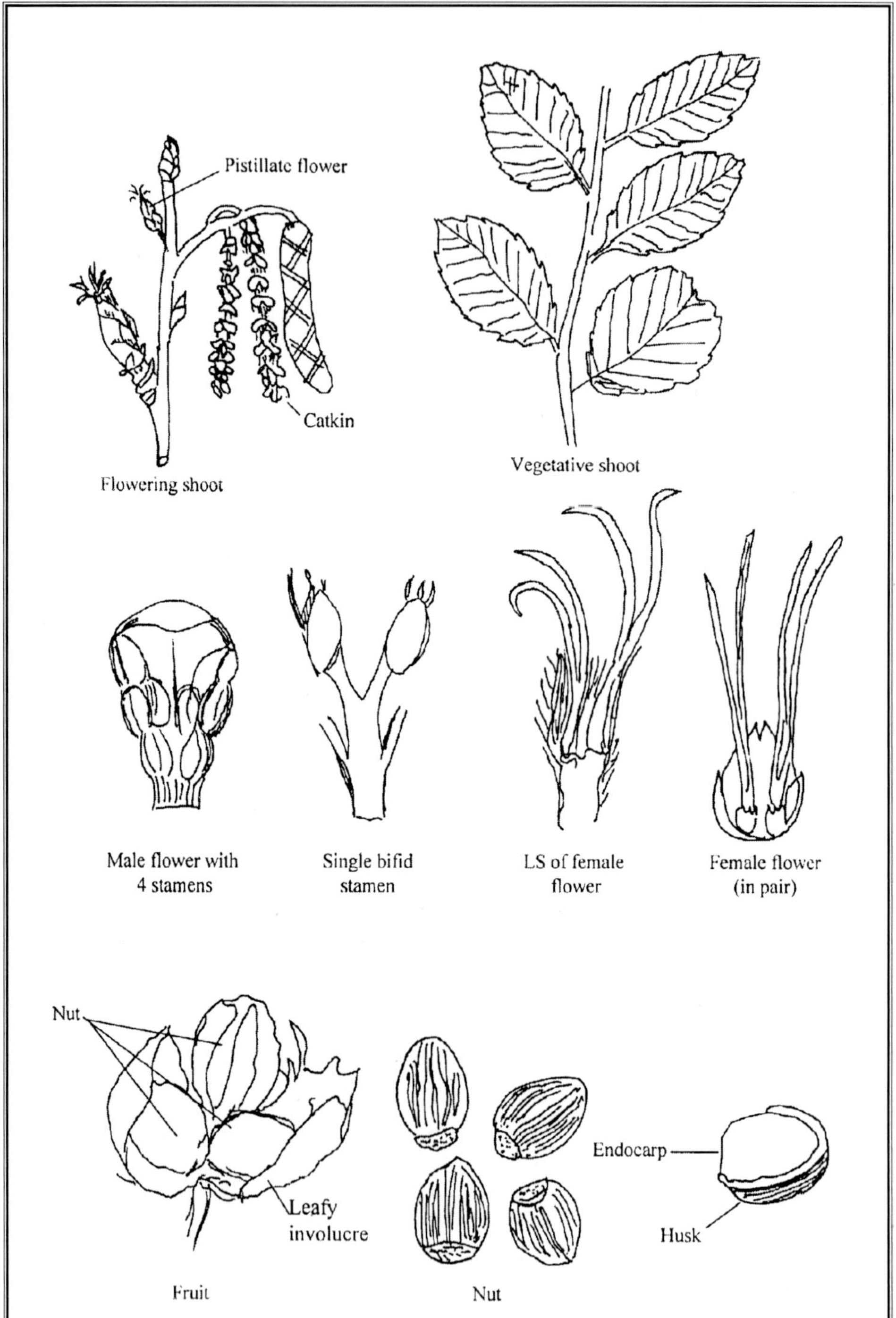
Pistillate flower
Catkin
Flowering shoot
Vegetative shoot
Male flower with
4 stamens
Single bifid
stamen
LS of female
flower
Female flower
(in pair)
Nut
Leafy
involucre
Fruit
Nut
Endocarp
Husk

is used in European and Filbert in Asian countries for and its other species. However in America, the term is used to refer to three native species *C. archericana, C. californica* and *C. cornuta*. Filbert is an important member of birch family, Betulaceae which has a long history of cultivation (8000 to 5500 B.C.). The plants of genus *Corylus* were commonly called as hazel nut or filbert at early times, but on the basis of variation in husk, distinction has been made between the two. Filberts from very early time were usually recognized as one having long husk whereas hazel with short husk. However this view is no more held true as *C. maxima* and *C. avellana* cross freely and the cultivars grown today is the mixture. *C. avellana* is the most important specie grown in Europe, whereas *C. maxima* is native to south eastern Europe and western parts of Asia.

Although it is widely distributed throughout the world but performs best in areas with moderate climates. The filbert plants are not classed as large trees, usually round topped and is multi stemmed. All species of filberts are deciduous, however in some species the leaves are seen on the branches during winter also particularly in areas that have moderate climate.

Origin and Distribution

Various members of birch family Betulaceae are widely distributed in northern Hemisphere especially in the temperate region, is regarded to be native to Europe and south western Asia. Ancient records indicate its existence in the glacial period (17000-18000 BC) in British Isles, northern Europe and Scandinavia and its cultivation area is bounded by Atlantic coast of Europe in West to North Africa, from Morocco to Norway and in East to Soviet Union. Southern boundaries spread from Morocco, Algeria, and Spain though Italy, Yugoslavia, Greece and Turkey to northern Iran and Transcaucasia.

Uses

The tasty edible nut is highly valued in commerce for their pleasant taste, high food value for its distinctive flavour and taste, kernel or seed extensively used in baked products, candies, ice-creams, pastries and whole lot of confectionary products.

Description

'Hazel' is used for nuts of all Corylus spp., collectively all over the world whereas the term 'filbert' originated in England and was employed for long husked hazel types (*C. avellana*) to separate them from short husk types. Various members (species) of *Corylus* are typical deciduous perennial shrubs or small trees 4-6 m. tall. The branches are some what drooping/willowy which ultimately form very dense zigzag thicket. Tree bears globose to ovoid buds about ¼ each long, terminal buds are missing. Leaves are thick some what rough, alternate to ovate heart shaped, slender or thin, about 10cm long with prominent ribs, margins toothed, veins are depressed on the surface some purplish beneath, very pubescent below more along the mid rib and veins. Male catkins are yellow about 6-10 cm long and appear laterally on previous season wood as drooping inflorescence. Female flower occur terminally on current season growth and are head like and enclosed in a small scaly bud with 3-4 red styles projecting outward. Fruit is an ovoid nut which is surrounded by leafy tubular involucre/husk borne in clusters and ripen in autumn. The nut is highly valued.

The more important species of filbert are *Corylus americana* Marsh, *C, avellana* L., *C. chinensis* Franch, *C. colurna L., C. cornuta, C. heterophylla, C. maxima, C. sieboldiana* and *C. ferot*,which are briefly described as.

Corylus americana, (Marsh). American filbert. Habitation ranges from Saskatchewan to Maine, Missouri, Minnesota, Oklahoma and Florida. Trees are shrub like-shrubby, fruits medium in size, kernel edible, nuts covered by husk which is double the length of the nuts.

C. avellana, L. Is distributed widely throughout Europe. The four major production areas are marked by large water bodies i.e. Turkey by Black Sea, Italy and Spain by Mediterranean and Oregon by the Pacific Ocean. Is known as European filbert and is the most commercially grown specie. Its crosses with other species resulted in wide variability and produced many present day cultivars.

C. chinensis Franch. Chinese hazel. Is native of Yunnan, Hupei and Szechwan and grows at an elevation of 2000 m. or more. Nuts are enclosed in husk which is tube like and deeply forked at the apex.

C. colurna L. Turkish hazel. Is native to south eastern Europe, northern Turkey, Caucasia, Iran and the Himalayas to China. Trees are tall, 20-40 m. in height. Young bark is light grey and deeply furrowed, still very soft, later turn rough and fall off in vertical plates. Nuts are small and thick shelled.

C. cornuta Marsh. Beaked filbert, is indigenous from Quebec to Georgia and through Saskatchewan to Missouri. Is very winter hardy tolerates - 50° C temperatures.

C. heterophylla Fisch. Known as Siberian or Japanese filbert, extends from Hupei, Shensi Kansu and Shansi to Manchuria, Japan and Korea, Plants are shrub like grows to 2-3 m. in height, leaves are truncated.

C. maxima Mill. Is native to southern Europe and Asia Minor, grows upto an elevation of 1300 meters. As the ranges of *C. maxima* and *C. avellana* overlaps with no genetic blocks, many of presently grown cultivars are the result of natural interspecific cross.

C. sieboldiana Bl. Manchurian or Japanese filbert. The name also suggests the range where they are native of Plants are shrub like, husk drawn out like a tube above the nut tip and for this character resembles *C. cornuta*.

C. ferox. Wall. Himalayan filbert is indigenous to northern India, Nepal and Sikkim at 3000 m. or more height. Leaves are typically ovate-lanceolate in outline, nuts are small quite thick shelled. Nut clusters are like that of chestnut burrs.

C. tibetica Batal. Tibetan filbert. Indigenous to China. Plants are shrub like or a bush of 3-7 m. in height. Husk markedly spiny like the burrs of chestnut.

Pomological description of some commercially grown important filbert cultivars are briefly described below.

Cultivars

Barcelona : Cultivar is regarded to be the most important as large plantation in Europe are under this cultivar. Is known as coutard hazelnut in France. Late maturing mid- september, damaged by frost produces good yields, sometimes erratic in bearing. If there is cold during June percentage of empty nuts is very high. Good pollinizer for Segorbe and is effectively pollinated by 'Segorbe', 'Butler' and 'Merveilte de Bollwiller' at start, full bloom and end of flowering respectively.

Butler : Is of seedling selection parentage not known. Cultivar is very productive and is highly compatible with 'Segorbe' and number of other cultivars but has pronounced biennial bearing.

Ennis : Parentage unknown, obtained from seedling selection. Plants are fast growing and resistant to cold. Low tendency towards biennial bearing compared to Barcelona, high cropper, nuts are large light brown each about 4-5 g, occur singly or in pairs, husk very small. Large quantity of defective pollen is produced by small catkins (male) which is incompatible with Barcelona.

Skinner : Cultivar is of hybrid origin, Dropmore x seedling of Italian Red. Tree very hardy late ripening-September end, produces nuts which are large more wide than long, shell quite thick of chestnut colour smooth, kernel sweet plumb, crisp resembles that of Barcelona, with very pleasant flavour.

❑❑❑

49 Pista

Genus : *Pistachia*
Species : *vera L.*
Family : *Anacardiaceae*
Chromosome No. : 2n = 30

Among the nut crops Pistachio nut (*Pistacia vera* L.) occupies a significant place for its delectable and nutty flavoured kernels which in large quantity is imported to India mainly from Iran and Afghanistan. On account of its demand for very exacting climatic requirements the crop has failed to establish successfully in number of new regions where efforts were made to introduce this crop, despite the fact that it resists drought and adverse soil conditions much better than more important fruits of the family Anacardiaceae. Pistachio trees are long lived but grow very slowly in early years of its plantation

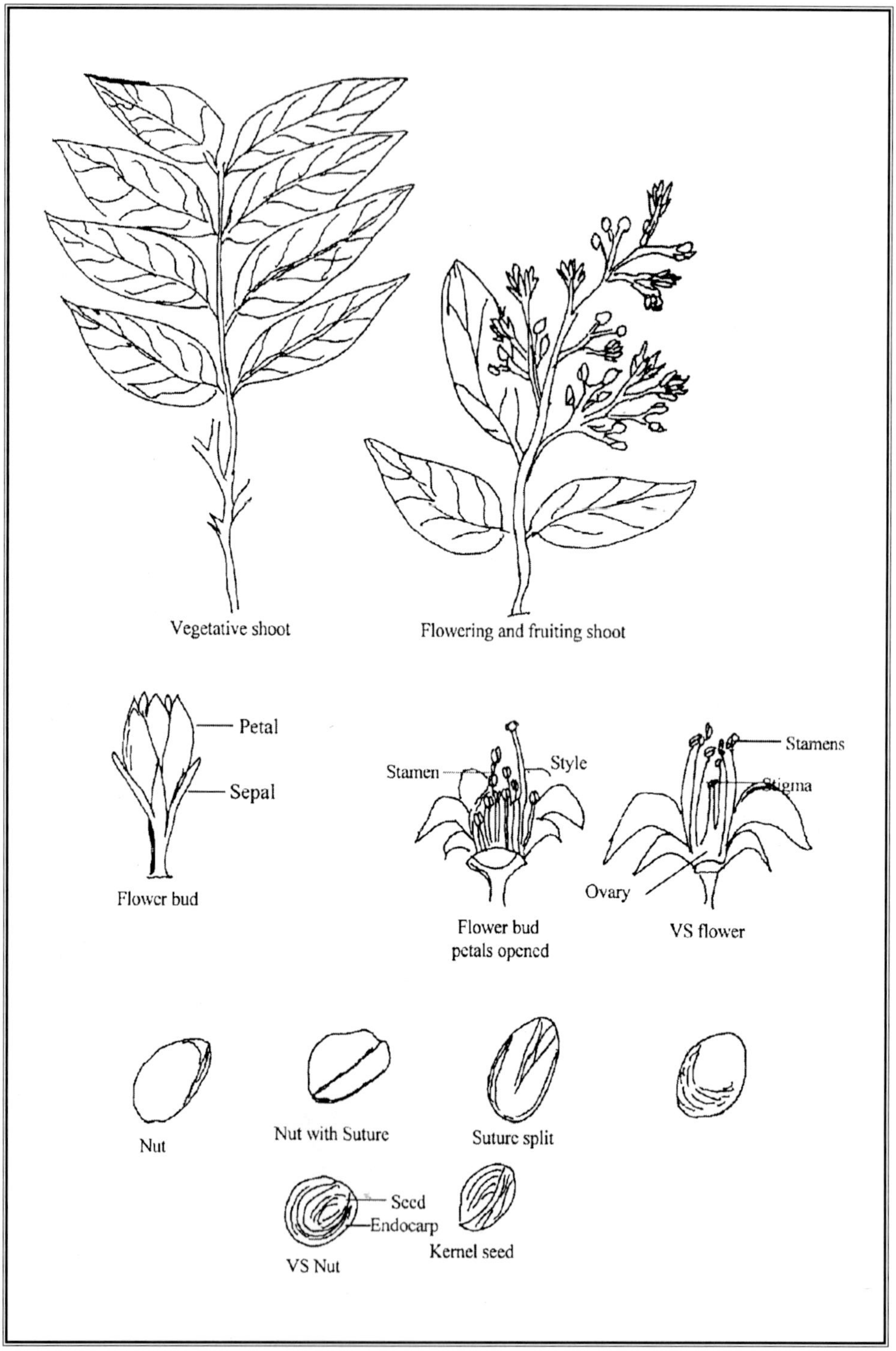
Vegetative shoot
Flowering and fruiting shoot
Petal
Sepal
Flower bud
Stamen
Style
Flower bud
petals opened
Stamens
Stigma
Ovary
VS flower
Nut
Nut with Suture
Suture split
Seed
Endocarp
VS Nut
Kernel seed

compared to many of the deciduous fruit trees. Alternate bearing is also a wide spread phenomena in this crop. Pista is a fruit of temperate region and can withstand wide fluctuation in temperature. The exact place of its origin though not definite, is often regarded to be indigenous to central Asia, where large stands of wild pistachio trees are found growing in abundance. Its cultivation started first in Iran, Turkey and Syria and was later introduced in Mediterranean region of Europe by the turn of the first century. Wild pistachio trees are also found in Afghanistan and Turkmenistan.

In the international market price determining factor for pistachios is primarily based upon the green colour and flavour of the nuts. Dark green, large nuts with pleasant flavour fetches higher price compared to small nuts which are used mainly by pastry and ice cream industry.

Origin and Distribution

Pista is regarded to be native of eastern Mediterranean region, Afghanistan, Iran and parts of central Asia. Presently it is cultivated commercially in Iran, Afghanistan, Pakistan, India, Syria, Turkey, Lebnon, and on small scale in California.

Uses

Pista nuts (kernels) are highly rated for its delicious flavour, sweet kernel usually used as dessert, roasted, salted, employed as an important ingredient in various confectionary and ice cream, husks made into marmalade and organic fertilizer. Pista oil used as spice oil and in medicines. Leave used to feed cattle and wood for furniture and ornamental work.

Description

Plants are dioecious trees or shrubs, grows upto 10 m. in height, spreading, branches also spreading, leaf imparipinnate compound, leaflets broadly ovate to lanceolate smooth green above light pale beneath, leathery/coriaceous, pubescent when young, glabrous later. Leaves of pista contain irregular gland which contain tannins that are used in dyeing and tanning industry. Young and mature leaves yield shikimic acid. Flowering occurs in lateral panicles, with the young leaves. Flowers are small sized, red unisexual, calyx small 5 lobed, petals absent/wanting, stamens 5-7, carpel syncarpous and three in number. Fruit is a compressed dehiscent drupe, oblong linear to globose, endocarp bony in texture, greyish white separates readily from the husk and splits along ventral suture, seed or kernel light yellow to deep green coloured with reddish covering on the seed.

No other specie, other than *P. vera* is important from horticulture point of view.

Description of some important cultivars is given below.

Kermen : Widely cultivated in California and is the backbone of pistachio industry there. Trees are vigorous in nature, branches upright high cropper, produces large and excellent quality nuts, kernel white crisp firm, but suffers from biennial bearing, blank nuts and low natural nut splitting.

Joley : Selected at California from open pollinated seedling. Nuts are of small size, tendency of blank nuts is low, nut splitting is very good compared to Kermen, but flowering and maturity are earlier than Kermen.

Sfax : Suitable for warm temperate places as its chilling requirement to break the winter dormancy is low compared to Kermen. Selected in Tunisia, is very early maturing, nuts are smaller in size, quality and shell splitting is quite high.

Lassen : Obtained from seedling selection at California from seeds which were brought from Rafsinjan, Iran. Late flowering helps to escape frost. Nuts are bigger in size, shell splitting is high and is opened by hand very easily, fairly high cropper, quality of roasted kernel is very good.

Being cross pollinated in nature some pollinizers have also been selected and are described briefly.

Peters : A well established and universal pistachio pollinizer. The cultivar produces large quantities of fertile pollen with the added advantage that the flowering coincoides with a large number of pistillate cultivars and that of Kerman too.

Chico : Obtained from seeds of *P. vera* at California which were brought from Syria. Like Peters it also produces pollen in abundance, flowers early in season, desirable pollinizers for Bronte, Red Aleppo and Trabonella but very early for Kerman. Chico and Peters in combination cover the flowering period of both early and late blooming cultivars.

□□□

50 Macadamia

Genus : *Macadamia*
Species : *integrifolia Maiden and Betch*
Family : *Protaceae*
Chromosome No. : 2n =28

Macadamia nut is known as 'Australian Nut' or 'Queensland Nut' because of its origin in the rainforests of New South Wales and Queensland. The crop is sensitive to frost hence the site which suits orange and lemon tree cultivation is also good for macadamia. Macadamia through cultivated commercially in Australia it is in Hawaii that maximum improvement and advancement in terms of selection of superior cultivars has taken place and in short time of about 100 years macadamia culture has become a very fruitful entrepreneur.

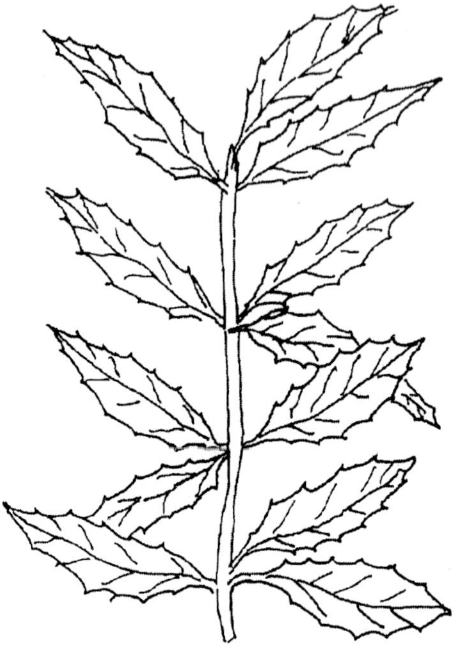

Vegetative shoot

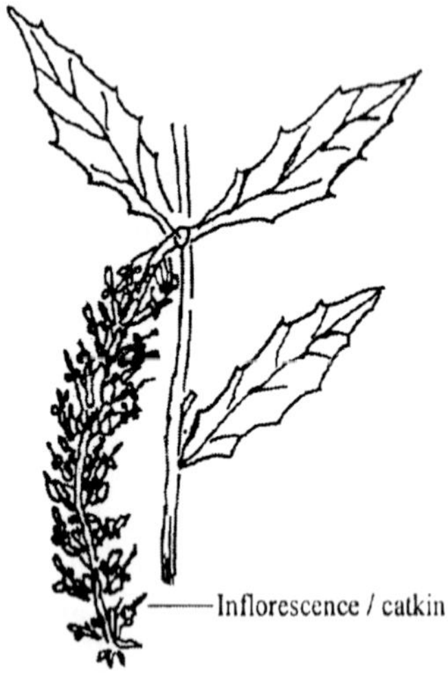

Flowering shoot

Mature nuts

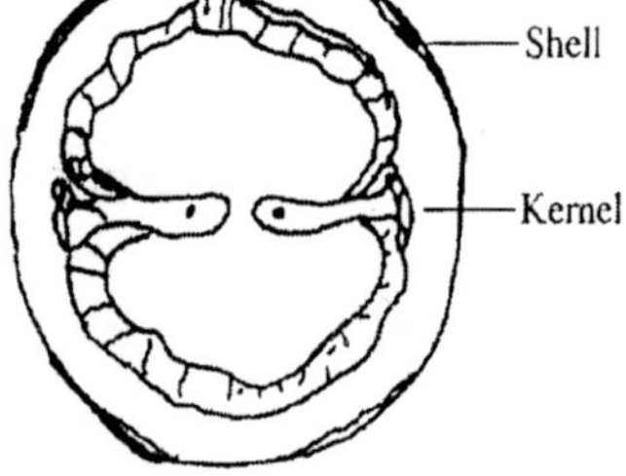

TS of nut

Presently there is huge gap between the macadamia nut demand and supply therefore efforts are being made to introduce macadamia into other tropical and subtropical parts of the world. Two species are commercially grown *Macadamia integrifolia,* the smooth shell type and *M. tetraphylla* rough shell type, former specie is considered to be commercially more important and most of the named cultivars belong to it. In texture and flavour, the finished macadamia product is rated as the finest confectionary nut.

Origin and Distribution

The crop has its origin in Australia, more precisely in the rain forests of the south-eastern Queensland and north-eastern New South Wales. The present day macadamia nut is credited to have developed in Hawaii, and from there the crop further moved to east Africa. Isolated trees of this nut are found in number of south-east-Asian countries.

Uses

Macadamia is regarded as one of the best dessert nuts on account of its flavour, crispy crunchy texture, dense cream colour. The nut attains high quality after roasting and tasteing done in coconut oil, used extensively in confectionary and processed food industries

The nut husk when decomposed adequately makes fine potting mixture.

Description

The macadamia plant in a large spreading tree (17x15 m), leaves generally glabrous, coriaceous, margins dentate, spiny during early stage of growth, later becomes smooth, each leaf axil has 3 buds of which the top sprout, making an acute angle with trunk. Inflorescence pendulous appear axillary on mature wood of current season growth, long (30 cm) with 100-500 flowers ,which are creamy white, occurs in groups of 2-4, measuring 1-1.2 cm in length, short pedicelled (3-4 mm), perianth tubular which is petaloid sepals, stamen 4 in number inserted above the base of the tube, pistil, long ovary, carpellate, 2 ovuled ,ovary superior. Fruit round or globose follicle (3-4 cm in diameter), pericarp thick (3 mm). Nut, thick hard, smooth, globular in shape, which encloses kernel, the edible part.

Species of Macadamia other than M. *integrifolia* and *M. tetraphylla* are

1. *M. terniifolia* F.J. Muell. Small fruited Queensland nut. Tree small 15 ft. tall and wide, new growth pink to red coloured. Leaves in whorls of 3 or 4, petioled, dense and serrate when young, lanceolate when old, 6

inch long with 8-10 teeth on each side. Flowers pink 50-100 flowered racemes, 2-5 inch long. Fruit pubescent, dehiscing when still on tree, kernel ¼ inch in diameter bitter inedible.

2. *M. tetraphyll* L.A.S. Johnson. Called by various names, Australia or Queensland Nut, rough shell. Tree 50-60 ft. high, branchlets dark coloured new growth pink to red rarely yellow green, leaves in whorls of 4 rarely 3 or 5, sessile serrate when young, at adult stage oblanceolate 4-20 inch long acute, finely serrate with 15-40 teeth on each side. Flower pink rarely white in 100-300 flowered racemes, 6-18 inch long. Fruit pubescent, dehisces when on tree, kernel ½ to 1.5 inch in diameter, sweet, edible.

Cultivars

Very few commercially grown macadamia cultivars are available but the ones described are commercially grown all over the world.

Kakea : An early selected cultivar, tree with moderate branching semi spreading, foliage clustered towards the terminals, less hardy under hot and dry condition, produces small nuts. Flowers early and for long duration. Nut medium sized brown round, shell somewhat thick, with few irregular spots, kernel white and round.

Keauhou : Very old Hawaiian cultivar. Tree large, branches well developed and spreading, leaves dark green, flowers early (2 weeks) than most of the cultivars, nut medium to large in size brown, spot buff coloured.

Keaau : Tall trees, upright suitable for high density plantation, crown rounded, branches medium sized. Produces excellent quality nuts, yields are high, shell rounded medium to thin, kernel white, round, nuts have tendency to mature and fall early.

Kau : Is a new cultivar like Keauhou but more hardy and upright produces 2-3 branches per node on young tree which requires thinning for good structure. In nut traits and productivity is similar to Keauhou but quality is better.

Mauka : A new cultivar from Hawaii suitable for higher elevations, resembles to Kau in tree characteristics, nut quality is good but with some kernel discolouration.

□□□

Section - IV
TEMPERATE FRUITS

51
Apple

Genus : *Malus*
Species : *domestica (Borkh.)*
Family : *Rosaceae*
Chromosome No. : 2n = 34, 51, 68

Like many other fruit crops, apple too is a very old crop in cultivation. Theophratus as early as third century B.C. has indicated that the Greeks and Romans knew apple culture and among the temperate fruit crops, apple is the most ubiquitous of all fruits and is extensively grown all over the world. Huge genetic variability present has permitted the better adapted types to be selected for plantation under different agro climatic conditions. Through continuous selection both from wild population and also from hybrid seedlings, varied types or varieties have been selected, some are suitable for cultivation in warm

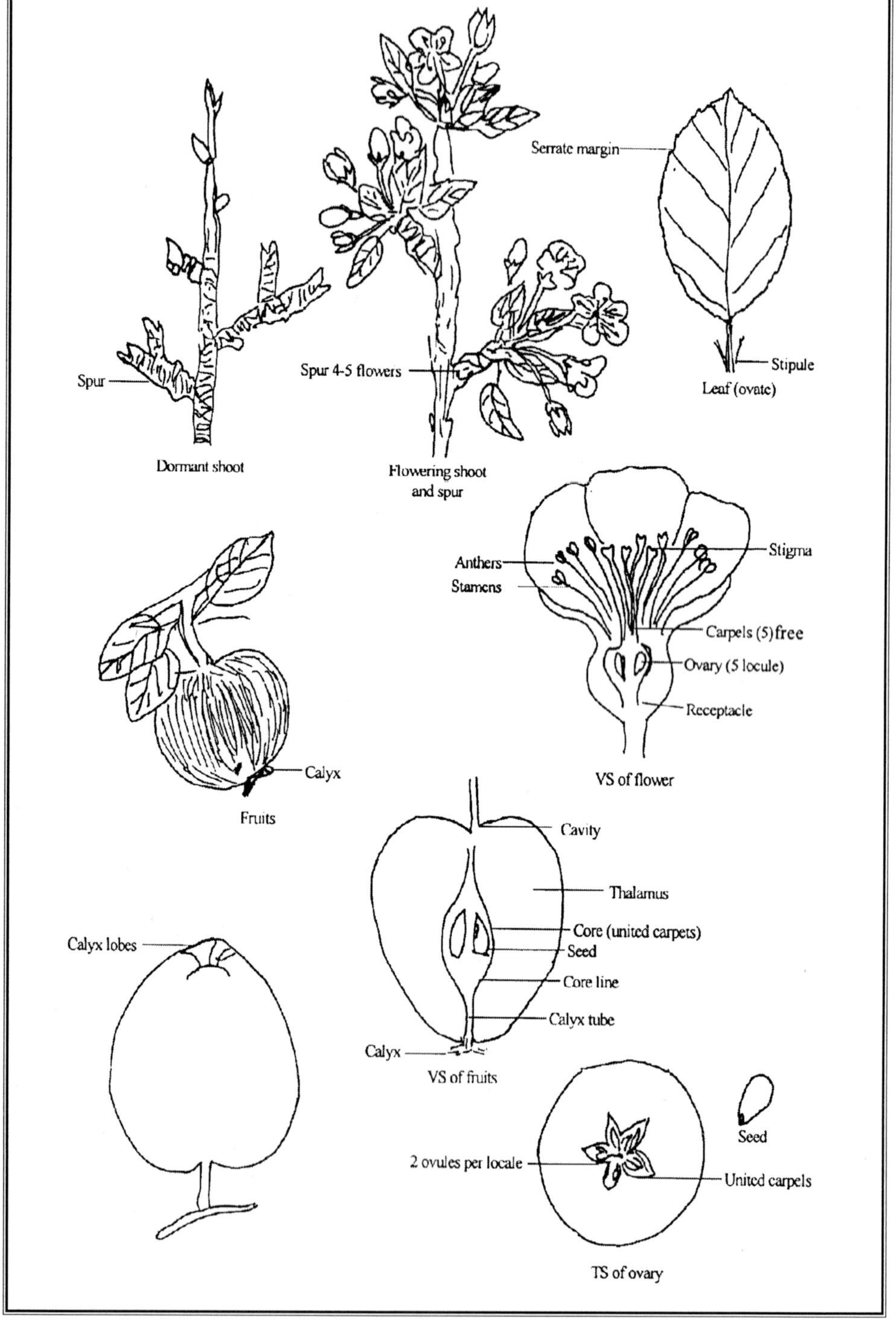
Spur
Dormant shoot
Spur 4-5 flowers
Flowering shoot
and spur
Serrate margin
Stipule
Leaf (ovate)
Anthers
Stamens
Stigma
Carpels (5) free
Ovary (5 locule)
Receptacle
VS of flower
Calyx
Fruits
Cavity
Thalamus
Core (united carpets)
Seed
Core line
Calyx tube
Calyx
VS of fruits
Calyx lobes
Seed
2 ovules per locale
United carpels
TS of ovary

temperate regions and others for extreme cold conditions. This has resulted/ added to the gene pool of desired types of apples enormously. At present cultivars are available destined for specific purposes like dessert, cooking, juice, cider and for large number of other uses.

The family Rosaceae is a very large diversified and comprises mainly of fruits like apple, pear, quince etc., collectively are referred to as pome fruits and are characterized by fruits having five carpels enclosed in a fleshy receptacle or thalamus and this forms the edible part of the fruit. For this reason fruit is also called as false fruit. The genus has about 25-30 species and also number of sub species which for their enormous blossom and small attractive fruits of various colours are infrequently called as crab apples. Apples are of interspecific origin and the cultivated apple is said to be derived from the specie *Malus pumila* Mill. Besides this specie other species that played important role in its evolution major being *M. sylvestris, Malus baccata* imparted winter hardiness, whereas *M. micromalus, M. prunifoli*a and *M. floribunda* were disease resistant sources.

Most of the present day cultivars have been derived from *M. pumila* or in other words it is the progenitor of apple cultivars and is found growing in abundance in south eastern Russia, Balkans, Transcaucasus, Iran, Turkistan etc. As a result of travels and invasions, the crop moved to other parts of the world, later the medieval religious houses took up its cultivation and by the end of the thirteenth century number of named cultivars were established along with their detailed characterization.

Origin and Distribution

De Candole after extensive work and survey indicated that in all likely hood apple is indigenous to the region south of the Caucasus from the Persian province Ghilan on the Caspian to Trebixand on the Black sea. Charred remains of fruit in Switzerland indicate apple cultivation from time immemorial in Europe. Later apple was introduced from Holland to America by the early settlers from western Europe.

Uses

Highly flavored and delicious fruits are eaten fresh, processed, juice making, mixed with other fruits to make jams, jellies for flavoring and preserves.

Description

Apple plants are medium size trees (10-12m tall), trunk short, low branching, crown broad rounded, bark thick dark brown-grey, which brakes into irregular flakes, sometimes, also in irregular longitudinal shallow fissures on trunk. Branches spreading, stout, twigs light green stout densely pubescent later turns to purplish or reddish-brown, old twigs later becomes smooth with dark-brown-grey in colour. Dormant buds, small, rounded at tip densely pubescent. Leaves are simple deciduous, alternate, measuring 4-10 x 3.6cm in size, smooth dark green above densely pubescent below, ovate, broadest at base or towards middle, apex short pointed, margin rounded (crenate) single or double, petiole pubescent 2-3cm long, thick (compared to pear). Flowering takes place on spurs and terminally on two or more years old wood, flowers produced in few flowered clusters usually 5, sometimes more 3-8cm across, pedicle, pubescent 2-5cm long. Calyx urn shaped, green pubescent light green fine free, alternate to petals attached to the thalamus, corolla 5 white or slightly pink tinged, orbicular (rounded at tip) free, stamens 20, carpels 5 usually free. Fruit rounded, ovate-oblong dented at both the ends, 2-8 cm in diameter, green, yellow, red, skin smooth waxy, glossy, flesh firm (at maturity) later softens, sweet juicy, white to creamish, calyx persistent, core fleshy, leathery 5 loculed each with 2 ovules enclosing small, plumb brown to black seeds.

Species

About forty species and a number of botanical varieties are classed under genus Malus. Important species which are of interest to pomologist are classified as.

A. Calyx lobes persistent on mature fruit

American species

i) *M. angustifolia* : Wild crab apple.

ii) *M. coronaria* : Garland crab apple.

iii) *M. ioensis* : Prairie crab apple.

iv) *M. soulardi* : Souland crab apple.

Old world species

i) *M. pumila* : Common apple.

ii) *M. spectabilis* : Chinese flowering apple.

B. Calyx lobes deciduous from mature fruit

i) *M. baccata*

ii) *M. pulcherrima*

Horticultural traits of some important species are summarized below.

Malus sylvestris Linn. Is native of western and central Europe, wild type with very less pubescence somewhat glabrous. Young/small branches and leaves both are glabrous with few scattered pubescence more so along the mid and lateral veins beneath. The petiole, pedicel and calyx tube with slight pubescence whereas the calyx lobes are pubescent from inside and glabrous from outside.

Malus pumila Henry. The specie is considered to be native primarily in south east Europe and in Asia and is also present in wild form in other temperate regions. Most of the cultivated varieties have been evolved from this specie. Has variable tree form, may be bush-like, small or a big tree. Plant parts including young shoots, branches, leaves, pedicels, calyx tube and calyx lobe surfaces both inside and outside are densely pubescent (tomentose). Leaves dull green, oval or ovate in shape.

Malus astracanica Dum. In all probability is an Asian specie. The specie is easily recognized by its large coarsely serrate or doubly serrated leaf margins which are densely pubescent below. Petioles are long.

Malus apetala Asch and Graebn. Bloomless apple. The specie is unique in that it is petalless and petals are in the form of rudimentary sepals which are bract like, sepals hence appear in 2 rows, without stamens, styles 10-15, ovary 6-7 or even more celled. Fruit similar to common apple except deep and open cavity. Besides these species *M. communis* has number of botanical forms for example var pendula Hort., branches, drooping or weeping type, var. aurea Hort., leaves are yellow variegated, var. plena Hort., Double flower type.

Malus sargentii Bean. Native to Japan. Plants are low growing bush, branches stout usually spiny, leaves variable, ovate to elliptic- oblong or ovate oblong 2-3 inch long, margin sharply serrate, petiole thin. Flowers are white about 1 inch across in 5-6 flowered clusters, pedicel shiny 1 inch long, calyx tube and calyx lobes are shiny outside and with soft pubescence inside. Calyx lobes ovate lanceolate with apex pointed, petals almost double the length of calyx lobes, short-clawed shiny, stamem 15-20, styles generally 4 united. Fruit dark red, round flat, small about ½ inch or less across.

M. seiboldii Regel (syn. *M. toringo* Sieb; *M. macrocarpa* var. Torringo Carr.). This specie is also native to Japan. Plants are shrub, leaves usually ovate or oblong-ovate in shape, pubescent. The leaves have a typical prominent notch on either side of the leaf at or below middle. The central lobe again is notched at the top. The remaining part of the leaf margin is dentate sharply. Flowers are small reddish in colour, petals rounded at base, oblong or obovate occur on thick shoots. Style, 3-4 connate at base, fruit very small, yellow or red about the size of pea or more, calyx deciduous.

The specie has some botanical forms i.e. *M. sieboldii* var. arborescens Rehd., which is more tree like and is less pubescent. Leaves less lobed and larger, flower usually white, fruit red or yellow. Indigenous to Japan, var. calocarpa (*M. sieboldii* var. calocarpa Rehd.), leaves are usually ovate oblong, margin crenate-serrulate but on rapid growing shoots they are 3 lobed. The lateral lobe broad and short. The flowers and fruits both are large and latter bright red in color.

M. spectabilis Ait (*M. sinensis*, Dum). Chinese flowering apple. Originated likely in China or Japan. Tree small, leaves oval to oval oblong, stalk thin shiny or slightly pubescent on both surfaces, margin closely serrate. Pedicels and calyx tube glabrous, flowers are coloured. Fruit reddish yellow round without a cavity at the base, sourish in taste.

M. prunifolia Willd. Indigenous to Siberia. Small shiny tree, but calyx and pedicel densely pubescent-tomentose. Leaves ovate to obovate or rounded, apex short pointed, margin unequally serrate. Flower on stalkless clusters of 6-10 flowers, white, 1-2 inch across, sepals lens like, petals round or oblong, style 5 connate below. Fruits yellow or red, cavity at base globose or ovoid about 1 inch in diameter.

M. micromalus Bailey. Cultivated in Japan and China, and is considered to be a hybrid of *M. spectabilis* and *M. floribunda*. Is a useful showy and profusely bearing plant, flowers are red coloured so are the calices and pedicels. It differs from *M. spectabilis* in that fruit has depression both at apex and base, calyx deciduous, pedicel and calyx both are densely pubescent.

M. baccata Linn. (*M. microcarpa* var. baccata, Carr. *M. baccata* var. sibirica, Schneid.) Siberian Crab. Tree found growing in abundance from Siberia to Manchuria and N. China. Tree small, compact, round topped, all parts almost glabrous. Leaves ovate to ovate lanceolate, shiny and thin, bright green, margin serrated evenly. Pedicels thin and long, flowers white, stamens shorter than styles. Fruit small to about 1 inch across, calyx deciduous at maturity, yellow or red in color, never mellows, specie has few botanical forms.

M. baccata var. mandschurica Maxim. (*M. baccata* var mandschurica, Schneid). Native of Japan, central China, Korea. Low growing with dense branches, leaves round or broad elliptic, margin entire, petiole, midrib, veins and flower stalk all are hairy, fruit round ½ inch across.

---------------- var. himalaica Maxin. (*M. baccata* var. Schneid). Found in abundance in the western Himalaya. Leaves hairy on the lower side, broad oval in shape margin coarsely serrated.

M. sikkimensis Hook. f. Indigenous to Himalayan region. Tree small, branchlets highly pubescent, leaves ovate to ovate-oblong, apex long pointed, base narrow, tomentose on underside margin sharply serrated. Flowers in corymbose cluster with 5-8 flowers and measure about 1 inch across, buds rose coloured, petals pinkish outside otherwise white. Fruits small dark red with white spots, calyx missing.

M. zumi Mats. (*Malus zumi*, Rehd.). Found in plenty in the mountains of central Japan. Low growing tree profusely branched, round topped. Leaves with long petioles, oblong to ovate oblong, apex sharp pointed, round or narrowed at base, margin crenate-serrate, light green below and yellowish green above. Flowers white or pink tinged about 1 inch across, pedicels 1 inch long, calyx lobes lanceolate. Fruit small ½ inch in diameter, red, globose, without calyx.

The total apple germplasm (cultivars) is very large and specific cultivars have been designated for specific purpose i.e. dessert, juice, culinary, green, yellow, red, early mid, late ripening and Cox type apples. Besides each growing area has its own preference for particular cultivars. At present the new cultivars being grown commercially have been obtained from very precisely planned crossing and selection programme. Description of some very important and commercial cultivars are detailed below

Cultivars

Red Delicious : Most widely grown cultivar all over the world, evolved as a chance seedling in N America, has number of strains which are equally important. Fruits are large, conical oblong calyx lobes very prominent and protrude outward. Skin smooth, shiny, with prominent red streaks on pale yellow background, flesh creamy white texture crisp, fine grained, later turns tender, sweet juicy and very aromatic, ripens in mid season. Tree moderately vigorous, semi spreading.

McIntosh : Tree vigorous, crown rounded, branches spreading, susceptible to scab. Fruit medium in size, oblate skin glossy smooth lenticels few conspicuous, skin pale yellow green overlaid with bright carmine. Flesh white with some red tinge after harvest, crisp, juicy tender, sweet with pleasant acid blend, flavour mild.

Golden Delicious : A very good pollinizer for Delicious group, tree moderate in vigour, branches spreading somewhat upright. Fruit medium large oblong skin golden yellow with prominent brown lenticels, russet prominent at the stalk end. Flesh creamish white firm crisp when mature, later turns soft, sweet with pleasant acidic blend.

Gloster- 69 : German cultivar, evolved from a cross between Glockenapfed x Richared Delicious. Tree vigorous, upright bears regularly with high yields, moderate preharvest drop. Fruit large, short conical uniform light red (Richard Deli.), flesh greenish yellow, free of bitter pit, sweet but acid content more than Delicious.

Red Chief : Bud mutant of Starkrimson Delicious, produces heavily, semi dwarf, vigour moderate spreading. Fruit, medium-large shape like Delicious with bright attractive cherry red stripes over pale yellow, flesh creamish white, quality excellent harvest midseason.

Royal Gala : Bud mutant of Gala. Tree of medium size, vigour moderate, precocious, productivity very high. Fruit medium sized smooth, skin shiny, golden yellow overlaid with bright scarlet, very attractive in form, flesh light yellow, texture firm crisp, tolerate bruising, free of bitter pit, sweet juicy, flavour and quality excellent ripens in mid season.

Red Fuji : Bud mutant of Fuji. Hardy variety, fruit medium in size round to conical with bright red stripes over a light red base, flesh firm texture fine, flavour aromatic late in maturity, stores very well over long period.

Anna : Low chill cultivar originated in Israel from cross between Red Hadassiya x Golden Delicious. Fruit large shiny pale yellow overlaid by bright red more so on cheeks, flesh white, sub acid to sweet in taste, flavour mild ripens by June end.

❑❑❑

52
Pear

Genus : *Pyrus*
Species : *communis L.*
Family : *Rosaceae*
Chromosome No. : 2n= 34

Pear like apple is also a prehistoric crop and unlike apple its cultivation in Europe started much before that of apple at an early time. According to Homer (1000 B.C.) the ancient Greek said that pears were one of the "gifts of god's". Romans also enormously added to the knowledge of pear cultivation. Pliny in Rome had described 35 cultivars (23-79 AD). The variation in fruit characters described earlier match to the range of fruit characters which are found in presently grown commercial cultivars. In Europe for its buttery soft texture and distinct aroma many consider European pear as the most detectable of

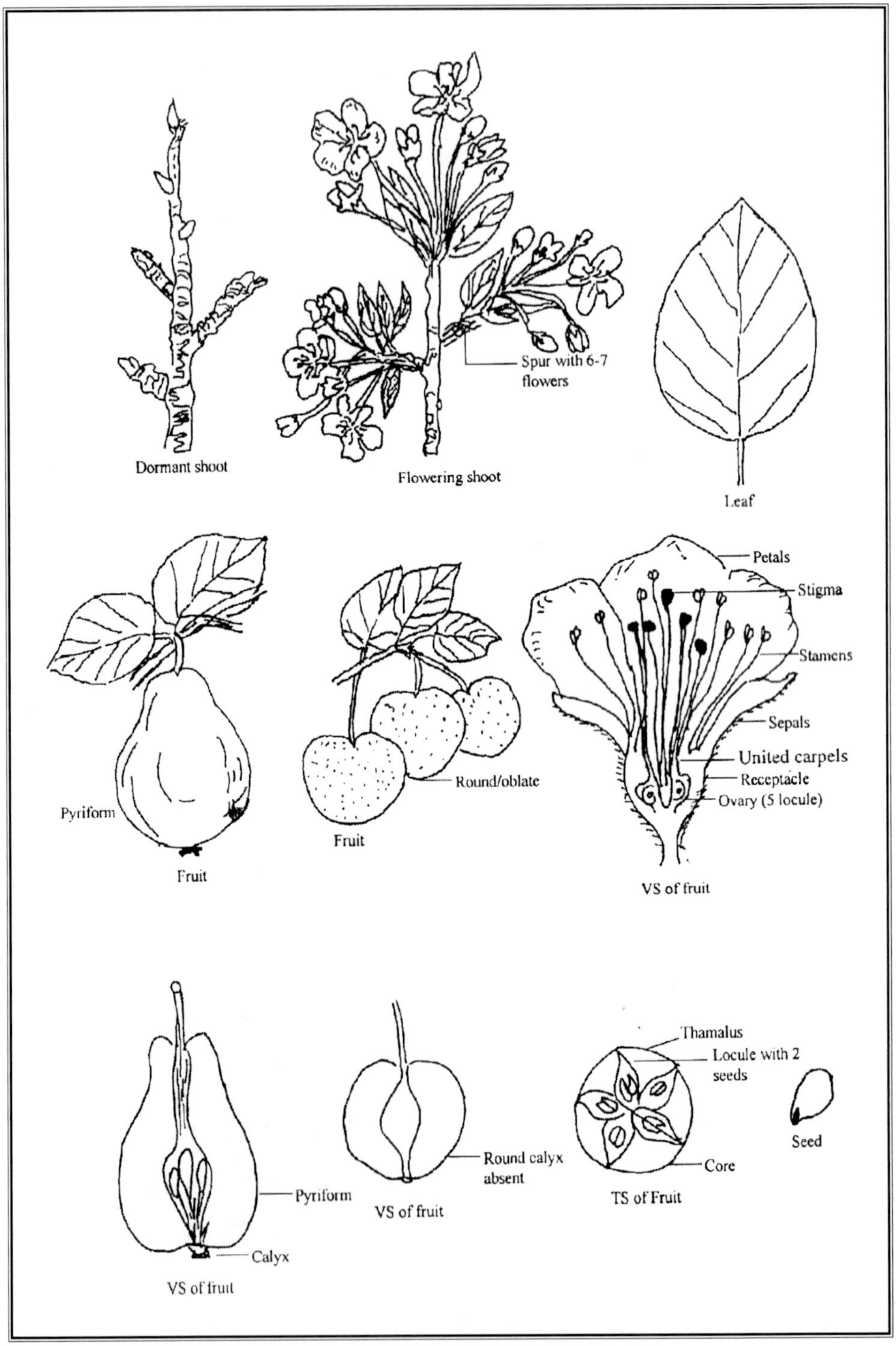
Dormant shoot
Flowering shoot
Spur with 6-7 flowers
Leaf
Pyriform
Fruit
Round/oblate
Fruit
Petals
Stigma
Stamens
Sepals
United carpels
Receptacle
Ovary (5 locule)
VS of fruit
Pyriform
Calyx
VS of fruit
Round calyx absent
VS of fruit
Thamalus
Locule with 2 seeds
Core
TS of Fruit
Seed

all temperate fruits, while the Asian pears for their firm crisp texture and juiciness are preferred more in China and Japan - their native place. Pears are readily consumed as fresh, dried, cooked or as preserves or to make perry and pear wine.

The genus Pyrus genetically is very diverse and belongs of family Rosaceae, sub family Pomoideae and has more than 20 species, most of which are indigenous to Europe and Asia. In Europe *P. communis, P. nivalis* are commercially more important, whereas *P. pyrifolia, P. ussuriensis, P. betulaefolia, P. bretschneideri,* and their hybrids are more important in China and Japan. On the basis of variability three centers of origin have been demarcated, Chinese Center (forms of *P. pyrifolia* and *P. ussuriensis* are grown), The Central Asiatic Center (*P. communis* is grown) comprise of vast areas of Tajikistan, Uzbekistan, western Tian-Shan, India and Afghanistan and the Near Eastern Center where *P. communis* in grown, includes Caucasus mountains and Asia Minor. The last center is regarded to be more important because the domesticated forms of *P. communis* from which the present day cultivars have been obtained originated there.

The pear history with which the pomologist for long has been involved is *Pyrus communis.*

Origin

Pear is a native of southern Europe and western Asia. For long it has been recognized as pear of Egyptians, the Assyrians, the Greeks and the Romans, and has rightly been nick named as pear of western civilization.

Uses

The highly flavoured and delicious fruits are usually consumed fresh, processed to make juice, juice blended with other fruits, made into jams, jellies, preserves, flavouring of products etc.

Description

Pear plants are tall (15-20m), crown cone shaped or rounded , low branching, trunk usually straight about 1 foot in diameter rarely more, bark thick smooth when young brown to grey, later becomes scaly with shallow to medium fissures (longitudinal). Branches upright, short and stout, twigs, smooth, reddish brown, lenticels light yellow. Dormant bud, long ovate (oval) pubescent. Leaves, alternate deciduous, thick, simple, leathery, smooth pubescent when young later turns shiny/ glabrous, green above, paler below, slightly pubescent underneath, cordate, broadest at base or middle, narrowing and ending in a sharp pointed apex, margin finely serrated, petiole thin long (sometimes as long as leaf blade), smooth, shiny. Flowering takes on spur and terminally on previous seasons wood, borne in a few to many flowered clusters (6-9), each flower pentamerous, calyx urn shaped, 5 lobed, pubescent light green, corolla 5, usually white sometimes pink tinged, rounded at tip, narrowing at base, stamens 20, anthers pink tinged, pistils/carpels 5 united at base. Fruit round, globose or pyriform,10 cm long, sometimes even more in pyriform pears, green, yellow or red in colour, hard at maturity, softens later, skin smooth, flesh white or creamish, soft juicy, grit cells very small concentrated at core in European pears, core fleshy leathery, with several large brown seeds, calyx persistent.

The genus Pyrus comprises of more than 30 species all being native to northern hemisphere of the old world. Non is native of Australia or north and south America. The important species are grouped into two native groups (geographical distribution) Occidental group and Oriental group.

1. *Occidental Group:* are characterized by persistent calyx, fleshy pedicels with pyriform shape. Native of eastern Europe, south western Asia including south Afghanistan, Asia Minor and Transcaucasia. Important species are *P. communis* and *P. nivalis* (snow pear).

2. *Oriental Group:* Has deciduous calyx, non fleshy pedicels and with apple shaped fruits. Native to eastern Asia including Japan, China and Manchuria. Important species are *Pyrus bretschneideri* Rehd. (Chinese white pear), *Pyrus pyrifolia* (sand pear), *Pyrus ussuriensis, P. serrulata, P. betulaefolia, P. calleryana and P. phaeocarpa.*

Pyrus is considered to be a polymorphous genus in the northern hemisphere. Important *Pyrus* species of potential value are described below.

Pyrus heterophylla Regel and Schmalh. Trees are small glabrous thorny, foliage variable, but usually ovate. Fruit almost globose, and ends flattened, very small 1 inch across.

P. nivalis Jacq. Snow Pear. Used for making perry or pear cider. Commonly grown in Europe especially France. Trees without thorns, short with dense greyish pubescence, leaves oval to oval obovate, short acute or obtuse at base, with dense greyish pubescence. Flowers white large and showy. Fruit roundish - short pyriform acidic later becomes sweet.

P. salicifolia Pall. Trees are relatively small about 30 feet tall, somewhat spiny, young branches with dense grey pubescence. Leaves lanceolate or linear lanceolate, tapering at both ends, pubescent below. Flower white on short pedicels in corymbs. Fruit small ¾ inch across yellow or greenish, stalk small, calyx persistent. A botanical variety pendula has drooping branches.

P. serotina Rehd. Plants are medium tall 30-50 feet, young branches not pubescent, leaves 3-5 inch long ovate-oblong, base round, apex long pointed and sharply serrate, under surface pubescent when young later turns glabrous, flowers white in umbellate racemose 6-9 flowered, pedicel thin. Calyx long, corolla short clawed, oval, stamens 20, styles 5. Fruit brown, sub globose calyx deciduous has one botanical variety culta-sand pear. Fast growing trees, shoots thick, leaves dark green broadly ovate, apex long pointed, margin sharply serrated, flowers white large appear before foliage. Fruit rounded hard, rough, depression at stalk, flesh gritty flavour fair, calyx fall prior to maturity.

P. bretschneideri Rehd. Native of China. Plants are medium in vigour. Leaves are ovate or elliptic ovate apex acuminate, leaf base cuneate rarely rounded, margin sharp serrated. Fruit globose-ovoid or subglobose with deciduous calyx, yellow pale spots, pedicel 1-1 ½ inch long.

P. ussuriensis Maxim. Oriental or Asian pear, wide spread in N. China, Manchuria. Plants somewhat spiny, branches with grey pubescence, branchlets light yellowish brown in colour. Flowers in dense clusters, stalk small, cluster therefore hemispherical, petals obovate narrowing towards the base. Fruit globose with short stalk.

P. betulaefolia Bunge. Has wide ornamental value. Tree not tall growing, 20 feet generally without thorns. The branchlets and buds are with grey pubescence, foliage light in colour, leaves ovate or rhombic ovate apex pointed, base cuneate, glabrous or tomentose below shiny above. Flowers white, fruit almost rounded brown and with dots. Calyx falls at maturity.

P. calleryana Decne. Native of China. The specie resembles Eurasian type in crenate serrate leaf margin and not the sharp serrate margins of Asian types. Leaves are small shiny usually ovate in shape base generally rounded, margin crenate. Flowers are small white pentamerous but with 2-3 styles. Fruit small rounded, stalk slender calyx deciduous.

P. pashia Buch-Ham. Widespread in foot hills of Himalayas and W. China. Trees are medium tall usually spiny. Young growth woolly in nature and tomentose. Leaves are ovate or ovate-lanceolate apex long pointed, margin crenate or serrulate when mature and growth becomes glabrous, usually three lobed and doubly serrate when young. Flowers are small appear on woolly corymbose clusters of 7-9, pedicel short, calyx lobes pointed and fall off as fruits mature. A close relative of this specie is *P. kumaoni*, Decne. Is considered to be native to Kashmir, Kumaon in western Himalaya and Yunnan in China.

Plants are tall growing 50 feet or more, usually thornless buds, young branchlets all are glabrous. Leaves ovate, narrow apex long pointed, subcordate and shiny. Flowers in shiny corymbs. Fruit rounded-globose about 1 inch across calyx on fruit fall off early in the reason.

The varietal worth of both European and Asian pears is well established. Salient features of some pear cultivars are given below.

Cultivars

Bose (Beurre Bosc) : Originated in Belgium. Tree large upright, vigorous but very susceptible to fire blight. Fruits are large, very long with tapering neck, skin dark yellow russetted, flesh creamish yellow, soft, sweet juicy with aromatic flavour, quality very good to excellent ripens late.

Clapp's Favourite : A very old cultivar selected by Thaddeus Clapp. Tree large upright very productive. Fruit medium-large skin pale lemon yellow mottled with bright red colour, flesh yellowish, grit cells present at core, soft melting very juicy sweet, quality very good, ripens late.

Flemish Beauty : An old and popular midseason Belgian cultivar. Tree vigorous, hardy, spreading and productive. Fruits are large, roundish, skin smooth shiny, yellow blushed with red, flesh white, firm, tender, sweet juicy aromatic with slight musky flavour, quality very good.

Max Red Bartlett : Bud mutation of Bartlett found in Washington. Tree characters are like of Bartlett but shoots and leaves slightly reddish when young. Fruit short globose skin dark cranberry red turns bright red-purple red in colour on ripening, flesh fine grained firm crisp white with more sugar, sweet juicy, ripens with Bartlett.

Regal Red Comice : Bud mutation of Doyenne du Comice, tree and fruit characters resemble that of the parent cultivar, but fruits are complete overall red with some purple tinge over greenish yellow ground colour when ripe, flesh usually white fine soft, melting, eating quality excellent.

Rogue Red : Originated in New Jersey from cross between Gorham x NJ 1. Tree fair in vigour, branches somewhat spreading. Fruit medium in size acute pyriform, skin straw yellow turns red at ripeness, flesh creamy white, fine without astringency, almost buttery, quality good, resistant to fire blight.

Chojuro : Asian pear, tree medium sized spreading and productive requires pollinator for high production. Fruit round oblate medium sized, greenish brown to brown russet, flesh white, mildly sweet, sourish at core, flesh firm crisp coarse due to grit cells, pulpy quality good, ripens mid season.

Hosui : Developed in Japan, Asian pear. Tree medium sized spreading very vigours and productive. Fruit medium large, round, skin as of Chojuro, flesh white, firm crisp, texture fine, very sweet and juicy quality excellent.

Kosui : High quality commercial Asian pear. Fruit small medium, light green to golden bronze, russetted, flesh white crisp tender juicy, very sweet, quality excellent. Tree vigorous and moderate in production.

❑❑❑

53
Quince

Genus : *Cydonia / Pyrus*
Species : *oblonga* Mill. *vulgaris Pers.*
cydonia L.
Family : *Rosaceae*
Chromosome No. : 2n= 34

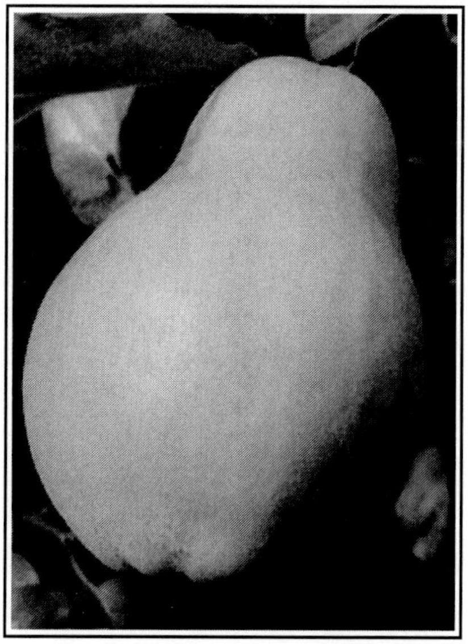

The fruit of pome class referred as *Cydonia oblonga* (also *Pyrus cydonia*) is an ancient fruit and yet its significance is not fully recognized and crop has received very less attention. There is hardly any mention of this old fruit in respect of its merit, its cultivation or in the form of some standard literature. One reason could be that amongst pome fruits where quince is placed, better fruit form, raw eating quality persisted, which markedly restricted quince improvement. Quince fruit lacks dessert class except when cooked but is commercially used for flavouring, in preparation of jellies marmalade and preserves. However

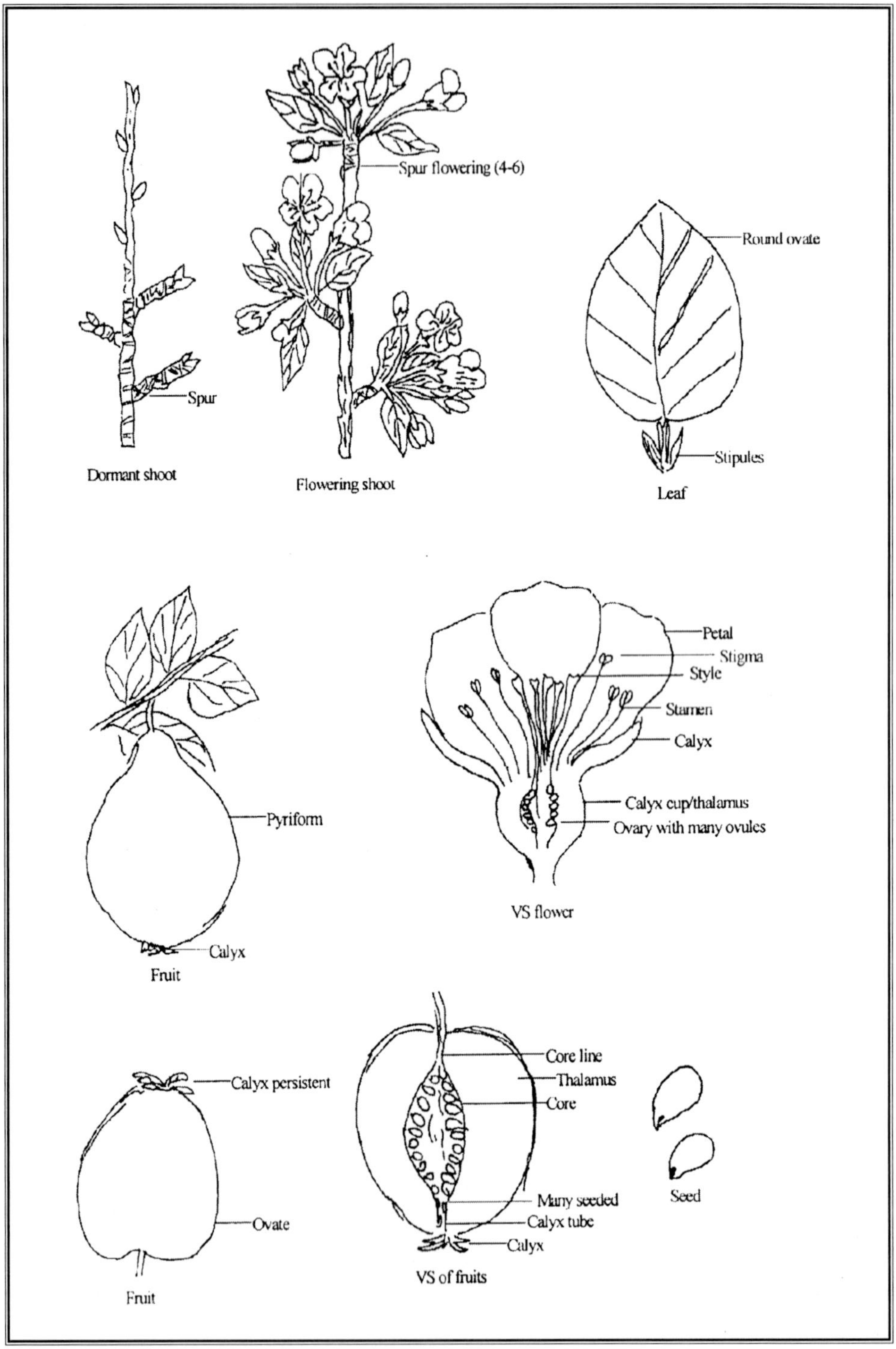
Spur flowering (4-6)
Round ovate
Spur
Stipules
Dormant shoot
Flowering shoot
Leaf
Petal
Stigma
Style
Stamen
Calyx
Calyx cup/thalamus
Ovary with many ovules
Pyriform
VS flower
Calyx
Fruit
Core line
Thalamus
Core
Calyx persistent
Many seeded
Seed
Ovate
Calyx tube
Calyx
VS of fruits
Fruit

in near East and Persia, quince cultivation is highly improved through the selection of better varieties which are eaten raw.

The quince plants are small and somewhat twisted in its growth. Its flowering and fruiting behaviour is different from other orchard fruits but resembles that of walnut. After the break of dormancy short vegetative growth takes place from the terminal bud and on its apex flower and fruits are produced during the same season. The fruit stalk is highly reduced or sessile and fruits appear to rest on short leafy branch. On account of this growth pattern quince has crooked appearance.

Origin and Distribution

Quince has its origin in Asiatic region, most likely the Transcaucasia and areas adjoining Asia Minor, as records indicate its long cultivation there dating back to about 4000 B.C. Another close related quince of Chinese origin is the chaenomeles flowering quince. From western Asia the crop moved westward and was very well established in Greece by 600 B.C. It is here that the genus name Cydonia was derived from Cydon a village located in the coast of Isle of Crete. In the early 18th century the genus reached north America.

Uses

The quince fruits usually being very acidic and sourish in taste are usually grown for preserves, for drying and rarely for fresh use. With little flavour and high pectin content, it is often mixed with other fruits to make jellies, various flavouring agents and preserves.

Description

The quinces are generally of two types the common or true quince *Cydonia oblonga* and the Oriental quinces of the Asiatic genus *Chaenomeles*, the latter usually recognized as ornamental flowering type. The quince plants are small (8-12 ft. high), deciduous, slow growing with irregular branches. Leaves are small round light green, highly pubescent, young growth also pubescent. Flowers appear terminally or solitary, with white or pink flowers at the end of the leafy shoots. Flowers are quite similar to apple but more rough and colourful. The individual flowers may be 2 inches across, calyx five light green pubescent, corolla five usually cup-shaped (white to scarlet, cultivar based), usually 20 or more stamens, carpel five, resulting in 50 or more seeds. The stamens and pistils are almost twice as large and thicker compared to apple. The fruit is an article of food, is a pome fruit usually apple shaped (round to ovate), may measure 5 inches or more in diameter, very pubescent. The golden yellow flesh is not fit for consumption when fresh, therefore, mainly used in preserves and cooking.

In the genus Cydonia, specie *oblonga* Pers; (*C. vulgaris* Pers.) is by far the most well defined and widely grown specie and has number of botanical varieties which are described below.

Cydonia oblonga var.pyriformis Rehd. (*C. vulgaris* var. pyriformis, Kirchn). Fruits of this botanical form are typically pear shaped.

---------------- var pyramidalis Schneid. The plant shape is like a pyramid.

---------------- var lusitanica, Schneid *C. lusitanica* Mill.). The fruit shape is like that of pear, fruits are markedly ribbed, leaves are larger and growth is more vigorus.

---------------- var. maliformis Schneid (*C. maliformis*, Mill.). The fruit shape is like that of an apple.

---------------- var. marmorata Schneid. The leaves of this botanical form are variegated with white and yellow.

Cultivars

In comparison to apple and pear, quince has very few cultivars which are grown commercially salient features of the most important is detailed below.

Arthur Coby : A very old cultivar named in honour of Dr Arthur AS Coby, Urbana Illinois, derived from *Chaenomeles lagenaria and C japonica*. Small bush like, grows upto 5-6 feet, upright and spreading, flower bright red in colour two or more inch across, showy with ornamental value. Fruit large round flat greenish yellow, smooth, late in ripening, good for making jelly.

❑❑❑

54
Peach

Genus : *Prunus*
Species : *persica Batsch*
Family : *Rosaceae*
Chromosome No. : 2n = 16

The peach, *Prunus persica* (L) Batch is unique among the temperate fruits as it is adaptated to varied climatic conditions and it is for this reason that peach and its close relative the nectarine, in a short period has spread rapidly to different parts of the world. Peach is a tree of temperate zone, but has attained higher quality (fruit colour, taste etc.) in areas where the summer temperatures are relatively warm to hot. With the discovery of low chill types 'Honey' and 'Peen to' forms from southern China, peach cultivation has been spread to warm temperate regions of the world. Normally peach requires from 500-1000 hours

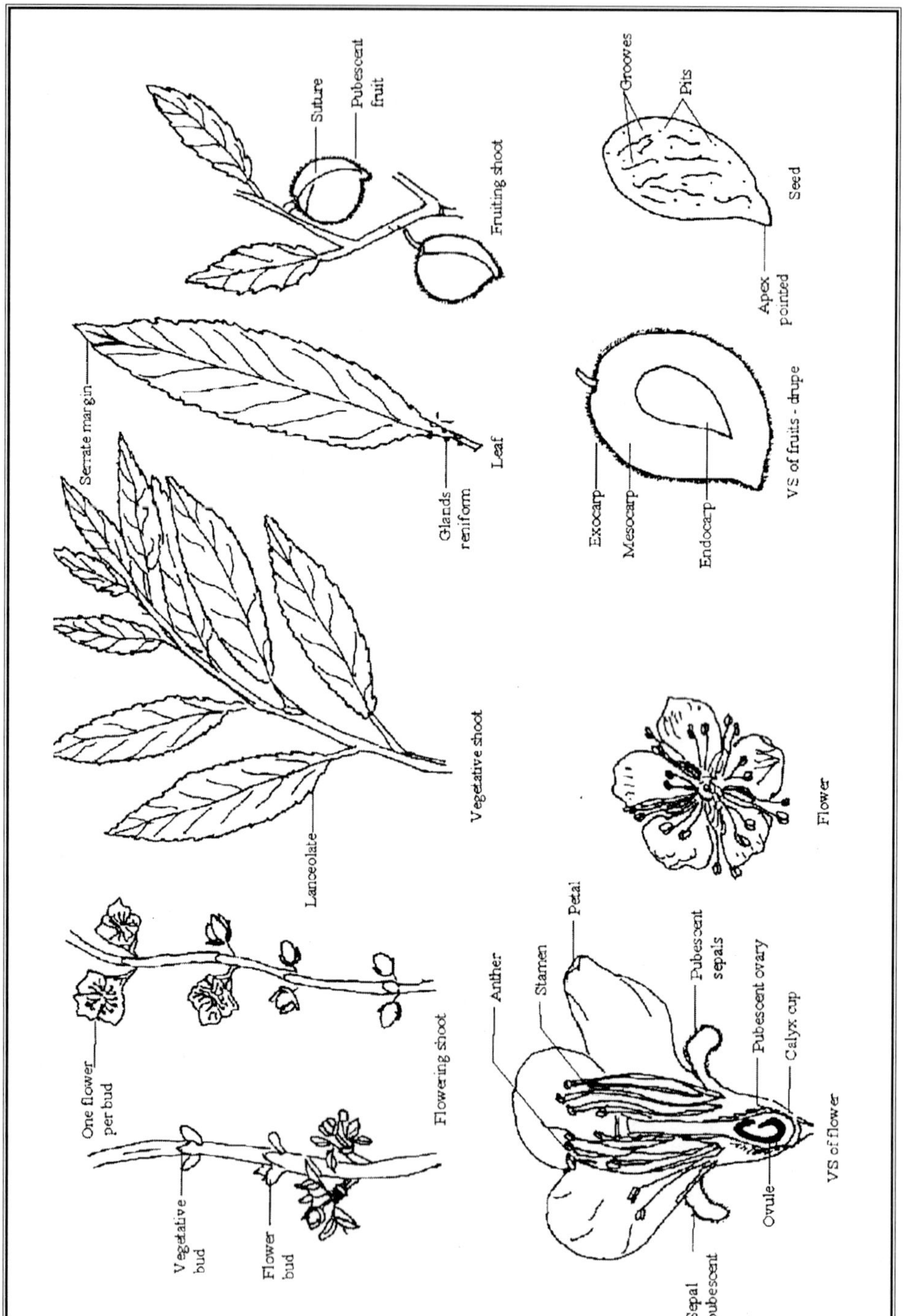
Vegetative bud
Flower bud
One flower per bud
Flowering shoot
Lanceolate
Vegetative shoot
Serrate margin
Glands reniform
Leaf
Suture
Pubescent fruit
Fruiting shoot
Sepal pubescent
Anther
Stamen
Petal
Pubescent sepals
Pubescent ovary
Calyx cup
Ovule
VS of flower
Flower
Exocarp
Mesocarp
Endocarp
VS of fruits - drupe
Grooves
Pits
Apex pointed
Seed

of cold/chill hours below 7.2°C for growth and flowering in the spring. Peach history as a garden crop in China is well documented, and forms varying in flesh colour (white, yellow or red), free stone or cling stone, flower colour, large and showy flower type seen presently were also present at very early time i.e. the beginning of the present century.

Origin and Distribution

Peach is native to China as suggested by old records (2000 BC) though some earlier writers indicated Persia to be the native place of peach. Alexander's expedition to the east and later return brought peach to Greece about 400-300 BC and to the Romans sometime later. From here the crop reached England in the twelth century. The Spanish introduced peach into America. At present it is grown all over the world in the warm temperate region.

Uses

Mostly used as dessert fruit, juice making and preserved as halves pulp used to prepare jam, mixed fruit jam, peach nectar or peach beverage, and wine, seed yields oil.

Description

Peach plants are small trees (7 m tall), crown round topped/broad, bark, smooth when young, rough and scaly later, reddish brown in colour. Leaves are simple, alternate, deciduous, dark green, shiny above pale below, broadly lanceolate, long (8-15 cm) tip long pointed, rounded at the base, margins sharpely serrated, petiole usually with glands which may be round, globose or reniform, 2-4 in number sometimes even more. The flowering usually takes place before leaf opening, occur laterally on previous season's growth, flowers usually solitary appear before leaves, receptacle cup-shaped calyx, 5 lobed, purplish red, pubescent, corolla 5, pink to white in colour large 8-20 mm long, usually rounded at tip, stamens, 20-30, pistil single. Fruits almost round to round oblong, apex pointed, 4-10 cm in diameter, distinct groove/suture, seen all along the fruit length from base to the tip, reddish pink to orange, epicarp has soft pubescence (hairy), flesh soft, sweet juicy, white or orange, enclosing a very hard, oval-oblong seed, with deep grooves and pits on endocarp, apex pointed.

Nectarine. The absence of fuzz (pubescence) on the nectarine fruit fundamentally is the only difference between peach and nectarine. Peaches and nectarines pomologically have been classified into five 'races' or 'groups' namely 1) Honey or south China 2) Persian or True peaches 3) Spanish or Indian 4) Peen-to or Flat peaches 5) North China or Chinese cling.

Peach has two well defined forms, the cling stone or pavies (*Persica vulgaris* Risso.) and the free stone (*Persica domestica* Risso.). Apart from these there are number of different forms that represent variation in leaf colour, flower colour and petal number etc.

Persica vulgaris flore alba plena. Flowers are white in colour.

------------------- roseo plena. Rose coloured flowers or petals.

------------------- folis purpureis. Group represents purple or red leaved peaches.

------------------- pendula. Branches are drooping.

------------------- pyramadilis. Tree canopy is like that of a pyramid.

Apart from these botanical forms, pomological descriptions of some peach species which are of significance from horticulture point of view are detailed as.

Prunus persica var. necturina. Maxim (syn. *Persica nucipersica,* Borkh *P. laevis* D.C. *P. persica* var. laevis). Nectarine. Nectarine is said to have originated both from bud variation and also through seed. In most of the characters it is more or less like peach, but the fruit is smaller and without pubescence or fuzz. Nectarine cultivation is very popular in N. America especially California.

Prunus persica var. platycarpa. Bailey (*Persica platycarpa,* Decne.) Flat peach native of China. 'Peen to' is said to be a form or variety of this and is grown in warmer states of southern China. It resembles closely to the cultivated peach and differs in that the fruit is highly compressed more so towards both the ends, stone compressed, rough, small flat and irregular in outline.

Prunus davidiana French (Persica Davidiana Carr. *Prunus persica* var. Davidiana, Maxim). Native to China used as stock for other stone fruits. Tree slender looks willow like. Leaves much smaller and narrower compared to peach, light green in colour and begins to taper near the base and ends in a long narrow apex at distil end, margin sharply serrated. Flowering takes place early in the season, flowers are relatively large 1 inch or more across single per bud, light pink in colour, calyx shiny. Fruits globular, pubescence grey or yellow with conspicuous suture. Stone small, free from flesh almost round.

Although peach has narrow genetical base, still cultivars show variation with respect to blooming, fruit shape, colour, flesh colour firmness, pubescence, leaf glands, chilling requirements etc.

Cultivars

Description of some well known established cultivars are summarized briefly.

July Elberta : Originated in California, parentage unknown, very fruitful, tree vigorous, chill hours 750. Fruit round, medium sized greenish yellow overlaid with dull red blush or streaks, pubescence dense flesh yellow firm faintly red at pit, stone small free, quality very good.

Flordared : Obtained from Southland x Hawaiian, very low in chilling requirment 50-100 hrs, flowers large showy, glands reniform. Fruit small medium round skin about 50% blushed with red, flesh white somewhat firm free stone.

Flordasun : Requires 300 chill hours to break dormancy, self fruitful, flowers showy. Tree medium tall, spreading, fruit small round yellow slightly blushed, flesh yellow red at pit, medium firm, ripens uniformly, almost freestone, quality not very good.

Sunhaven : Tree vigorous, large self fruitful productive and ripens early, high chill cultivar - 900 hrs. Fruit medium large uniform round, bright red over yellow, pubescence soft short, flesh yellow, texture firm, fine, does not brown, free stone, flavour excellent, good for fresh market.

Nectarine : Requires 300 hrs of chilling. Fruit small round surface almost completely red coloured, flesh yellow, firm, semi free stone, dessert quality excellent ripens early mid May.

Starking Delicious : Bud mutant of July Elberta. Most of the tree, flower, leaf characters are similar to the parent cultivar. Flesh yellow, freestone, ripens almost 15-20 days earlier than July Elberta, flowers are pink and small.

Maya and Vesna : Evolved at Beograd, Yugoslavia from self pollination of Glohaven, self fruitful, tolerant to peach mildew, scab and brown rot. Plant vigorous than standard. Maya is more productive 20 tons/ha. Fruit large roundish firm 60-70% red blushed over yellow ground colour, flesh yellow firm resists browning, free stone, flavour very good. Maya has excellent table quality and also when processed, whereas Vesna is good to very good for fresh and superior as processed fruit.

55
Apricot

Genus : *Prunus*
Species : *armeniaca L.*
Family : *Rosaceae*
Chromosome No. : 2n = 16

Of all the temperature fruits belonging to tribe pruneae, apricots after cherries are more exacting in their climatic requirements. For this climate specific, apricots are often described as a fruit that fails where other fruits grow successfully. Apricots lavish continental climate, which is characterized by low rainfall (50 cm per year), warm summers and without much variation in winter temperatures and this winter temperature in temperate regions remain close to -33° C. Apricot trees are said to be drought resistant and can sustains low atmospheric humidity but is very sensitive to low soil moisture content.

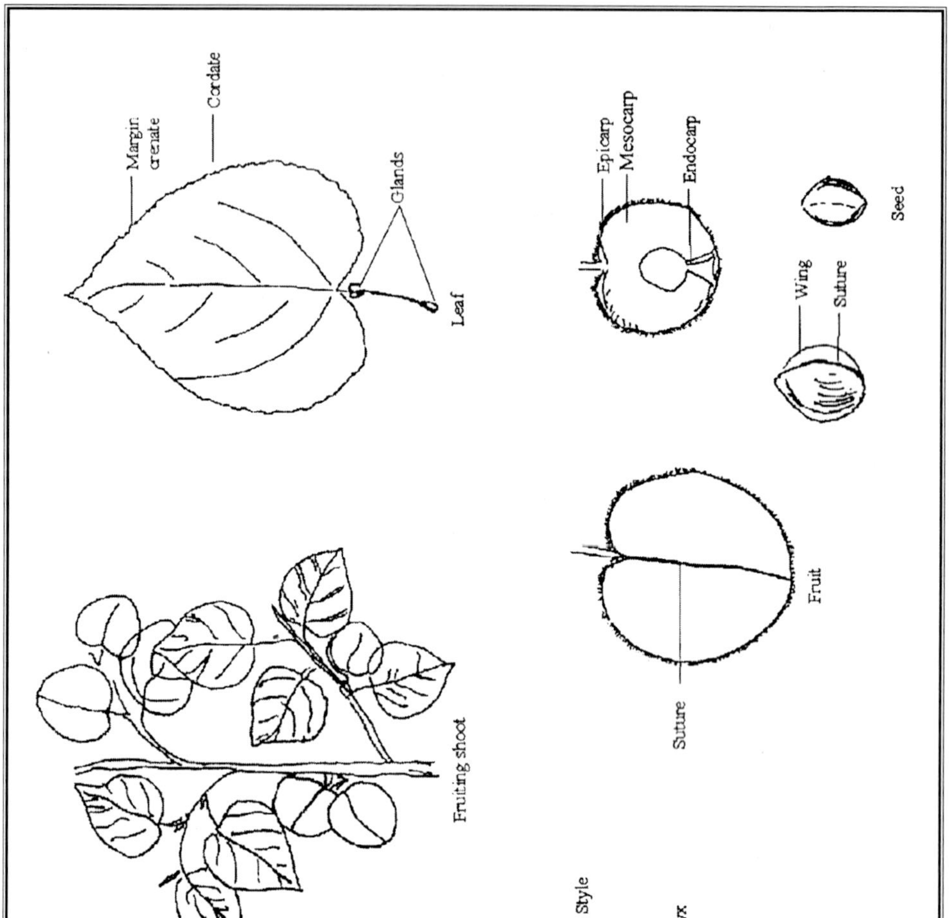
Margin crenate
Cordate
Glands
Leaf
Epicarp
Mesocarp
Endocarp
Wing
Suture
Seed
Fruiting shoot
Suture
Fruit
Stigma
Style
Pubescent calyx
Pubescent ovary
Ovule
Flowering shoot
Dormant shoot
Calyx cup
VS of flower

Apricot is a very desirable detectable fruit and compared to other stone fruits, grows successfully with minimum of cultural care. It is for this reason that despite growing for thousands of years on the mountain slopes of temperate region of China and central Asia, the crop failed to extend successfully beyond the ecologically adapted few areas.

Origin and Distribution

Is native of China, as it has been cultivated since ancient times. Apricot belongs to genus *Prunus*, sub genus prunophora and apricot of commerce belong to the specie *Prunus armeniaca*. Three centres of origin have been enlisted, 1) Chinese Centre - the mountainous regions of northern and western China, 2) Central Asian Centre - the mountainous area extending from Tien-Shan including Hindu-Kush to Kashmir, and 3) Near Eastern Centre mountainous regions extending from North Eastern Iran to Caucasus and Central Turkey later was introduced to western Asia and reached North America (California) in 18th century.

Uses

Generally consumed as fresh, apricot halves can be preserved, used for drying to make various products, pulp utilized to make apricot jam, chutney, squash, appetizer, nectar, ready to serve drinks, to prepare alcoholic beverages.

Description

Apricot plants are small to medium trees (5-8m), crown round topped, bark reddish brown, branches stout stiff. Young growth (leaf, petiole, shoot), purple tinged. The leaves are simple deciduous, shiny green, broadest at base with a typical notch (cordate) 5-10 cm long, pointed at tip, petiole 2-3 cm long, margin crenate. The flowers are produced laterally on previous seasons shoot / wood, generally singly, rarely in cluster of 2 or 3. Flower pentamerous, calyx cup shaped, 5 lobed, pubescent, purplish red in colour, corolla 5, pinkish to white, stamens 20-30,all placed on the rim of the calyx cup, pistil single, ovary inferior. Fruits are round or nearly so, 2-3 cm sometimes more in diameter in better cultivars, orange to yellowish or white coloured, epicarp slight to densely pubescent, flesh, sweet, orange or white, enclosing a smooth slightly flattened stone, distinct suture on the dorsal and slightly winged on ventral side.

The horticultural description of some apricot species important from pomology point of view are summarized below.

Prunus sibirica Koch. Siberian Apricot. Plants are bush or small tree 10-12 feet tall or even more planted as ornamental, leaves are small, ovate to rounded, apex long pointed, pink or white, sepal pubescent. Fruit small half inch in diameter, inedible, yellow, cheek reddish, stone pointed and smooth.

P. mandschurica Maxim. Wide spread in Manchuria. Leaves roundish, subcordate at base, apex acute margin doubly toothed. Fruit small roundish or globular, yellow with red spots, flesh soft juicy, stone small and smooth.

P. mume Sieb. and Zucc. Japanese apricot. Tree characters are like those of common apricot, but bark grey or greenish, leaves dull green lighter below, small round to small ovate - cordate, apex long pointed, margin serrated finely, petioles with glands. Flowers almost sessile with characteristic aroma, fruit small, yellow or greenish, flesh dry adhere to pit.

P. ansu Maxim. Leaves are broadly elliptic, wedge/cuneate at base, apex pointed, shiny, margin crenate-serrate, flower two per bud, fruits subglobose sulcate/grooved, highly pubescent, red, flesh greyish brown, sweet, free stone, stone sharp pointed.

P. dasycarpa Ehrh. Purple or Black apricot. Plants are usually small in general same as that of common apricot, leaves elliptic ovate, smaller and narrow compared to apricot, thin, dull green, margin serrated closely with or without glands on petioles. Flowers large, showy attractive and with long stalks, fruits plum like globular, dark purple (hence the specie name), pubescent, flesh soft but sour in taste, stone fuzzy.

Cultivars

Apricot although requires very exact climatic conditions for its successful culture, however it represents wide variation at varietal level. Some commercially grown cultivars along with their important pomological traits are detailed below

New Castle: This cultivars is very popular especially for its consistent regular bearing habit and also for its wide adaptation more so to the warmer temperate locations. Fruit ripens early – May end, fruits are roundish medium in size, suture prominent, pubescence present, yellow in colour, flesh yellow to orange, sweet juicy, stone partially free.

Shipley Early: Has performed exceptionally well in the foot hills, regarded as an excellent dessert cultivar, early ripening – mid May. The fruits are creamish white and overlaid by light pink – red blush on cheeks, flesh firm, white to creamish, very sweet and juicy, kernel however is bitter in taste.

Royal : A very popular cultivar of mid hills, in most of the characters it resembles Blenheim, tolerates high heat. Fruits ripen by mid June, large, yellow in colour, flesh firm yellow, sweet and juicy.

Nugget : This cultivars has gained popularity for its high fruit quality, regularity in bearing heavy crops, and is a highly adapted cultivar of mid-hill regions. Fruits are medium large, round in form, pubescence thin and soft surface shiny, dark yellow to orange blushed with bright red colour. Flesh firm, texture fine, deep orange, sweet and juicy, quality very good, kernel also sweet.

Charmagz : Cultivar has high chill requirements hence suitable for cultivation at higher altitudes, shy in bearing. The fruits are rounded, medium in size, skin straw yellow in colour surface pubescent attractive, flesh pale yellow, flavour aromatic and high, very sweet in taste, used both as dessert and for drying, being rich in sugar content is a very prime commodity for drying purpose.

Harogem : Evolved in Canada. Tree medium in vigour crown open, spreading self fruitful produces consistently, flowers white bloom mid season, long petiole with number of glands. Fruit round-ovate, sides compressed, bright red blush on attractive orange background, flesh orange firm, free stone with moderate juice, flavour good, keeping quality very good, stores easily at room temperature for several weeks.

❑❑❑

56
Plums

Genus : *Prunus*
Species : *domestica institia* L.
saliciana Lindl.
Family : *Rosaceae*
Chromosome No. : 2n - 48

Of all the stone fruits, plums are the most varied. They belong to the genus *Prunus* of the subfamily *Prunoideae* of the family Rosaceae. All plums have a basic chromosome number of x=8. Plums and related species have been placed in the subgenus Prunophora section Euprunus and Prunocerasus. Six species have been placed in the section Euprunus. Their chromosome number range from diploid to hexploid and are native to the land extending from Europe to China. Thirteen species native to the North American continent are all diploid in nature and have been placed in the sections *Prunocerasus.* Separation of the plums into the two sections does not appear to have any genetical basis since the species appear to be closely related. *Prunus domestica,* the European plum, is believed to be the most recent crop species to have arisen within the section *Prunophora. Prunus insititia. Prunus cerasifera* and *Prunus spinosa* are thought to be involved in its ancestry.

Plums are widely grown throughout the world. Europe produces about four times as many plums as prunes in the United Slates. Early United States settlers found native plums which were used by the Indians for food, growing in most parts of the country. The settlers brought with them the European plums *(Prunus domestica* L.) which were far superior in fruit characters to the native plums. These became well established except in the south and in colder regions of the country. About 1870, the Japanese plums *(Prunus salicina* Lindl.) were introduced, and soon became popular because of their size, productivity, and shipping qualities. About one-fourth of the plum crop in the United States are used for fresh consumption. With the exception of small quantity used for canning most of the remaining crop is dried to make prunes.A prune is a firm-fleshed plum fermenting at the pit. All prune cultivars belong to the species.

P domestica. Fruits of several prune type plums may be dried, and the product is known as dried plums. Some plums may be dried when halved, and the product is known as dried plums.

Plums belong to genus Prunus and include many species. Plums resemble other common fruits in the genus - e.g., peach , nectarines, apricots, cherries, and almonds - but differ from these fruits, except the cherry, in that the fruits are borne on stems. Peaches, nectarine, apricots, and almonds have short peduncles. Plums differ from cherries by having a flattened pit, usually longer than broad, cherries have globular pits. Plums inter-hybridize with apricots and with cherries. Plums inter-graft with other stone fruits with varying degrees of success.

Plums are placed within the *Prunoideae* subfamily of the Rosaceae. which contains all of the stone fruits such as peach, cherry, and apricot. The subgenus *Prunophora* contain plums and apricots. Hybrids between plums and apricots have been produced recently which are said to be finer fruits than either parent. A "Plumcot" is 50% plum, 50% apricot, an "Aprium" is 75% apricot, 25% plum, and most popular hybrid, the "Plutot" is 75% plum 25% apricot.

The cultivated plums belong to two species, viz., *Prunus domestica* L. (the European plum) and *Prunus salicina* (the Japanese plum). Other species like

Prunus spinosa (the black thorn sloe), *Prunus cerasifera* (the myrobalan plum). *Prunus ussuriensis* (the Ussurian plum), *Prunus nigra* (the Canada plum) and *Prunus americana* (the American plum) are of interest to the plum breeders for imparting winter hardiness to the cultivated plums or for use as rootstocks. Cultivars could be classified on the basis of (i) temporal relationship between flower bud burst and leaf burst (ii) the number of flower/bud, (iii) the diameter of the corolla at the beginning of the peak flowering stage, (iv) the spacing of petals in the corolla, (v) petal shape,(vi) calyx shape, (vii) sepal shape, (viii) pedicel length, and (ix) degree of pedicel pubescence.

Plums comprise of number of species but three are important horticulturally brief description is as follows.

1. European plums - *Prunus domestica* L. Worldwide this is one of the main specie grown. Fruits are generally oval, smaller, and more variable in color than Japanese plum. In the USA, *P. domestica* is used for prunes or fruit cocktail or other products, and rarely eaten fresh.
2. Japanese plums - *P. salicina* Lindl. and hybrids. These are most common fresh eating plums in USA.They are larger, rounded (or heart shaped), firmer than European plums and are primarily grown for fresh market.
3. Damsons, Bullace plums, St. Juliens, and Mirabelles- *P. insititia L.* These are small, wild plums native to Europe, cultivated there prior to the introduction *of P. domestica.* The 'St. Julien' types are used as dwarf rootstocks for plums. Fruit are small and oval (1 inch.) purple and clingstone for Damsons and yellow and freestone for Mirabelles, with heavy bloom. They are used primarily for jams/ jellies/ preserves.

Prunus domestica - European plum

The wild form of the European plum, which comprises the bulk of the cultivated plums, has never been found. Many botanists and horticulturists believe it to be a hybrid of the diploid myrobalan plum *(Prunus cerasifera)* and the tetraploid black thorn *(Prunus spinosa)* both of which occur wild in Europe and western Asia. The triploid hybrid so generated could have doubled its chromosome number resulting in the hexaploid, *Prunus domestica.* Recently it have been demonstrated that both *Prunus domestica* and *Prunus spinosa* carry the *Prunus cerasifera* genome suggesting thereby a more complex origin for *Prunus domestica.* This origin is supported by the fact that in *Prunus domestica* fruits exhibit colour and blue anthocyanin with very limited variation, *Prunus domestica* fruits exhibit both yellow and green ground colours and also both red and blue skin colours, with an unlimited range of variation. Fruit characteristics of size shape and flavour also lend support to the above origin.

European plum is a strong growing shrub or a strong growing short trunked tree, about 26-30 feet tall, with thin greyish to almost black bark, twigs pubescent.

The leaves are large and thick alternate, deciduous broadest near or above the middle, 5-10 cm long, pointed at the lip. Tapering at the base, ovate or obovate, coarsely and some what irregularly toothed along the margin, dull dark green in colour, much reticulated, paler and hairy below. The white flowers are large, occur singly or in clusters of two and three. The fruits are variable but usually egg-shaped, 2-3 cm long, sometime even more, firm in texture, not depressed blue or bluish black, with a sweet flesh enclosing a large stone which is slightly round and pitted.

The cultivated varieties of European plums have been classified into the following groups.

Prunes

This is commercially the most important group in the eastern world especially in north America continent. Blue or purple in color, the fruit is oval with bulging ventral side bilaterally. Flesh is firm and thick. The distinguishing feature of this group is high sugar content which makes them suitable for drying without removal of pit, thus giving rise to age old adage that all prunes are plums but all plums are not prunes.

Reine Claude or green gage plums

These are perhaps hybrids of *Prunus domestica* with *Prunus insititia.* The fruit is greenish yellow round and of high quality.

The fruit is characterised by yellow skin and fresh. Golden Drop is an important cultivar of this group.

Lombard plums

The fruit is purplish red and of good quality. Important cultivar of this group are Lombard, Bradshaw, Pond and Victoria.

Prunus salicina, Japanese plum

Also known as the Japanese plum, it is next in importance to the large fruited European plums, it is actually believed to have originated in China. It was introduced in Japan some 200 to 400 years ago from where it got disseminated around the world. Hence the name Japanese plum.The species is more vigorous, productive, precocious and resistant to diseases than the *Prunus domestica.*

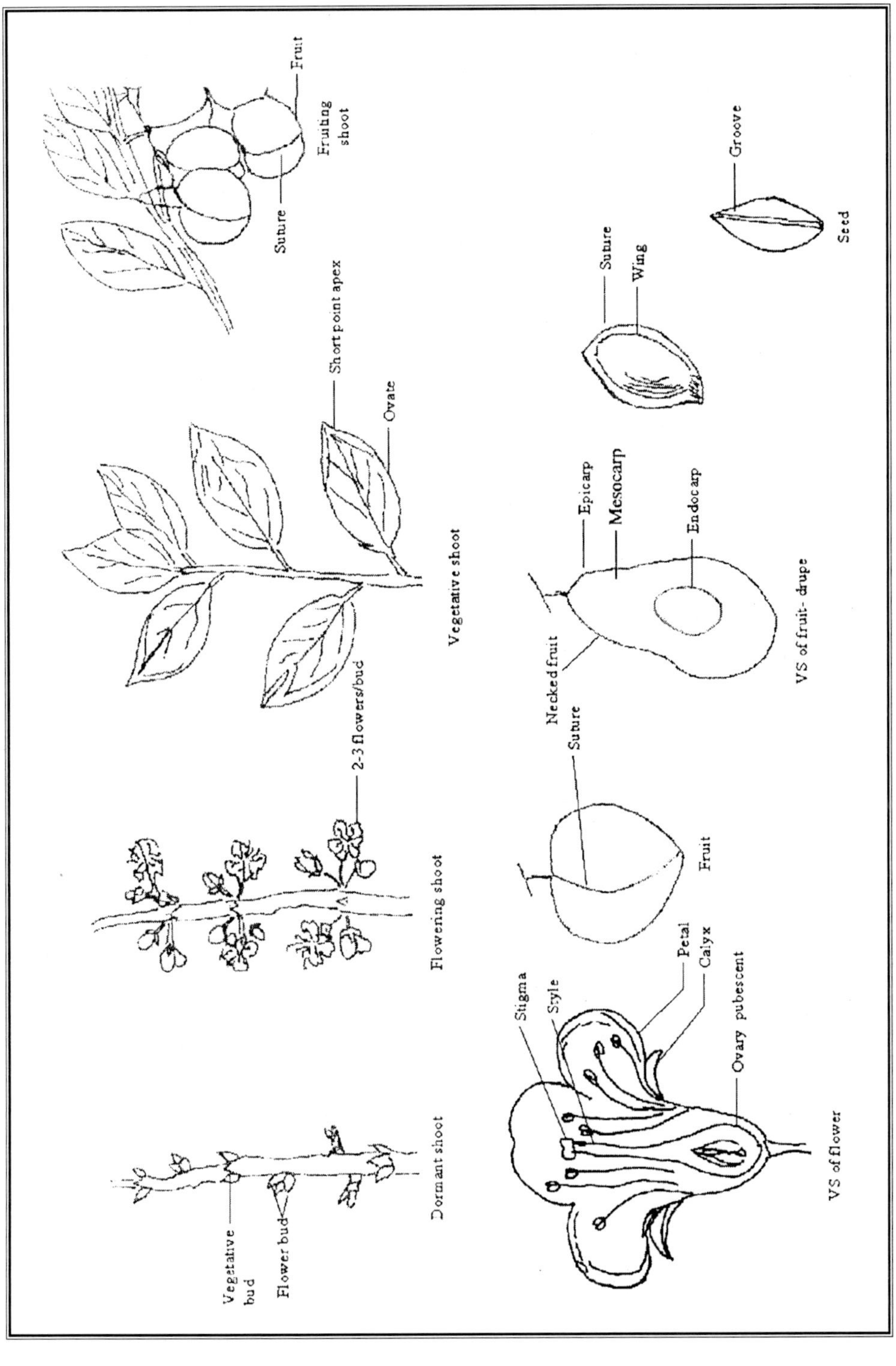
Fruit
Fruiting shoot
Suture
Short point apex
Ovate
Vegetative shoot
2-3 flowers/bud
Flowering shoot
Vegetative bud
Flower bud
Dormant shoot
Groove
Seed
Suture
Wing
Epicarp
Mesocarp
Endocarp
Necked fruit
Suture
VS of fruit- drupe
Fruit
Stigma
Style
Petal
Calyx
Ovary pubescent
VS of flower

It is characterized by medium sized sharp pointed leaves, that are free from hair on the lower surface. It comes into bloom quite early and thus is a leader for early market, but this feature also makes it very susceptible to spring frosts. The fruits of Japanese cultivars are usually large and heart shaped. Often with a pronounced apex which distinguishes them from other types of plums. A few cultivars are however, oblate or round. Tree growth habit varies considerably amongst cultivars. Some are low and spreading, while others exhibit an extremely upright growth. Flowers are produced three in a bud on either one year old shoot or on spur. Fruit set is heavier and more consistent on spurs.

Description

Japanese plum has great adaptability, is a hardy small strong growing tree, bears laterally on one year shoot and spurs, more productive, early to stand, bearing requires large amount of heat. It is free from most of the diseases except ground rot, can be easily identified by its rough bark, which with age develops deep longitudinal incisions. Young shoot (current growth) is shining smooth cinnamon-brown, later turns reddish in colour. Leaves, shining green above pale below, pubescence slight more on the axils of the vein on lower surface, shape oblong obovate, pointed at lip, margin serrated but rounded (obtuse), the veins distinctly looping near the margin. Pedicels smooth and glabrous. Flower white showy, pedicel thin/slender, green without pubescence. At each node there are three buds two lateral are floral each with 1-3 flowers per bud, central bud is vegetative,calyx 5, light green, slightly pubescent, corolla 5, white round, stamens 30 or more arise from the rim of calyx cup. Ovary inferior, apubescent, style single stigma solitary simple. Fruit variable usually large and firm with distinct cavity at stalk end and suture on the fruit surface, round with short apex (slightly heart shaped), yellow, light red or red in colour never bluish or blackish. Seed compressed, stone compressed longer then broad grooved on dorsal side and winged on the ventral suture.

Uses

Consumed fresh, sometimes dried, pulp is utilized to prepare different processed product like plum squash, beverage, appetizer, plum sauce, jam, chutney, wine etc., Plum pulp mixed with pulp of apple, apricot and other fruits to make mixed fruit jam. Oil extracted from seed is low in quantity and used for industrial purposes.

Prunus insititia L. Bullace Plum/Damson plum.

Origin and Description

Damson plums are native of western Asia and Europe and was introduced in north-eastern North America and frequently spotted along roadside, waste lands and fencerows.

Uses

Fruits are very bitter, eaten by animals and birds, reddish brown wood is hard, heavy and close grained, has little commercial value.

Description

The damson plum plants are bushy shrubs or small trees about 6 m high, tree compact, crown rounded. The branches are short, stiff and typical of its dense spiny branching system. The leaves are simple alternate, deciduous, dark green, paler beneath, broadest near or above middle, 3-5 cm long, tip pointed, tapering at base, margin doubly serrated. Flowering occurs singly or in small clusters of 2-3 laterally on previous year growth. Receptacle, cup/ bell shaped, calyx 5 lobed green, corolla 5, white appear in spring before leaves appear, broadly rounded 20-30 stamens emerging from the rim of calyx cup, style one. Fruit small (10-12 mm), rounded, dark blue or black, flesh thin bitter, encloses small flattened stone.

Prunus americana, American plum

This is a small fruited European plum. It is hexaploid like the *Prunus domestica* to which it appears to be related. It grows wild through Europe and western Asia. Plums of this species are commonly known as the Damsons, Bullaces and Mirabelles and extensively grown in western Britain. Damsons are believed to be native of the area around Damascus, where seeds of this species have been found in ancient ruins. The variability seen in this species is much less compared with *Prunus domestica*. The fruits are small and purple (in damsons) or yellow (in mirabelles). The damsons are highly prized for preserves and jam while mirabelles have an excellent flavour. The flavour is retained even after preservation for several years if necessary. The plants of this species are compact and form excellent hedgerows or shelter belts. Their silver grey blossoms and fruits are very ornamental in spite of the fact that this is a very hardy species, it has remained largely neglected.

Description

Plants are shrubs or small tree (11 m tall), trunk straight, crown broad and spreading, bark thin to thick, smooth when young, becomes rough with age into small plates dark brown to reddish brown in colour. Branches, spreading, dark reddish brown, twigs/branchlets stout stiff ending with sharp spine like apex. Dormant buds are small (3-8mm long), broadest at the base, apex pointed, covered with overlapping scales, light brown in colour. Leaves smooth dark green above, pale beneath, simple alternate, deciduous, 6-10cm long, 3-5 cm wide, broadest at or above the middle, apex sharply pointed, tapering or rounded at the base, margins doubly serrated, petiole short with two conspicuous large glands at the base of lamina and on the petiole. Flowering takes place laterally on previous seasons growth, flowers appear with or before leaf opening in cluster of 2-5, flower stalk is thin slender, flower pentamerous, receptacle is cup or bell shaped, sepals 5 green, petals 5 while, rounded (9-12 mm long), stamens 20-30, pistil single. Fruit occur singly or in clusters of 2-4 rounded or slightly elongated, orange to red, suture faint, more evident at base less at apex, flesh bright yellow, juicy enclosing a flattened stone.

Important species under American plum are

A. American group

1. *P. americana* - Common wild plum
2. *P. maxicana* - Big tree plum
3. *P. nigra* - Canada plum

B. Chickasaw wild goose group

1. *P. hortulana*
2. *F. angustifolia* - Chickasaw plum
3. *P. munsoniana* - Wild goose plum

C. Beach or shore plum group

1. *P. marithima* - Beach plum
2. *P. gracilis* - Oklahoma plum

D. Pacific coast plum

1. *P. subcordata* - Specific coast plum.

Species of American plum

P. subcordata Benth. Pacific plum, Small tree, bark is blackish, fruit about 1 inch in diameter inferior in quality, fresh sub-acidic, cling stone.

P. mexicana Wats, Big tree plum. Has number of syn. *P. reticulate, P.polyandra.*

P.arkansana. Plants are tree like, leaves oblong-obovate,3-5 inches long, base rounded apex abruptly pointed pubescent more beneath, margin doubly serrated, petiole with 1 or more glands, flowers about 1 inch across white, fruit globose, purple red, bloom bluish, stone globular, smooth and turgid.

P. angustifolia Marsh. Chickasaw plum. Mountain cherry. Small tree 10ft. or more, top spreading, branches zigzag thin, red in color, leaves lanceolate to oblong lanceolate small sized about 2 inch long, base narrow apex short pointed, soft pubescence on veins below, margin finally serrated. Flower small white, fruit small cherry like yellow or red shiny, fresh soft juicy clingstone, stone rough.

P. munsoniana. Wight and Hedr. Wild Goose plum. Plants medium sized 20-50 feet tall. Leaves long lanceolate to oblong lanceolate 3-4 inch long base obtuse apex pointed, margin serrated, upper surface shiny slightly pubescent on veins, petiole with glands 2 or more, flower white ½ inch across, fruit oval or globular, yellowish or bright red, stone oval surface rough.

Besides commercially significant species described other species that hold substantial promise are described below.

1. *P. cerasifera* Ehrh. Cherry plum. Plant are shrub like or small tree, thorny, leaves slightly pubescent, later turns shiny, thin, small, light green ovate in form, apex with short tip, margin serrated finely. Flower small white, pedicle short. Fruit also small, round or globular, usually yellow or red, flesh yellow, soft juicy, sweet with characteristic flavour.
2. *P. divaricata* Bailey. Branches are spreading and start low on the trunk. Leaves obtuse at base, fruit not depressed near the stone, yellow in colour.
3. *F. spinosa* Linn. Blackthorn. Stiff thorny growth acts as effective barrier. Plants with thorny growth, branches low on trunk and spreading, young growth pubescent. Leaves small elliptic ovate or oblong obovate, margin finely serrated. Flowers white, solitary or in pairs, fruit small deep blue shiny, quality poor. Wickson plum is a hybrid of *P. salicina* and this specie.
4. *P. simonii* Carr. Simon or Apricot plum. Fast growing plant, leaves thick dull green in colour, veins conspicuous beneath, shape lance-obovate or long oblanceolate. flower white, stalk short, fruit small 1-2 inch in diameter smooth, red- maroon in colour with deep suture, flesh yellow adhere to the seed.

The other species, which are of less significance, as they do not produce fruit of desirable fruit quality is-

P. thibetura, French Ornamental tree, flowers are pink blushed.

Some very popular cultivars of each species is briefly described.

Santa Rosa : (*P. salicina*). Prolific and regular bearer, self fruitful with characteristic flavour, trees are upright, fruits are large purplish red in colour with whitish bloom, flesh golden yellow, red near the skin, sweet, juicy, acidic at seed. Skin thick.

Mariposa : Japanese plum : Originated as chance seedling, tree grows upright, vigorous, needs cross pollination for good fruit set. Fruit large, heart shaped, skin greenish yellow to maroon, thick flesh red firm juicy flavour sweet, stone free, quality good.

Methley : Japanese plum : Tree vigorous, upright, hardy, low chilling productive, self fruitful. Fruit small to medium, round, skin purple blushed with red, flesh red soft very juicy, sweet, flavour pleasant, quality good, ripens early in the season.

Kelsey : Japanese plum. Plants with moderate growth, upright and low in chill requirement, bears moderately and regularly. Fruit large roundish oval, apex distinct, suture deep towards stem end. Skin greenish with bright red blush and greenish dots, flesh yellowish with firm texture, juicy, sweet sub acidic, quality medium, keeping quality good, semi cling stone.

Greengage : European plum. Famous cultivar of Reine Claude group. Tree medium in growth, round top, flowers medium to late in season. Fruit small round with vivid greenish yellow, yellow skin, mottled with red has finest flavours amongst plums, flesh firm, greenish yellow very juicy tender, sweet joy to eat, stone partly clingstone, round and thick.

Early Transparent Gage : Reine Claude European plum cultivar. Tree dwarf, growth compact, leaves oval intitally brownish, flowers early, self compatible, regular bearer. Fruit small and roundish, ends flat, green with primrose bloom with red dots, flesh yellow, juicy very sweet characteristic of gage flavour, very good as dessert and excellent for jam and canning.

Victoria : European plum. Tree small, branches spreading, chilling requirement high bears regularly with medium to heavy yields. Fruit oval, suture deep, apex round skin colour strawgreen, pink to red when ripe with green dots, flesh deep red firm juicy, sweet acidic, free stone, stores well, quality good canning quality excellent.

❑❑❑

57 Cherries

Genus : *Prunus*
Species : *avium cerasus, mahaleb* L.
Family : Rosaceae
Chromosome No. : 2n = 16, 24, 32

Records indicate that cherries were in cultivation ever since the beginning of civilization. Cherries exhibit vast genetic variability as is evident from the different species which vary in their ploidy level. Sweet cherries (Prunus avium L) are diploid while sour cherries *(P. cerasus,* L.) are tetraploid, and commercially more important than the other species. Duke cherry has resulted from hybridization between sour and sweet cherry with intermediate tree and fruit characters. Successful cherry cultivation revolves around climatic factors, prime being the extent and spread of rainfall, variation in day and night temperature, occurrence of frosts, high heat day largely determine the type of cherry that will grow successfully in a particular areas. Hence cherry improvement around the world has so far been localized to few selected areas where different cherries have been successfully grown. Sour cherries being hardy succeed in climate which are considered to be too severe for sweet cherries. Concerted efforts are required to develop cherries with wide adaptability so that they can be grown in more extensive areas.

Brief pomological description of sweet, sour and mahaleb cherries is briefly described:

Origin and Distribution

Sweet Cherry - *Prunus avium* L.

Uses

Sweet juicy fruits are eaten fresh, used for canning pulp used to make Jam, beverages, wine etc., seeds yield oil in small quantities.

Description

The sweet cherry plants are wide spreading trees, may reach height of 20 meters, bark smooth grey to reddish brown. The leaves are simple alternate, dull green pubescent below, broadest at or near the base 6-12 cm long, short pointed at tip, margin doubly serrated. Flowering takes place laterally on two or more year old shoots or spurs, flowers solitary or in clusters of 3-5, pentamerous, calyx 5 green lobed, urn shaped, corolla 5 white rounded, stamens 15-25, pistil single. Fruits are round to heart shaped, 2-2.5 cm in diameter, yellow, red, dark red to black colour, flesh red juicy, sweet firm to tender with distinct flavour, enclosing round turgid stone.

Sour Cherry - *Prunus cerasus* L.

Description

Sour cherry plants are less tall and spreading (15 m), short trunked, bark with thin scales and it splits horizontally, greyish to brown coloured. The leaves are simple, alternate, deciduous, light green, shiny round to cordate (3-5 cm long) apex acute, finely serrated along the margin. Flowering occur laterally on two years wood and on spurs, flowers are white sometimes pink tinged produced in cluster of 2-5 flowers, which are narrow and elongated. Flower pentamerous, calyx, light green 5 lobed, urn shaped, corolla 5, white which are broadest and rounded at tip, stamens 20-30, pistil, single. Fruit small (0.8 – 1.2 cm. in diameter) reddish to reddish-black, flesh light coloured, juicy, firm, sour in taste, stone small round turgid.

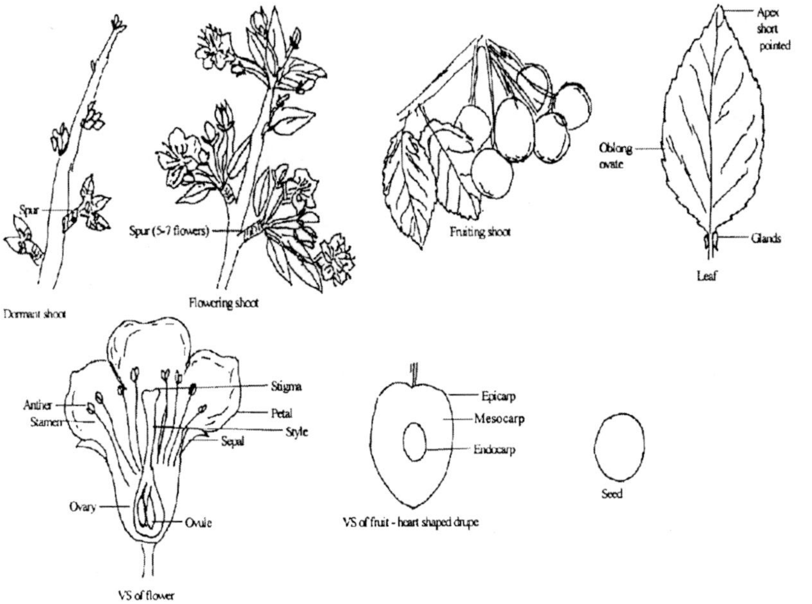
Spur
Dormant shoot
Spur (5-7 flowers)
Flowering shoot
Fruiting shoot
Apex short pointed
Oblong ovate
Glands
Leaf
Anther
Stamen
Stigma
Petal
Style
Sepal
Ovary
Ovule
VS of flower
Epicarp
Mesocarp
Endocarp
VS of fruit - heart shaped drupe
Seed

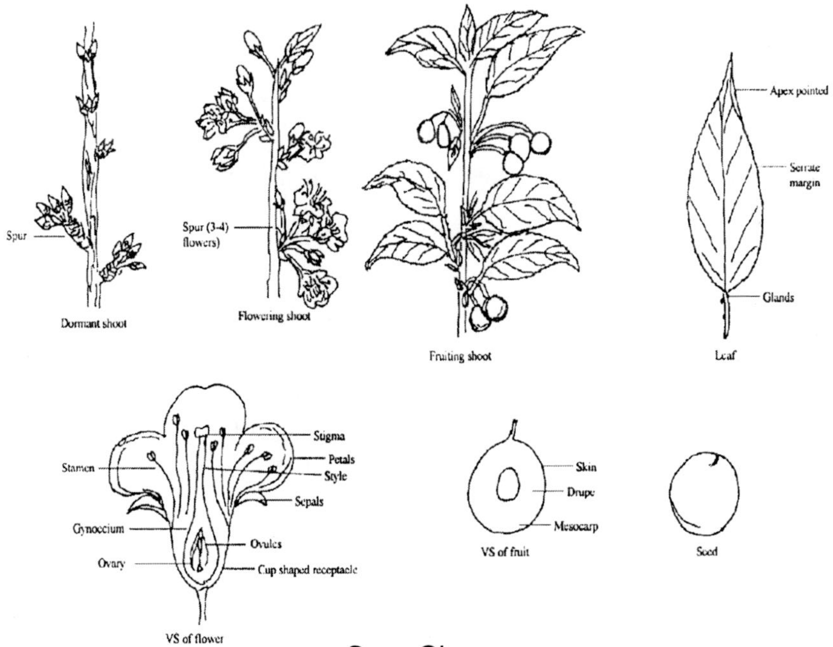
Spur
Dormant shoot
Spur (3-4) flowers)
Flowering shoot
Fruiting shoot
Apex pointed
Serrate margin
Glands
Leaf
Stigma
Petals
Style
Stamen
Sepals
Gynoecium
Ovules
Ovary
Cup shaped receptacle
VS of flower
Skin
Drupe
Mesocarp
VS of fruit
Seed

Sour Cherry

Mahaleb Cherry *Prunus mehalab* L.

Origin and Distribution

Is native of Europe and western Asia.

Description

Mahaleb cherry is a small tree (8 m) or a shrub, crown broad and rounded, bark dark grey, thin. The leaves are simple, alternate, deciduous, light green in colour, round (2.5-5 cm) long and wide, cordate at base with short pointed tip, margins are finely serrated. Flowering takes place laterally on previous seasons growth and spurs in clusters of 6-10 flowers, flowers are pentamerous, calyx 5 lobed light green, corolla 5 white, stamens 20-30, style, one. Fruit oval to round, small (5-8 mm), reddish black, skin smooth, flesh, enclosing a small rounded stone.

Cerasus, the common or fascicled cherries are generally characterized as having flowers is cymes or fascicles, fruit glabrous or with faint pubescence very juicy, stone may be smooth or rough.

The specie *Prunus avium* has number of ornamental or botanical types important being.

P. avium var. variegate Hort; the foliage has characteristic yellow and dull spots.

P. avium var. plena Hort; double flowers.

----- var. salicifolia Dipp. leaves are very narrow.

----- var. pyramidalis Hort; the crown of the tree is in the form of a pyramid.

----- var. pendula Hurt; the branches are drooping type.

Prunus mahaleb Linn. (*Prunus odorata Lan. Padus mahaleb, Borkh.*). Mahaleb cherry is wide spread in central and south Europe and the Caucasus region. Plants are small thin, branchlets shiny, leaves round ovate or rounded light green, 2-3 inch long, apex short acute, slightly cordate at base, margin serrate, flowers are white, small in terminal umbels, fragrant. Fruit dark red very small and inedible. Used as root stock. Has number of botanical forms.

Prunus mahaleb var. variegate. Foliage is variegated.

----- var. pendula. Branches are drooping or weeping.

----- var. campacta. Tree top compact.

----- var. chrysocarpa. Fruits are generally yellow in color.

----- var. marginata. Leaves are with white edges.

Other well established cherry species along with their distinctive features are described as below.

Prunus pumila Linn. Sand or Dwarf Cherry. Has ornamental value. Early growth erect reddish in colour and shiny but becomes spreading with age, dull green on ventral and pale green beneath, narrowly oblanceolate, apex short pointed or rounded. Flowers in umbels in 2-5 flowered clusters, thin pedicels. Fruits round purple black in colour, fruit stalk slender.

P. besseyi Bailey. Western sand cherry. More spreading in growth than *P. pumila,* leaves thick broad, generally elliptic oval or elliptic lanceolate, stipules green, larger and serrate, fruits are on short stalks, sweet in taste about 1 inch in diameter.

P. tomentosa Thunb. Common in Japan, Manchuria and N. China. Plants are small and compact trees wide spreading, branches are very close and young growth very pubescent, leaves broad oval to short obovate, pedicle short, margin serrate, thickly pubescent below. Calyx bright red in color, petals white flowers appear before leaves. Fruit globular, stalk very small, small in size and light red in color, with soft pubescence, fruit edible.

P. cerasus. Linn. Sour Pie or Morello Cherry. Native to Asia Minor and to south east Europe, the specie sprouts from roots, forms dense thickets. Two pomological groups are well established, one with coloured juice (Morellos) and with uncoloured (Amarelles). A number of botanical types derived from Latin names have been applied to the cherries in this group, pronounced form are as.

P. cerasus var. typica, Scheid. Tree forms are similar to those of the cultivated types.

---------- var. frutescens Schneid (P. *acida,* Koch). Bushy plants with small fruits, usually wild types.

---------- var. persiciflora Koch. Double flower type petals pink or rose coloured.

---------- var. variegate Hort. Leaves are variegated with white and yellow colour.

---------- var. globosa Spaeth. Bush with rounded top, leaves small.

This specie is also said to be divided mainly into two groups Eucerasus and Acida. Eucerasus consists of three types, branches erect strong and later drops, petioles with or without gland. Fruit globular, bigger in size seed round Acid group comprises of ornamental bush like plants, branches are drooping petioles short with glands. Fruits globular, small, seed ovoid upto half inch in length.

P. cerasus var. semperflorens Loud. Ever blooming cherry. This cherry flowers throughout entire summer, is considered as an important ornamental type. Bush or a small tree with drooping branches. Leaves oval to oblong obovate, apex short pointed, margin dentate texture firm, flowers occur on axillary or terminal long pedicels. Fruits are with long pedicels small in size (Pie Cherry) dark red in colour.

P. cerasoides D. Don (*P. Puddum,* Roxb). Native of temperate Himalayan region and found in abundance at an elevation of 3000-8000 feet height. Is considered to be a member of *P. pseudo cerasus.* Plants are large with attractive shiny green and pink or rose coloured flowers. The young shoots however are slightly pubescent, leaves ovate lanceolate or oblong lanceolate, apex sharp pointed, margin serrate shiny, petiole with 2-4 glands. Flowers appear singly, fascicled or in umbel cluster, calyx tube bell shaped, petals linear or obovate. Fruit ellipsoid or oblong with both apices round, stone furrowed.

--------- var. campanulata Maxim. (*P. cerasoides var. campanulata, Koidz.*). Himalayan specie closely resembles *P. cerasoides.* Ornamental widely cultivated in Japan. Tree medium tall, leaves ovate to elliptic ovate, glossy, margin double serrate. Flowers are hanging campanulate, dark rose in color, calyx purple coloured. Fruits red ovoid.

P. microcarpa C.A. Meyer. Indigenous to western Asia and Baluchistan. Shrub, branches are long, leaves shiny margin sharply serrate, petiole smaller than blade. Flower, ½ inch across, calyx tube cylindrical wide at base, fleshy drupe, ovoid about ¼ inch long.

P. rufa Wall. Native to Sikkim, Nepal and Tibet, grow at an elevation of 9-12000 feet. Small tree, young growth pubescent, leaves elliptic-lanceolate, margin serrated sharply, petiole and veins pubescent beneath. Flowers pink single or fascicled, pedicels about 1 ½ inch long. Fruit red and fleshy.

P. Jacquemontii Hook f., found growing in Baluchistan and W. Himalayan region. Small shrub, leaves deeply and sharply serrate, leaf blade 1-2 inch long Flower single pink, pedicel very small, calyx tube funnel shaped, fruit bright red.

Another group of cherries the Padus or racemose cherries in which flowers appear in raceme is well recognized under Laurocerasus which is characterized by small globular inedible fruits, flowers usually white and small occur in racemes relatively late in season, plants are relatively big trees. Important species of this group are described as.

P. serotina Ehrh. (*Padus serotina*, Agardh). Wild black cherry. Indigenous to N. America. Tree stout and straight growing, bark dark brown bitter, leaves oblong or oblong ovate, firm thick, shiny above, margins with incurved teeth. Flowers occur on long and loose racemes and appear when leaves are fully grown. Fruit very small, bitter in taste, purple black in color ripens late in summer upto september. Specie has number of botanical forms.

P. serotina var. pendula Dipp., branches are dropping.

---------- var. variegate, Hort., leaves are with yellow mark or pattern.

---------- var. cartilaginea Diff., foliage very attractive distinctly with long and shiny leaves.

---------- var. pyramidalis Zabel. Plant canopy is in form of a pyramid.

P. padus Linn. (P. *racemosa, Lam. Padus vulgars, Borkh. Cerasus padus,* DC). European Bird Cherry. Native of Europe and Asia. Small tree bark rough and specked with strong odour, leaves elliptic to oblong ovate to oval shiny, base broad, apex sharp pointed, margins serrated sharply, glands present on petioles. Flowers are larger and appear after leaves. Fruit inferior in quality stone round and rough. Has many varied botanical forms.

P. padus var. pendula Dipp., drooping branches.

---------- var. variegate Hort., has different colour pattern on leaves.

---------- var. bracteosa Sers., at the base of raceme, large leaves are present.

---------- var. plena Hort., double flowered form.

---------- var. cornuta Henry. Native to the Himalaya region, leaves bluish green on the dorsal side with reddish pubescence along the veins, base round, racemes hairy, fruit small about ½ inch across stone smooth.

Cultivars

Cherries represent vide variation for number of characters like tree form, leaf morphology, flower number and color, fruit shape, size color, taste, quality etc. Some salient features of important cultivars are described below.

Sweet Cherry

Early Rivers : A very old cultivars but still is in cultivation due to number of desirable characters like resistance to cracking, bears regularly though in fair quantities. One of the finest early ripening cultivars, ripens early May, fruit large oval black in colour, surface smooth and glossy, stone small, rounded, flesh pink to reddish highly flavoured, texture firm, sweet in taste.

Bing: A very popular and commercially grown cherry cultivar in almost all cherry growing areas of the world. Tree upright and spreading, bears moderate to profusely. Fruits are large to very large, roundish, skin red black coloured, flesh dark red very firm and crisp, sweet in taste, very flavourful and aromatic, juice dark red, export cultivar because flesh is firm, may be pollinated by Burlat, Van etc. Major defects are fruit cracking in rain and flowers are very tender to low temperature, self sterile.

Compact Lambert : Has been obtained from Lambert through induced radiation mutation. The plants are very dwarf i.e. 80-90 percent of the standard cherry plant on a vigorous rootstock, foliage is dense, flowering occurs late hence escapes frost damage, bears heavily and fruit cracking is also less. Fruit medium sized, black in colour.

Compact Stella : Cultivar is radiation induced mutant of Stella, developed at Summerland, Canada, is suitable pollinizer for number of cultivars, fruit cracking due to rain is moderate. Trees are semi dwarf and about 50% of the size of parent cultivar Stella, precocious in bearing, self compatible, bears heavily. Fruits are medium to large, oval, heart shaped, black in colour, flesh moderately firm, texture partially firm, fine, flavour also moderate. Highly rated cultivar both as dessert and for canning purposes.

Van: Another very useful cultivar from Summerland, Canada, is an open pollinated seedling of 'Empress Eugenie' self sterile. Trees are much hardy and vigorous than many of the other sweet cherry cultivars. Starts bearing when young, very productive. Fruits are large, but on heavy bearing they are small and round, attractive with bright glossy red skin, not too thick, not very juicy and pale red in colour, fruit stalk very small, picking therefore is difficult cracking is less than Bing, flesh red, fine, texture firm with very good flavour.

Celeste : Self fruitful cultivar selected at Summerland Research Station in British Columbia, Canada. Plants are vigorous, growth compact, not very early in bearing, very productive, fruit matures early, large, stalk small, flesh dark, firmness medium, flavour good.

Sunburst : Self fruitful, selected at Summerland, Canada. Tree medium in vigour, branches spreading, precocious, mid to late in flowering fruit large, slightly heart-shaped, spheroidal. Skin shiny red, flesh not very thick, pale red, juicy, taste good aromatic ,stone adhere to flesh.

Lapins : Evolved from a cross between Van x Stella at Summerland, Canada. Tree upright, grows fast, productive. Flowers early in season. Fruit large-very large, round to heart shaped, cracking is less. Skin purple red to dark red, flesh firm juicy, sweet, somewhat acidic.

Sour Cherry

The sour cherry cultivars are classified into two groups on the basis of colour of fruit juice. Sour cherries with light red flesh and juice are known as Amarelle. Fruits are light red in colour mostly flattened at the ends and are less acidic. Montmorency and Early Richmond belong to the group. The others group Morello, comprises of cultivars which have very dark coloured fruits, juice usually reddish in colour. Fruits are more spherical or are heart shaped compared to Amarelles. Cultivars English Morello, Olivet and Ostheim belong to this group.

Brief description of well know sour cherry cultivars is summarized as below.

Montmorency : A very old French cultivar originated at Montmorency, and is perhaps the only cultivar which is grown commercially. Trees are very hardy and productive. Self fertile cultivars responds to good cultural practices.

Early Richmond : The trees are more hardier than Montmorency, fruits ripen early however the quality is inferior or just fair as compared to Montmorency.

English Morello : Plants of this cultivar are somewhat smaller, hardy in nature, produces regularly and moderately. Fruits are of medium size roundish dark red in colour, juice dark red in colour.

North Star : Originated as seedling selection from cross between English Morello x Serbian Pie No 1. Highly rated for its hardiness and self fertility. Plants are slender very small with thin drooping branches very productive bears regularly with moderate to high yields. Fruits are round medium sized. Both fruit skin and juice are intense red. Is also being tested as a dwarfing interstock for sweet cherries.

❏❏❏

58 Olive

Genus : *Olea*
Species : *europaea L.*
Family : *Oleaceae*
Chromosome No. : 2n=2x=46

Amongst the fruit crops, Olive is undoubtedly one of the world's oldest cultivated crop and its cultivations dates back to 3000 B.C. and till date is recognized as a symbol of peace and prosperity. Olive is an evergreen subtropical plant, but for flowering and fruiting it does require chilling as most of the temperate fruits do. Although slow to grow, olive is a very long lived plant and with its typical dense foliage of olive green and silvery leaves also serves as an ornamental type. The centre of origin includes Palestine, Lebanon, north west Syria and Cyprus. The Greeks and Roman were well aware of the significance of olive oil during 900 to 600 B.C. and oil was considered as a luxury item in body anointment. From its native place olive established itself in the Mediterranean region especially Italy, Spain and from here to other parts of the world. Today majority of the olive producing countries lie in the Mediterranean basin and Spain is the leading olive oil producing country. Other important olive oil and fruit producing countries are Italy, Greece, Turkey Tunisia, Portugal, Morocco, France, Syria, Algeria, Jordan, Cyprus, Israel, Argentina and united states of America.

Ripe olive fruits contain substantial amount of proteins, sugars, minerals and vitamins B, C and A. Very few cultivars are suitable for table purposes and in majority oil is extracted, fruit is pickled, used in salad and is important from medicine point of view. Olive oil is used commercially in cooking, perfume and soap industry, as textile lubricant, sulphonated oils and as laxative. The precious olive oil is very nutritious, tasty, rich in polysaturated fatty acids resists cholesterol formation, regulates high blood pressure and is safe for people suffering heart and other ailments. Is a potential baby oil very effective for human skin, muscular pain, messaging and provides warmth in cold weather. Mature olive fruits are astringent and the substance responsible for bitterness is 'Oleuropein'. The bitterness is usually removed by treating the fruits with 1.5 to 2.0 percent sodium hydroxide.

The genus Olea comprises of about 30 species of which only *europaea* produces edible fruits and belongs to the family Oleaceae. Other species which do not yield edible fruits are indigenous to New Zealand, India, Afghanistan and S. Africa. Mostly all of the cultivated olives are diploid with 2n=2x=16, but triploid and tetraploid forms are also present.

Origin and Distribution

Olive is considered to be an ancient and one of the most characteristic plant of Mediterranean and originated as hybrid swarm in mountains of eastern Mediterranean region. It is one of the relicts of the Tropical mid Tertiary flora of Mediterranean, with two botanical varieties being recognized, cultivated sativa and wild oleaster. Olea has about 35 species, distributed from Africa to New Zealand. In this regard three species are important which are *O. laperrinii* and its related forms are found in the Macoronesian archipelago, and north Africa, Red Sea and Persian Gulf. *O. Africana* is distributed from far southern Africa, Sudan and Egypt border and *O. ferruginea* is native from Iran to Himalayas. From eastern Mediterranean, the crop moved westward resulting in a secondary centre of diversity in the Aegean area. The crop was also known to the Romans by 600 B.C. Chance seedlings have resulted in supply of new clones.

Flowering shoot

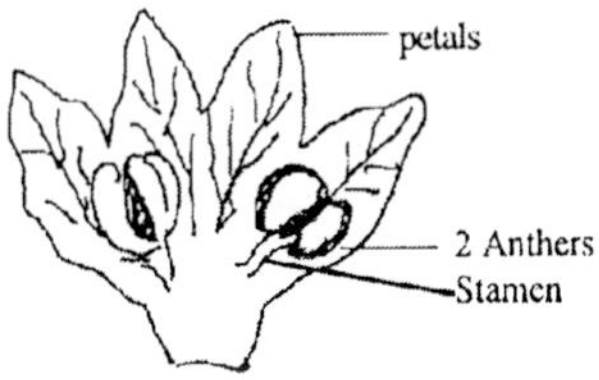

VS flower showing petals and stamens

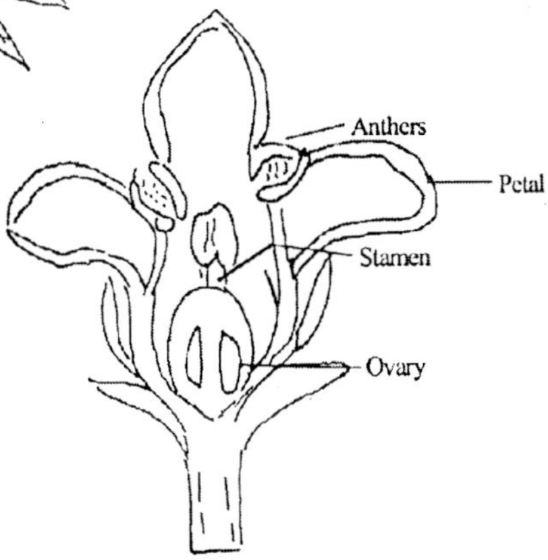

VS of flowers

Fruit

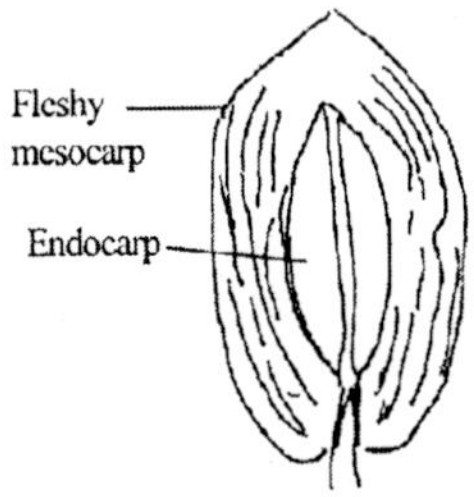

VS of fruit

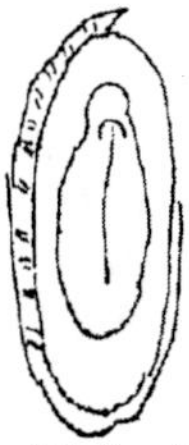

VS of seed

Uses

Ripe olives contains sugars, minerals, proteins, vitamins B,C, and A. Fruits are mainly utilised for oil extraction, for table purposes, pickled and used as salad, is also important medically. Oil is used in cooking and in perfume industry. Oil is tasty nutritious, free from cholesterol has polyunsaturated fatty acids, regulate high blood pressure, is good for human skin and massaging, releaves muscular pains, used in textile lubricant, sulphonated oil and as laxative.

Description

Plants are generally trees or woody shrubs, sometime may be scandent or twining, leaves simple or compound, dark green above, silvery beneath, margin entire, obovate–oblanceolate, exstipulate. Flowers generally bisexual, hypogynous, sometimes may be unisexual, actinomorphic, sepals 4, light green–pale yellow small, petals 4-5 small, creamish white, stamens 2 rarely 4 and gynoecium 2, in terminal or axillary cymes, panicles, anthers bilobed. Ovary superior 2 celled with 2 ovules in each ovarian chamber, style simple, stigma bilobed. Fruit is a typical drupe, green later turns purple to purple black in colour, flesh soft, white to creamish at maturity, round ovate to oblong shaped. Endocarp thick hard, shape oblong.

It is often expressed that the Euro Mediterranean olive that is *O. sativa* or *O. oleaster(wild)* the *laperrina* and Asiatic *cuspidata* are the three main species of *O. europaea*.

Brief horticultural description of some species is summarized below:

Olea chrysophylla. Found in abundance in Asia and N. Africa. The specie is highly polymorphic and number of its geographic types are common in its native place. Plants are small trees, branches thin and somewhat upright, leaves are long lanceolate (5-10 cm), petioles relatively very short. Fruit is a true drupe, globose sometimes long oblong purple to black in colour.

O. laperrini. Wide spread in Morocco and central Sahara. Is regarded as an intermediate taxa between *O. europaea* and *O. chrysophylla*.

Olea europaea var. communis (cultivated olive). Plants are handsome with typical olive green silvery foliage, medium tall, young shoot green later turns grey in colour, branches wide angled, leaves lanceolate, fruits are oblong or globular.

O. europaea var oleaster or var. sylvestris (wild olive). Plant medium tall, 20 feet sometime even more, thorny, branches angular, leaves oblong elliptic, fruit small round inedible, used mainly, as rootstock.

O. cuspidata (Indian olive) syn *O. ferruginea*. Plants of this specie are found in abundance in the foot hills of the Himalayas from Kashmir to Kumaon hills of Uttaranchal upto an elevation of 2400 meters. Tree medium tall, leaves oblong lanceolate, fruit ovoid dark purple to almost black when ripe, unpleasant in taste.

O. glandulifera (wild olive). Plant small bushy glabrous tree with dull grey bark, branches spreading and alternate, leaves elliptic-lanceolate somewhat bigger in size, fruits usually inedible, small round to oblong, endocarp crustaceous, flower white bisexual in terminal and lateral compound pyramidal panicle, petal round.

Cultivars

On the basis of use the olive cultivar are classified as oil type and pickle type. Some important cultivars along with their distinct characters are described below.

Frontoio : Tree are medium tall, crown rounded, branches drooping regular and heavy bearer. Leaves medium sized, lanceolate, bright green above, silvery below margin entire. Inflorescence pale yellow, quite large and sparse, flowers paniculate. Fruits are purple black, oval elongate, size medium, ripens late. Oil content varies from 23-25 percent. Stone large, oval elongate, surface furrowed.

Ascolano : Originated in Italy at Merche. Plants are fast growing, upright in stature, crown dense and compact. Leaves elliptic of medium size, bright green above, silvery below, margin entire. Inflorescence few and small, light yellow, flowers borne in paniculate thyrse. Fruits, large, spheroid, somewhat asymmetric, almost black when ripe. Stone medium, elliptic-elongate, surface rough, asymmetric, pulp stone ratio relatively high with oil percentage varying between 22-24. Resistant to cold and olive fly.

Leccino : Originated in Italy at Toscana. Resistant to low temperature but susceptible to sooty mould. Tree vigorous, branches spreading crown open. Leaves elliptic-lanceolate, light green, silvery below. Inflorescence few and small, flowers occur as racemose. Fruits spheroid, medium sized apex round purple black at picking. Stone shape as that of fruit, asymmetric and rough. Bears regularly, yield fair with 25-30 percent oil.

Coratina : Italian cultivar. Tree semi vigorous, crown open, branches somewhat drooping. Leaves dark olive green silvery white beneath, elliptic. Inflorescence short, flowers with paniculate thyrse. Fruits black bigger in size, oval somewhat asymmetric, ripens late. Oil varies from 23-25 percent, heavy cropper stone shape like that of the fruit, surface furrowed. Resistant to cold and drought susceptible to olive fly, knot etc.

Picholine : French cultivar, tree moderate in growth, upright, crown open. Leaves pale green, elliptic-lanceolate, entire, tomentose below. Inflorescence compact, large with racemose flowers. Fruits, green spheroid-elongate, apex pointed stone small, oval, surface smooth late in ripening. Medium in yield with 16-18 percent oil.

Pendolino : Tree with moderate growth, crown open, branches drooping, leaves bright green lanceolate, entire tomentose below. Inflorescence like that of picholine. Fruits, oval medium sized, asymmetric oval, apex rounded, black when ripe. Stone spheroidal, medium in size, surface smooth. High cropper, oil percent ranges from 23-25, susceptible to cold and cycloconium.

❏❏❏

Glossary

Horticultural Terms

Term	Meaning
Abaxial	The side of an organ away from the axis or center of the axis, dorsal.
Abortive	Defective, barren, not developed.
Abrupt	Changing suddenly rather than gradually, as a leaf that is narrowed quickly to a point or as a pinnate leaf that has no terminal leaflet, not tapering.
Accessory fruit	Fruit in which fleshy parts are not derived from the pistil i.e. strawberry edible flesh is the enlarged receptacle the real fruits (achenes) are embedded in its surface
Achene	One chambered, small dry one chambered and seeded fruit
Achlamydeous	Flowers without perianth.
Acropetal	Developing in a longitudinal plane from a lower toward a more apical position. The opposite of basipetal.

Actinomorphic	Regular radial symmetry star shaped flower.
Acute	Sharp, ending in a point.
Adaxial	The side toward the axis, ventral.
Adherent	A condition existing when two unlike organs or parts touch each other but are not grown or fused together.
Adnate	Grown together, united with another part as the stamens on the corolla. Fusion of unlike parts.
Adventitious	Appearing but not from usual place like roots from cuttings of leaves or stem.
Affinity	A plant or part closely related to another or much resembling it in structure.
Albuminous	Presence of endosperm or albumen.
Androecium	Collective term for stamens.
Androgynous	Male and female present in the same inflorescence.
Angiosperms	Plants in which seeds are enclosed in the ovary.
Anther	The pollen-bearing part of the stamen, borne at the top of the filament or sometimes sessile.
Anthesis	Expansion or opening of a flower.
Apetalous	Without petals.
Apocarpous	Carpels separate, not fused .
Apomict	A plant produced without fertilization or fusion of egg and sperm.
Aposepalous	Sepals distinct.
Appendage	An attached secondary part as a projection or hanging part.
Appressed	Closely and flatly pressed against, adpressed.
Aril	A fleshy juice from or around the hilum of a seed sometimes partially or completely covering the seed.
Armed	Provided with any kind of strong and sharp defines, as of thorns, spines, prickles, barbs.
Asexual	Sexless,without sex, propagation by vegetative means.

Attenuate	Showing a long gradual taper.
Axil	Angle extended between the axis (stem) and any organ like leaf.
Axillary	Borne or appearing from the axil.
Axis	Main or central line of development of a plant or organ from which lateral parts arise.
Baccate	Berry-like pulpy or fleshy.
Barbed	Said of bristles or awns provided with terminal or lateral spike-like hooks which are reflexed backwards.
Base	The bottom or lower end of a part of structure or organ, even though this part may be uppermost as the organs hang on the plant.
Basipetal	Developing in a longitudinal plane from an apical or distil point toward the base.
Beak	A long prominent and substantial point, applied particularly to prolongations of fruits and carpels, especially drupaceous fruits.
Berry	A fleshy fruit, many seeded usually with one or more carpels.
Biennal	Refers to plants that complete their life cycle in two years, vegetative growth, followed by flowers, fruits which are formed in the second year.
Bifoliate	With two leaflets to a leaf.
Bilocular	Two celled, refers to any ovary or an anther.
Bisexual	Having both sexes present and functional in the one flower. Hermaphrodite.
Blade	Flat, expanded part of leaf.
Bract	A leaf – like structure present at the base of a flower.
Bracteole	A small bract present on flower stalk.
Branch	A lateral divisions of the stem or axis of growth.
Bristly	Bearing stiff strong hairs or bristles.
Bud	The rudimentary or beginning state of a stem, A thickened and condensed resting stage of a shoot or a flower or leaf before expanding.

Bud sport	A branch, flower or fruit that differs genetically from the remainder of the plant, usually arising spontaneously.
Bulb	Generally short thick stem, leaves also thickened to store food.
Caducous	Falling very early of sepals or petals.
Calyptra	A cap, hood or lid like structure.
Calyx	Referring to outermost usually green part of flower.
Calyx tube	Sepals when fused to form a tube as in gamosepalous, also used for receptacle of epi or perigynous flowers
Cambium	The growing or nascent tissue lying between the xylem and phloem of the fibro vascular bundle, and therefore on the outside of the woody trunk between wood and bast, or in trees and shrubs between wood and "bark". Its function is to increase the stem in diameter.
Campanulate	Bell shaped
Cap	In the grape, the cohering petals falling off as a cap. In the strawberry, a colloquial term for the calyx.
Capitate	Aggregated into a very dense or compact cluster.
Capsule	Fruit with two or more carpels generally with many seed, a dry dehiscent fruit.
Carpel	One of the foliar units (leaf-like structure) of a compound pistil or ovary containing ovules, a simple pistil contains one carpel. Consists of a basal portion, the ovary, a mid portion, the style and an apical portion, the stigma.
Carpellate	Composed of carpels.
Cartilaginous	Hard and flexible like cartridge.
Caryopsis	An achene like fruit, with the thin pericarp or covering grown fast to the seed, it is the characteristic fruit of the several grains and other grasses.
Catkin	An inflorescence usually a spike bearing unisexual flower generally fall off after maturity.
Caudate	Bearing tad like appendages.

Cavity	The depression at the bottom or stem-end of an apple or similar fruit.
Cell	One of the ultimate compartments or units of which plants are composed or made up, also a cavity or compartment or locule of an ovary or anther.
Chimera	A plant or plant organ having tissues of more than one genetic composition and origin.
Ciliate	Pubescent or hairs in the form finely divided hairs.
Circumscissile	Opening or dehiscing by a line around the fruit or anther.
Cirrhus, cirrus	A tendril.
Clasping	Partially or completely surrounding the stem, or bases of leaves
Cleistogamous	Self-pollination in a closed bud, prior to flower opening
Clone	A group of individuals resulting from vegetative propagation of a single plant. Sometimes spelled clone.
Collateral	Placed side by side.
Complete flower	All parts present.
Complete leaf	Having blade, petiole, stipules.
Compound leaf	A leaf with two or more separate leaflets,in some cases (as in citrus) some of the leaflets may be obsolete and the compound leaf have only one leaflet.
Compound pistil	Of two or more carpels united.
Compressed	Flattened especially flattened laterally.
Conduplicate	Two parts folded together lengthwise.
Connate	Fused or united.
Connolute	Twisted or rolled length wise, leaves or petals in the bud
Cordate	Heart shaped.
Coriaceous	Leathery, soft, smooth texture, leaf of mango.
Corm	A solid bulb-like underground asexual reproductive structure composed of an enlarged fleshy stem with rudimentary leaves or scales.

Corolla	Coloured, thin usually papery segment of the flower, appear next to calyx.
Corona	Crown, coronet, any appendage or intrusion that stands between the corolla and stamens, or on the corolla, as the cup of a daffodil, or that is the outgrowth of the staminal part or circle, as in the milk weeds.
Corymb	A racemose inflorescence, lower flowers with longer stalks than the upper ones so that all flowers reach at same level.
Cotyledon	Seed-leaf, the primary leaf or leaf of the embryo.
Creeper	A trailing shoot that takes root in the ground throughout its length.
Crenate	Margin with broad round teeth.
Cross	The offspring of any two flowers that have been interpollinated and fertilized. A cross-breed is a cross between varieties of the same species.
Cross pollination	Transfer of pollen to the stigma of a flower of different horticultural variety or species.
Crown	Corona, also that part of the stem at the surface of the ground, also a part of a rhizome with a large bud, suitable for use in propagation.
Cuneate	Wedge shaped.
Cuticle	The external rind or skin of a plant or part,usually applied to the thin waterproof membrane overlying the epidermis.
Cutting	A portion of a plant used in vegetative propagation, as a cutting of root,of stem or a leaf.
Cyclic	Whorled, the opposite or spiralled.
Cyme	Inflorescence determinate, apical or central flower opens first.
Deciduous	Not persistent, which falls off.
Decussate	Present in pairs, alternatively at right angles.
Definite	Said of a constant or known number, not exceeding twenty, contrasted with indefinite, above twenty, when the parts are usually not counted in systematic descriptions.

Deltoid	Triangular, Delta-like.
Dentate	With shape, spreading, rather coarse indentations or teeth that are perpendicular to the margin.
Depressed	More or less flattered endwise or from above.
Diadelphous	Stamens united into two sets.
Dicotyledonous	With two cotyledons.
Dimorphic	Having two forms
Dioecious	Male & female flowers present on different individuals.
Disk	An out growth (usually glandular) of hypanthium or receptacle.
Distinct	Separated, not united with parts.
Divergent	Spreading broadly but not too far apart.
Dorsal	Attached to the back or outer surface i.e. away from the axis.
Double	Said of flowers that have more than the usual number of floral envelopes, particularly of petals.
Downy	Covered with very short and weak soft hairs.
Drupe	A stone fruit like peach, plum, cherry in which pericarp differentiates into three layers, the epicarp thin and membranous, mesocarp, thick fleshy and juicy and endocarp stony with a seed. Fruit develops from single carpel having two ovules, one ovule degenerates.
Druplet	Small drupe, usually a component of an aggregate fruit e.g. black berry.
Ecology	Study of habits and modes of life of animals and plants.
Elliptic	Oval in outline, being narrowed to rounded ends and widest at or about the middle.
Elongate	Lengthened,stretched out.
Emarginate	With a shallow notch at the apex.
Embryo	The rudimentary plantlet contained within the seed, usually developing from a zygote as a product of fertilization.

Endocarp	Innermost layer of the pericarp.
Endosperm	Starch or other nutritive tissue outside or around the embryo formed within embryo sac.
Entire	Margin not indented, whole/continous.
Epicalyx	Segment or involucre, within or outside the calyx and resembles it.
Epigynous	Borne upon or top of the ovary or gynoecium
Epipetalous	Borne on the petals or the corolla tube referring to stamen.
Essential organs	Stamens and pistils.
Evergreen	Remaining green through the year, including the dormant season. Term applied to plant and not properly applied to leaves, but due to the persistence of leaves.
Exfoliating	Coming off in thin layers, as the bark of birch, pear, apple, and other plants.
Exocarp	The outside part of a pericarp or fruit wall.
Exserted	Stamens projecting beyond the petals.
Exstipulate	Without stipules.
Fascicle	A concerned or close cluster, as flowers of cherries.
Fibrous	Fiber-like containing fibers or thread like parts.
Filament	Thread like stalk of the stamen which bears the anther distally.
Flaccid	Limb flabby.
Flocose	With tufts or flocks or soft or woolly hair that usually rub off readily.
Foliaceous	Structure resembling a leaf.
Follicle	Fruit resulting from a single carpel opening from the ventral suture.
Forked	Branching or divided into nearly equal parts of members.
Free	Not joined or adnate to the organs as petals free from the stamens or calyx.

Fruit	The ripened ovary with its associated parts or seed bearing organ.
Funiculus	The stalk or stipe of a seed where it is attached to the placentae.
Furrowed	With longitudinal channels or grooves.
Gametophyte	The generation that bears the sex organs. In angiosperms reduced to the 3-nucleate pollen tube and the 8-nucleate embryo sac.
Gamopetalous	Petals united.
Gamophyllous	Perianth segments united.
Gamosepalous	Calyx of one piece, sepals united.
Genotype	The characteristics of a plant determined by its genetical constitution i.e. genes.
Germination	Resumption of growth of the embryo. Term used in reference to the sprouting of seeds and also the development of pollen tubes from pollen grains.
Glabrous	With pubescent or hairy out growth.
Gland	Properly, a secreting part of prominence or appendage but often used in the sense of a gland like body.
Glandular	Gland like structure present on leaves, ovary, fruit rind etc.
Glaucous	Covered with waxy bloom, variable in colour, can be easily removed.
Globose	Almost spherical.
Glutinous	Mucilaginous or sticky in nature - axil of many fruits.
Granular	Covered with very small grains, minutely or finely mealy.
Gymnosperms	Plants with naked seed.
Gynoandrous	Stamens when used with pistils.
Gynodioecious	Bisexual flowers on some plants and only pistillate flowers on others.
Gynophore	Projetion of the torus that forms the stalk of pistil.

Habit	The looks, appearance, general style or mode of growth, as an upright, open decumbent.
Habitat	Locality in which a plant grows wild.
Head	A short dense spike,capitulum.
Herb	Herbaceous or semi-woody plants used for culinary, ornamental or medicinal purposes.
Hermaphrodite	Androecium and gynoecium in same flower, bisexual.
Hesperidium	The fruit type of the orange. A special type of berry with a leathery outer rind.
Heterogamous	With two or more kinds or forms of flowers.
Heterophyllous	Leaves of different forms.
Husk	An outer covering of some fruits (nut crops) derived from perianth or involucre.
Hybrid	A plant resulting from a cross between two or more parents that differ genetically.
Hypanthium	Receptacle tube on which calyx, corolla, stamens and pistil are borne.
Hypocotyl	The axis of an embryo lying below the cotyledons and above the roots.
Hypogynous	Flower parts when borne below the ovary.
Imbricate	Overlapping - sepals and petals in the bud
Imperfect flower	Unisexual flower i.e. lacking either stamens or pistils.
Incised	A condition between toothed and lobed.
Incomplete	Lacking some of its parts, as a flower deficient in stamens, calyx.
Indehiscent	Fruits which do not split/open on their own.
Indeterminate	Growth of the main axis is not stopped by flower opening.
Indigenous	Native of the region, not introduced from some other country.
Inferior ovary	Beneath, lower, below, as an inferior ovary, one that is below the calyx.

Infinite	Two large, many in number.
Inflorescence	The arrangement of flowers on an axis. The flower or flower cluster of a plant.
Inserted	Attached, as a stamen growing on the corolla.
Internode	The part or space of stem between two nodes or joints.
Involucre	A whorl of small leaves or bracts standing close underneath a flower or flower cluster.
Irregular flower	Some parts different from other parts in same series.
Laciniate	Cut into narrow lobes.
Lamina	Expanded portion of a leaf or petal.
Lanceolate	Lance-shaped, much longer than broad widening above the base and tapering to the apex.
Lateral	On or at the side.
Latex	Coloured or colourless fluid produced by some plants
Leaflet	One part of a compound leaf, secondary leaf.
Leaf-stalk	The stem of a leaf, petiole, foot stock.
Linear	Long and narrow, the sides parallel or nearly so.
Lobe	Any part or segment of an organ,specifically a part of petal or calyx or leaf that represents a division to about the middle.
Locule	Cavity, cell or chamber of an anther, ovary or fruit.
Macrospore	A megaspore which gives rise to the embryo sac in plants.
Meristem	Undifferentiated tissue capable of developing in different organs/ tissues.
Mesocarp	The central or middle layer of succulent fruit.
Micropyle	Opening in the integument of an ovule through which pollen tube enters
Midrib	The main rib of a leaf or leaf like part. A continuation of the petiole.
Monoadelphous	Stamens united by their filaments to form a tube or a column.

Monochlamydeou	Presence of single whorl of perianth.
Monoclinous	Stamens and pistils present in the same flower.
Monocotyledonous	A plant producing seeds containing an embryo with a single cotyledon.
Monoecious	Staminate and pistillate flowers in separate flowers but present on same plant
Monogynous	Flower with single carpel or gynoecium.
Monopetalous	One-petaled, all the petals united to form one body or organ, as a gamopetalous corolla.
Multiple fruit	A fruit formed from fusion of several or many flowers with a common axis as in mulberry and pineapple.
Mutation	A sudden variation that is inherited.
Naked flower	With no floral envelops, without calyx and corolla.
Netted	Marked with reticulated lines or nerves that project somewhat above the surface.
Neutral	Without sex organs, sterile such flowers lack both pistils and stamens.
Node	A joint or point on a stem where a leaf is borne or may be borne, also incorrectly the space between two joints, which is properly an internode.
Nucellar embryo	An embryo developing from a cell in the nucellus rather than from fertilized egg.
Nut	An indehiscent 1-celled and 1-seeded hard and bony fruit, even if resulting from a compound ovary and in general partially or wholly enclosed in an involucre or husk.
Nutlet	Diminutive nut similar to an achene with thicker and harder wall.
Obcordate	Heart shaped with notch at the apex.
Oblanceolate	Inversely lanceolate, with the broadest part of a lanceolate body away from the point of attachment.
Oblique	At the base of leaf the two sides of the blade are unequal.

Oblong	Longer than broad sides nearly parallel for most of the length.
Obcordate	Inversely cordate.
Obovate	Inverted ovate.
Obsolete	Not apparent, vestigial or rudimentary.
Obtuse	Blunt, rounded.
Offset	A plant arising close to the base of another plant.
Orbicular	Almost circular or nearly so.
Ovary	Part of gynoecium bearing ovules.
Ovoid	A solid that is oval in flat outline.
Ovule	The body that after fertilization results in seed.
Paleobotany	The study of prehistoric plants.
Palmate	Lobed or divided in a palm-like or hand-like fashion.
Panicle	A compound racemose inflorescence.
Papilla	Soft, superficial projection.
Parietal	Placenta arising from the ovary wall.
Parthenocarpic	Producing fruits without fertilization.
Pendent	Hanging down.
Pedicle	Stalk of a flower
Peduncle	Stalk of an inflorescence.
Pentamerous	5-merous having five members in a cluster.
Perennial	Living for more than two years.
Perfect	Bisexual i.e. stamens and gynoecium in same flower.
Perianth	A collective term for the floral envelopes i.e. calyx, corolla or both.
Pericarp	Wall of a ripened ovary that is differentiated into outer layer (exocarp), middle layer (mesocarp) and inner layer (endocarp).
Perigynous	Borne around the ovary and not beneath it, as when calyx, corolla and stamens are borne on the edge of a cup-shaped hypanthium,such cases are said to exhibit perigyny.

Persistent	Remaining attached, not falling off.
Phyllode	A flattened petiole that performs the function of a leaf.
Phyllotaxy	Leaf arrangement on a stem.
Phylogeny	Evolutionary development of a taxon (genus/species) or part or organ of a member of a given taxon.
Pilose	Covered with a fine, soft hair.
Pinna	Primary unit or division of a pinnate leaf.
Pinnate	Compound leaf where the leaflets are placed along the sides of the rachis.
Pinnately	In pinnate fashion may not be compound, as pinnately lobed, pinnately divided or veined.
Pistil	Female organ of a flower.
Pistillate	Having pistils and no stamens, female flower only.
Pith	Soft spongy central part of stem composed of parenchyma cell in most of angiospermous plants.
Placenta	Ovary tissue by which ovules are attached.
Placentation	Arrangement of the placentae within the ovary
Pollination	Transfer of pollen from anther to stigma.
Poly	In Greek combinations signifying numerous or many.
Polygamous	Unisexual and bisexual flowers borne on the same plant.
Polymorphic	Appearing is several different forms i.e. leaves of different shape on a single plant or other morphological features.
Polypetalous	Petal separate not united.
Polyploid	A plant with a chromosome complement of more than two basic sets.
Polysepalous	Sepals/Calyx separate
Pome	Edible part is the enlarged flower axis and not the ovary as in apple, quince etc.
Posterior	At or on the back towards the axis.

Primocane	The first season's shoot or cane of a biennial woody stem, as in many brambles, usually without flowers.
Protandrous	Flower in which stamens mature before pistil.
Protogynous	Pistil(s) mature before the stamens.
	flower
Pubescent	Hairy i.e. covered with short, small fine hairs.
Punctate	Dotted with
Pyriform	Pear shaped
Raceme	An inflorescence where flowers are stalked along the axis. The oldest flower is at base and youngest at the top
Racemose	Bearing receme or receme like.
Rachis	The axis of an inflorescence or of a compound leaf.
Radical	Arising from near the root.
Radicle	Rudimentary root of the embryo.
Receptacle	Part of the flower axis which is greatly expanded to support flower or flower parts like calyx, corolla, stamens and carpels.
Recurved	Bend backward or downward.
Reflexed	Abruptly recurved/bent/back/downwards.
Refuse	Slightly notched almost round.
Reniform	Kidney shaped.
Repand	Slightly uneven or wavy margin.
Resinous	Producing or having resin/gummy coating or covering.
Reticulate	In the form of network.
Rib	In leaf any prominent nerve or vein.
Rosette	An arrangement of leaves radiating from a point.
Rotund	Nearly circular, orbicular, inclining to be oblong.
Rudimentary	Incomplete, very little developed.
Rugose	Wrinkled.

Runner	An elongated lateral shoot that roots at specific intervals.
Sagittate	Like an arrow-head in form.
Samara	One or two seeded dry indehiscent fruit e.g. maples.
Sapaloid	Resembling a sepal.
Saprophyte	Plant that survive on dead organic matter.
Scabrous	Small rough projections/ rough to touch.
Scale	Thin scarious body.
Scandent	Climbing.
Scarious	Membrane thin, dry and scale like.
Segment	One of the parts of a leaf, petal, calyx or perianth that is divided but not truly compound.
Self-fertilization	Secured by pollen from same flower,close-fertilization.
Sepal	Segment of a calyx.
Septate	Partitioned, division by partition or septa.
Septum	Partition in a compound ovary.
Serrate	Descriptive of a leaf margin when saw-toothed with the teeth pointing forward.
Sheath	The basal tubular part of leaf that encircles the culm (grass).
Sigmoid	Doubly curved to form the letter 'S'.
Siliqua	Fruits of cruciferae in which the two valves fall away from the frame, the replum on which the seeds are borne and across which a false partition is formed.
Simple pistil	Of one carpel.
Sinuate	Margin deeply wavy
Sinus	An indentation in a margin between two lobes or division of a leaf
Smooth	Said of surfaces that have no hairiness,roughness or pubescence, particularly of those not rough.

Solitary	Borne simply.
Spadix	Succulent axis that support an inflorescence.
Spathe	Leaf like coloured bract encircling the inflorescence.
Spike	An elongated unbranched, indeterminate inflorescence bearing sessile flowers.
Spikelet	Minute spike.
Spiral	Arrangement in a winding fashion, leaves or floral organs.
Spurious	False.
Stamen	Pollen bearing organ of a flower with filament and anther.
Staminate	Bearing only stamens or male reproducing organ.
Staminode	Sterile or abortive stamens.
Sterile	Non-functional i.e. sterile stamens or without functional sex organs.
Stigma	Part of pistil that receives pollen for fruit setting.
Stipule	A leaf like appendage at base of petiole, usually in pairs.
Stolon	A shoot that bends to the ground and takes root,more commonly, a horizontal stem at or below surface of the ground that gives rise to a new plant at its tip.
Stoloniferous	Producing/bearing stolons.
Stool	A clump of roots or rootstock that may be used in propagation.
Strobilus	A cone like structure sporophylls arranged on central axis
Stylar	Pertaining to style.
Style	Elongated part of pistil between stigma and ovary.
Succulent	Rather thick fleshy and juicy.
Sucker	A shoot arising from the roots or beneath the surface of the ground.
Sulcute	Grooved or furrowed length wise.

Superficial	On the surface.
Superior	Said of an ovary that is free and above the calyx or perianth.
Suppressed	Not being evident superficially.
Suture	Line of junction of two united organs, a line in dehiscent along which the splitting occurs.
Syconium	Fig fruit, hollow globose, receptacle open at one end, heavily set with flowers reduced to essential organs.
Symmetrical	Capable of dividing into similar halves.
Syncarpous	Carpels when united.
Tap root	More or less thickened, permanent root.
Tapering	Gradually becoming smaller or diminishing in diameter of width toward one end.
Taxonomy	Systematic botany involving identification naming and classification of plants
Tendril	A rotating or twisting thread-like process or extension by which a plant grasps an object and clings to it for support, morphologically it may be stem or leaf.
Tepal	A segment unit of a perianth not clearly differentiated into typical corolla and calyx.
Terminology	The subject dealing with names.
Ternate	Present in threes.
Terrestrial	A land plant.
Testa	Outer coating of seeds.
Thorn	A sharp woody spinelike out growth from the wood of a stem or reduced modified branch.
Tomentum	Pubescence covering of densely matted short woolly hairs.
Torus	Receptacle of a flower.
Tricolpate	Pollen grain having three grooves.
Trifoliate	Three leaved.

Tubercle	A small rounded/pointed projections on an organ.
Turgid	Swollen.
Twig	A shoot of a wooly plants that represents current growth.
Umbel	An umbrella like inflorescence in which pedicels radiates from a single point at the top of the peduncle.
Umbellate	In the form of umbel.
Umbilicate	With a navel like central depression.
Unarmed	Without sharp appendages like prickles or spines.
Undulate	Wavy surface.
Unifoliate	With one.
Unilateral	One sided.
Unilocular	Having one locule as in an ovary.
Urticle	A small bladder like structure i.e. Cheno-podiaceae.
Valvate	Dehiscing by valves.
Valve	One of the section into which the wall of a capsule or legume splits at maturity .
Velutinous	Velvety with firm erect hairs.
Venation	Arrangement of the veins or nerves of the leaf.
Ventral	Attached to the front or inner surface or part of organ facing towards the axis.
Vernation	Mode or arrangement of leaves in the bud
Verrucose	Warty, covered into wart like projections
Versatile	Attached by the middle, so as to swing freely, anther on the filament
Vertilliate	A whorled cluster of leaves or flowers.
Villous	With long shaggy hairs.
Viscid	Sticky.
Viviparous	Plants in which seeds or buds germinate while still attached to it.

Whorl	Organs clustered around on and axis i.e. flowers, leaves on the stem..
Wing	A thin dry or membraneous expansion of an organ.
Woolly	Covered with soft, long matted hairs.
Xerophyte	Plants that live on dry places, desert plant.
Zygomorphic	Flowers that can be divided into two equal halves through a single plane only.

Annexure-I

Fruit Crops, Their Common, Scientific Names along with Family and Type of Fruit

S.No.	English Name	Botanical Name	Family	Types of Fruits
1	Apple	*Malus pumila* Mill. (Syn. *M.sylvestris* Mill.; *M. communis* DC. ; *Pyrus malus* L.) *Malus*× *domestica* Borkh.	Rosaceae	Pome
2	Apricot (common)	*Prunus armeniaca* L.	Rosaceae	Drupe
3	Apricot (Japanese)	*P. mume* Sieb &Zucc.	Rosaceae	Drupe
4	Apricot (Black)	*P. dasycarpa* Enrh.	Rosaceae	Drupe
5	Annonas • Custard apple • Bullocks heart • Atemoya • Cherimoya • Sour sop • Pond apple	 *Annona squamosa* L. *A. reticulata* L. *A. atemoya* Hort. *A. cherimola* Mill. *A. muricata* L. *A. glabra*	Annonaceae	Etario of berries
6	Almond	*Prunus amygdalus*	Rosaceae	Drupe

7	Aonla (Indian gooseberry)	*Emblica officinalis* Gaertn.(Syn *Phyllanthus emblica* L.)	Euphorbia-ceae	Drupe
8	Avocado	*Persea americana* Mill.	Lauraceae	Drupe
9	Bael	*Aegle marmelos*(L.) Corr	Rutaceae	Amphisarca
10	Breadfruit	*Artocarpus altilis* (Park.) Fosb.	Moraceae	Sorosis
11	Banana	*Musa acuminate* Colla. *Musa balbisiana* Colla. *Musa sapientum* L.	Musaceae	Berry
12	Barbados cherry	*Malpighia glabra* L.	Malpighia-ceae	Drupe
13	Ber • (Indian) • (Chinese) • (Jharber)	 *Zizyphus mauritiana* Lamk. *Zizyphus jujube* L. *Zizyphus nummularia* (Burm.f.)	Rhamnaceae	Drupe
14	Blueberry	*Vaccinium symplocifolium* Alston.	Ericaceae	Many celled berry
15	Bilimbi	*Averrhoa bilimbi* L.	Oxalida-ceae	Berry
16	Cape gooseberry	*Physalis peruviana* L.	Solanaceae	Berry
17	Cashewnut	*Anacardium occidentale* L.	Anacardia-ceae	Nut
18	Carambola	*Averrhoa carambola* L.	Oxalidaceae	Berry
19	Chestnut • European • Indian • American • Japanese	 *Castanea sativa* L. *C. indica* *C. dentata* *C. crenata* Sieb.et Zucc.L.	Fagaceae	Nut
20	Cherry • Sweet • Sour • Duke • Wild • bird	 *Prunus avium* L. *P. cerasus* L. *P. avium*× *P.cerasus* *P.cerasoides* (syn. *P.puddum*) *P.cornuta* (syn *P. padus*)	Rosaceae	Drupe
21	Chinese gooseberry	*Actinidia chinensis* Planch.	Actinidia-ceae	Berry
22	Coconut	*Cocos nucifera* L.	Palmae	Fibrous drupe

23	Cluster fig	*Ficus glomerata* Roxb.	Moraceae	Syconus/ Syconium
24	Chinese laurel	Antidesma bunius (L.) Spreng.	Euphorbia-ceae	
25	Common mulberry	*Morus australis* Poir.	Moraceae	Sorosis
26	Cranberry	*Vaccinium macrocarpon* Alt.	Ericaceae	Many celled berry
27	Ceylon raspberry	*Rubus lasiocarpus* Sm.	Rosaceae	Eatrio of drupelets
28	Coromandel ebony (persimmon)	*Diospyros melanoxylon* Roxb.	Ebenaceae	Berry
29	Currants		Saxifraga-ceae	Many celled berry
	Black	*Ribes nigrum*		
	Garden	*R. vulgare*		
30	Citrus fruits		Rutaceae	Hesperidium
	Citron	*Citrus medica* L.		
	Calamondin	*C. madurensis* Lour.		
	Grapefruit	*C. paradisi* Macf.		
	Karna Khatta	*C. karna Raf.*		
	Kumquat	*C. japonica*		
	Lemon	*C. limon* (L.)Burm.f.		
	Lime	*C. aurantifolia (Chrism.)*		
	Mandarin	*C. reticulata Blanco.*		
	Pummelo	*C. grandis* Osbeck.		
	Rangpur lime	*C. limonia* Osbeck.		
	Rough lemon	*C. jambhiri* Lush.		
	Sour orange	*C. aurantium* L.		
	Sweet lime	*C. limettoides* Tanaka.		
	Sweet orange	*C. sinensis* Osbeck.		
	Trifoliate orange	*Poncirus trifoliata*		
31	Datepalm	*Phoenix dactylifera* L.	Palmae	One seeded berry
32	Dillenia	*Dillenia indica* L.	Dilleniaceae	Special
33	Durian	*Durio zibethinus* Murr.	Bombaceae	Berry
34	Dheoo(monkey jack)	*Artocarpus lakoocha* Rox.	Moraceae	Sorosis
35	Ebony persimmon	*Diospyrus ebenum* Koenig.	Ebenaceae	Berry
36	Fig		Moraceae	Syconus
	• Common	*Ficus carica* L.		
	• Wild	*F. glomerata* Roxb.		
	• Wild	*F. roxburghii* Lour.		

37	Filbert	*Corylus avellana* L.	Betulaceae/ Corylaceae	Nut
38	Feijoa	*Feijoa sellowiana* Berg.	Myrtaceae	Berry
39	Grapes • European • Fox • Muscadine	 *Vitis vinifera* L. *V . labrusca* L. *V .rotundifolia* Michx.	Vitaceae	Berry
40	Guava • Common • Strawberry • Guinea guava	 *Psidium guajava* L. *P.calttleyanum* Sabine. *P. guineense* Sw.	Myrtaceae	Berry
41	Himalayan mulberry	*Morus serrata* Roxb.	Moraceae	Multiple fruit
42	Hog plum	*Spondias pinnata* (L.f) Kurz	Anacardia-cea	Drupe
43	Houtbois strawberry	*Fragaria nilgerrensis* Schlecht.ex J. Gay	Rosaceae	Etario of achenes
44	Hazelnut (filbert)	*Corylus avellana* L.	Betulaceae/ Corylaceae	Nut
45	Indian Olive	*Olea ferruginea* Royle.	Oleaceae	Drupe
46	Indian almond	*Terminalia cattapa* L.	Combreta-ceae	Nut
47	Indian persimmon	*Diospyros peregrina* Gaertyn.	Ebenaceae	Berry
48	Jackfruit	*Artocarpus haterophyllus* Lamk.	Moraceae	Sorosis
49	Japanese persimmon	*Diospyros kaki* L.f.	Ebenaceae	Berry
50	Jamun	*Syzgium cuminni* (L.) Skeels	Myrtaceae	Berry
51	Karonda	*Carrisa carandus* L.	Apocynaceae	Berry
52	Kainth	*Pyrus pashia* L.	Rosaceae	Pome
53	Kokam butter tree	*Garcinia indica* (Thouars) Choisy	Guttiferae	Berry
54	Loquat	*Eriobotrya japonica* (Thunb.) Lindl.	Rosaceae	Pome
55	Litchi	*Litchi chinensis* (Gaertn.) Sonn.	Sapindaceae	Nut
56	Longan	*Euphoria longan* (Lour.) Steud.	Sapindaceae	Nut

57	Macadamia	*Macadamia integrifolia* Maiden & Betche.	Proteaceae	Nut
58	Mango	*Mangifera indica* L.	Anacardia-ceae	Drupe
59	Milk tree(khirnee)	*Manilkara hexandra* (Roxb.) Dub.	Sapotaceae	Berry
60	Mohua	*Madhuca indica* J.F. Gmel.	Sapotaceae	Berry
61	Mountain papaya	*Carica candamarcensis* Hook.f.	Cariaceae	Berry
62	Mammey apple	*Mammea Americana* L.	Guttiferae	Berry
63	Malay apple	*Syzgium malaccense* (L.) Merr & Perry	Myrtaceae	Drupe
64	Mysore raspberry	*Rubus niveus* Thunb.	Rosaceae	Etario of drupelets
65	Mauritius raspberry	*Rubus rosaefolius* Sm.	Rosaceae	Etario of drupelets
66	Natal plum	*Carissa grandiflora*	Apocynaceae	Berry
67	Olive	*Olea europea* L.	Oleacea	Drupe
68	Peach		Rosaceae	Drupe
	• Common	*Prunus persica*(L.) Batsch.		
	• Nectarine	*P. persica var nectarine* (Alt) Maxim		
69	Pear		Rosaceae	Pome
	• Soft	*Pyrus communis* L		
	• Hard	*P. pyrifolia* var *Culta Nakai.*		
	• Manchurian	*P. ussuriensis* Maxim.		
70	Phalsa	*Grewia subinequalis* DC.	Tiliaceae	Berry
71	Pineapple	*Ananas comosus* (L.) Merr.	Bromeliaceae	Sorosis
72	Pecan	*Carya illineonsis* Koch.	Juglandaceae	Nut
73	Pistachionut	*Pistacia vera* L.	Anacardia-ceae	Nut
74	Plum		Rosaceae	Drupe
	• Japanese	*P. salicina* Lindl.		
	• European	*Prunus domestica* L.		
	• American	*P. americana* Marsh.		
	• Wild goose	*P. munsoniana* Wight n Hedr.		
	• Damson	*P. institia* L.		

75	Plumcot	*Plum×apricot* hybrid	Rosaceae	Drupe
76	Pomegranate	*Punica granatum* L.	Punicaceae	Balausta
77	Passionfruit	*Passiflora edulis* Sims.	Passifloraceae	Berry
78	Papaya	*Carica papaya* L.	Caricaceae	Berry
79	Quince	*Cydonia oblonga* Mill.	Rosaceae	Pome
80	Rambutan	*Nephelium lappaceum* L.	Sapindaceae	Nut
81	Rose apple	*Syzgium jambos*(L.)Alston	Myrataceae	Berry
82	Rasp berry • Yellow • European • Red	 *Rubus ellipticus* *R. idacus* *R. strigosus*	Rosaceae	Etaerio of drupelets
83	Sapota	*Achras sapota* L.	Sapotaceae	Berry
84	Siberian crab apple.	*Malua baccata* (L.) Borkh.	Rosaceae	Pome
85	Strawberry	*Fragaria ×ananassa*	Rosaceae	Etario of achenes
86	Surinam cherry	*Eugenia uniflora* L.	Myrtaceae	Drupe
87	Tamarind	*Tamarindus indica* L.	Caesalpinia-ceae	Pod / Lomentum
88	Walnut	*Juglans regia* L.	Juglandaceae	Nut
89	Wild date	*Phoenix sylvestris* (L.) Roxb.	Palmae	One seeded berry
90	Wood apple	*Limonia acidissima* L. (Syn.*Feronia limonia* (L.) Swingle	Rutaceae	Amphisarca

❑❑❑

Annexure-II

Genera and species in different families

1. Actinidiaceae: 4 genera and 285 spp.

Actinidia: 36 spp.

A. arguta, A. burbidgei, A. chinensis, A. coriacea, A. kolomikta, A. lanceolata, A. purpurea, A. valubilis

2. Anacardiaceae: 45 genera and 200 spp. **Sapindales**

a) Anacardium: 15 spp.

A. occidantale

b) Mangifera: 40 spp.

M. indica

c) Pistacia: 10 spp.

P. chinensis, P. integerrima, P. lentiscus, P. mexicana, P. mutica, P. simaruba, P. texana, P. vera

3. Annonaceae: 75 genera

Annona: 100 spp. **Ranales**

A. cherimola, A. diversifolia, A. glabra, A. montana, A. muricata, A. paludosa, A. reticulata, A. squamosa

4. Betuleaceae: 6 genera and 100 spp.

Corylus: 10 spp.

A. americana, C. avellana, C. atropurpurea, C. californica, C. chinensis, C. colurna, C. cornuta, C. ferox, C. maxima, C. rostratex

5. Bombaceae: 27 genera

Duria: 30 spp.

D. zibethinus

6. Hippocastanaceae: 2 genera and 15 spp.

Asculus: Horse chestnut

7. Rosaceae: 100 genera and 200 spp. **Rosales**

a) Cydonia: 2 spp.

C. oblonga, C. sinensis

b) Fragaria: 12 spp.

F. annanasa, F. californica, F. chiloensis, F. indica, F. vesca, F. virginiana

c) Malus: 25 spp.

M. angustifolia, M. astracanica, M. baccata, M. brevipus, M. calocarpa, M. cerasifera, M. communis. M. coronaria, M. crataegifolia, M. dolgo, M. domestica, M. florentina, M. floribunda, M. formosana, M. fusca, M. glabrata, M. halliana, M. hillieri, M. hupehensis, M. kaido, M. sylvestris, M. pumila, M. sargentii, M. spectabilis, M. sikkimensis, M. sieboldii, M. prunifolia var. rinkii

d) Prunus: 400 spp.

P. acuminata, P. alleghaniensis, P. americana, P. amygdalus, P. andersonii, P. angustifolia, P. ansu, P. apetala, P. armeniaca, P. autumnalis, P. avium, P. besseyi, P. buergerana, P. companulata, P. communis, P. commulata, P. davidiana, P. cerasifera, P. cerasoides, P. cerasus, P. divaricata, P. domestica, P. dulcis, P. institia, P. integrifolia, P. japonica, P. lanata, P. mahaleb, P. mandschurica, P. myrobalan, P. nana, P. nigra, P. padus, P. pendula, P. persica, P. salicina, P. sargentii, P. serotina, P. sibirica, P. spinosa, P. tomentosa, P. virginiana.

e) Pyrus: 20 spp.

P. adamsii, P. americana, P. angustifolia, P. aucuparia, P. communis, P. dawsoniana, P. coronaria, P. florentina, P. floribunda, P. pashia.

f) Rubus: 250 spp.

R. albescense, R. allegheniansis, R. arcticus, R. argutus, R. australis, R. coronarius, R. ellipticus, R. flagellaris, R. henryi, R. hispidus, R. idaeus, R. mirus, R. occidentalis, R. odoratus, R. parviflorus, R. pubescens, R. spectabilis, R. stellatus.

g) Eriobotrya

E. buisanensis, E. deflexa, E. grandiflora, E. hookeri, E. japonica, E. koshunensis

8. Bromeliaceae: 45 genera and 200 spp. Commelinales

Annanas: 9 spp.

A. annanasoides, A. bracteatus, A. comosus, A. nanus, A. proteanus, A. sativus

9. Moraceae: 53-75 genera and 1400-1850 spp. Urticales

a) Artocarpus: 50 spp.

A. altilis, A. elasticus, A. heterophyllus, A. hirsutus, A. hypargyraes, A. incisus, A. integer, A. odoratissimus.

b) Ficus: 800 spp.

F. afzelli, F. altissima, F. aspera, F. aurea, F. auriculata, F. australis, F. belgica, F. benghalensis, F. carica, F. citrifolia, F. comosa, F. montana, F. philippinensis, F. ulmifolia, F. utilis.

c) Morus: 10 spp.

M. alba, M. australia, M. indica, M. laevigata, M. macrophylla, M. multicaulis, M. nigra, M. pendula, M. rubra, M. tatarica, M. tropicana

10. Apocynaceae: 130 genera

Carissa: 35 spp.

C. acuminata, A. arduina, C. bispinosa, C. carandus, C. edulis, C. grandiflora, C. longiflora, C. minima, C. spectabilis, C. spinarum.

11. Musaceae: 2 genera and 42 spp. Sectaminales

a) Ensete: 7 spp.

E. superbum, E. ventricosum

b) Musa: 25 spp.

M. acuminata, M. arnoldiana, M. balbasiana, M. gigantean, M. nepalensis, M. paradisica, M. nana, M. zebrine.

12. Ebenaceae: 6 genera **Ebenales**

Diospyros: 200 spp.

D. kaki, D. montana, D. discolor, D. malabarica, D. lotus, D. virgianiana

13. Burseraceae: 20 genera and 600 spp.

Canarium pimela: Chinese black olive

14. Dilleniaceae: 11 genera and 275 spp.

Dillenia indica: Elephant apple (Chulta in India)

15. Elaegnaceae: 3 genera and 45 spp.

***Hippophae*: 2 spp. (Seabuckthorn)**

H. rhamnoides, H. salicifolia

16. Myrtaceae: 80 genera and 3000 spp. **Myrtales**

a) Eugenia: 100 spp.

E. aggreta, E. apiculata, E. aromatica, E. australis, E. coronata, E. monticola, E. oblanceolata

b) Feijoa: 2 spp.

F. coolidgei, F. sellowiana

c) Psidium: 100 spp.

P. guineense, P. lettorale *var.* longipus, P. chinensis, P. guajava, P. lucidum, P. molle

d) Syzygium: 400-500 spp.

S. aqueum, S. aromaticum, S. australe, S. cumini, S. grande, S. javanicun, S. oblatum

e) Myrica: 50 spp.

M. asplenifolia, M. californica, M. cerifera, M. faya, M. gale, M. pensylvenica, M. rubra

17. Sapindaceae: 150 genera and 2000 spp. **Sapindales**

a) Euphoria: 15 spp.

E. longan, E. malaiensis

b) Litchi: Several spp.

L. chinensis

c) Rambutan

Nephelium lappaceum

18. Caricaceae: 2 genera **Parietales**

Carica: 25 spp.

C. papaya, C. candamariensis, C. pubscence

19. Euphorbiaceae: 283 genera and 7300 spp. **Euphorbiales**

Phyllanthus: 650 spp.

P. acidus, P. angustifolius, P. arbuscula, P. nivosus, P. speciosus

20. Ericaceae: 70 genera and 1900 spp. **Ericales**

a) Arbutus: 14 spp.

A. canariensis, A. menziesii, A. texana, A. xalapensis

b) Vaccinium: 150 spp.

V. angustifolium, V. arctostaphylos, V. ashei, V. atrococcum, V. australe, V. hirsutum, V. macrocarpon, V. occidentalis, V. parvifolium, V. virgatum

21. Oleaceae: 29 genera and 600 spp. **Loganiales**

Olea: 20 spp.

O. africana, O. chrysophyla, O. communis, O. europea, O. fragrans, O. illicifolia, O. lanceolata, O. montana.

22. Oxalidaceae: 7 genera and 1000 spp. **Geraniales**

Averrhoa: 2 spp.

A. bilimbi, A. carambola

23. Palmae: 210 genera and 2780 spp. **Spathiflorae**

a) Phoenix: 17 spp.

P. obyssinica, P. acaulis, P. andersoni, P. canariensis, P. cycadifolia, P. dactylifera, P. rupicala, P. sylvestris, P. tomentosa

b) Cocos: 18 spp.

Cocus nucifera

24. Punicaceae: 1 genus **Myrtales**

Punica: 2 spp.

P. granatum, P. nana

25. Polygonaceae: 40 genera and 800 spp.

Cocoloba: 150 spp.

C. diversifolia, C. floridana, C. laurifolia, C. pubscens, C. uvifera

26. Rhamnaceae: 55 genera and 900 spp. **Rhamnales**

a) Rhamnus: 150 spp.

R. alpina, R. californica, R. canariensis, R. catharicta, R. insula, R. integrifolia, R. japonica

b) Zizyphus: 40 spp.

Z. jujuba, Z. mauritiana, Z. mucronata, Z. obtusifolia

27. Solanaceae: 90 genera and 2000 spp. **Solanales**

a) Cyphomendra: 30 spp.

C. betaceae

b) Physalis: 80 spp.

P. edulis, P. alkekengi, P. bunyardii, P. franchtii, P. ixocarpa, P. lobota, P. peruviana, P. pruinosa, P. pubscens, P. subglabrata

28. Lauraceae: 47 genera and 2000-2500 spp. **Ranales**

Persea: 150 spp.

P. americana, P. breviflora, P. indica, P. japonica, P. leiogyna

29. Juglandaceae: 6 genera and 60 spp. **Juglandales**

a) Juglans: 20 spp.

J. ailantifolia, J. californica, J. cinerea, J. hindsii, J. neotropica, J. japonica, J. major, J. mandschurica, J. microcarpa, J. nigra, J. regia

b) Carya: 20-25 spp. **Juglandales**

C. alba, C. tomentosa, C. aquatica, C. cathayensis, C. cordiformis, C. glabra, C. illinoensis, C. laciniosa, C. microcarpa, C. pallida

30. Fagaceae: 6 genera 200 spp.

a) Castanea: 12 spp.

C. alnifolia, C. americana, C. crenata, C. dentata, C. henryi, C. japonica, C. mollissima, C. sativa, C. sequinii, C. vesca

b) Fagus: 10 spp.

F. americana, F. crenata, F. orientalis, F. sylvatica

31. Guttifereae: 40 genera and 1000 spp. Guttiferales

Garcinia: 200 spp.

G. dulcis, G. livingstonei, G. mangostana, G. spicata, G. tinctoria, G. xanthochymus

32. Passifloraceae: 12 genera Parietales

Passiflora: 400 spp.

P. alata, P. alba, P. biflora, P. bryonioides, P. caerulea, P. cinnabarina, P. coriaceae, P. edulis, P. foetida, P. gracilis, P. grandiflora, P. incarnata, P. jamesonii, P. lutea, P. manicata, P. mixta, P. mollissima, P. morifolia, P. quadrangularis, P. violacea, P. viridiflora, P. vitifolia

33. Saxifragaceae: 80 genera and 1200 spp.

Ribes: 150 spp. (Currants)

R. alpestre, R. alpinum, R. americanum, R. aureum, R. bracteosum, R. californicum, R. cereum, R. curvatum, R. diacanthum, R. floridum, R. japonicum, R. lacustre, R. malvaceum, R. nevadense, R. nigrum, R. odoratum, R. orientale, R. pinetorum, R. rotundifolium, R. rubrum, R. sativum

34. Rutaceae: 150 genera and 1600 spp. Rutales

a) Citrus: 16 spp.

C. acutifolius, C. albidus, C. algarvensis, C. corberiensis, C. criticus, C. crispus, C. hirsutus, C. hibridus, C. incanus, C. laurifolius, C. parviflorus, C. vagintus.

b) Aegle: 3 spp.

A. marmelos, A. sepiaria

c) Aeglopsis: 2 spp.

A. chevalieri

d) Fortunella: 5 spp.

F. japonica, F. margarita

35. Sapotaceae: 40 genera and 800 spp. **Ebenales**

a) Achras or Manilkara: 85 spp.

A. zapota, A. bidentata, A. kauki, A. roxburghiana

b) Pouteria: 50 spp. (Egg fruit)

P. campechiana, P. domingensis, P. hypogluca, P. sapota

36. Vitaceae: 12 genera **Rhamnales**

Vitis

V. acerifolia, V. aestivalis, V. arizonica, V. argentifolia, V. Antarctica, V. baileyana, V. californica, V. capensis, V. cortifolia, V. himalayana, V. labrusca, V. monticola, C. palmate, V. riparia, V. rotundifolia, V. rupestris, V. vinifera, V. vulpine

37. Malphighiaceae: 60 genera and 850 spp.

Malphighia: 30 spp.

M. clarensis, M. coccigera, M. glabra, M. punicifolia, M. suberosa

38. Tilliaceae: 50 genera **Malvales**

Grewia: 150 spp.

G. asiatica, G. biloba, G. damine, G. flavescens, G. occidentalis, G. orientalis, G. rotinervis, G. similis, G. tilifolia, G. trifoliolatum

39. Proteaceae: 55 genera and 1200 spp. **Proteales**

Macadamia: 10 spp.

M. alternifolia, M. integrifolia, M. ternifolia, M. tetraphylla

40. Rubiaceae: 400 genera and 4800-5000 spp.

Genipa: 6 spp. (Marmalade box)

G. americana

41. Dilleniaceae: 11 genera, 275 spp.

Dillenia: 60 spp.

D. indica, D. ovata, D. philippinensis, D. suffruticosa

42. Flacourtiaceae: 84 genera and 850 spp.

Dovyalis: 22 spp.

D. abyssinica, D. caffra, D. hebecarpa.

□□□

Annexure-III

List of tribes and generas of some horticulturally important families.

Family: Tiliaceae

Tribes:
1. Brownloweae
2. Apeibeae
3. Tiliae
4. Greweae

Genera:
1. *Grewia*
2. *Corchorus* (Fibre yielding)
3. *Etaeocarpus* (Rudraksha)

Family: Rutaceae

Genera:
1. *Citrus* (with no. of species)
2. *Aegle marmelos* (Wood apple)
3. *Murraya exotica, M. koenigii* (Meetha neem)
4. *Ruta*
5. *Feronia elephantum* (Hathi bael)
6. *Limonia acidissima*

7. *Skimmia*
8. *Toddalia, Toddalia asiatica*

Sub-families:
1. Rutoideae
2. Flindersoideae
3. Spatheliodeae
4. Toddaloideae
5. Aurantoideae (Citrus, Feronia)

Family: Rhamnaceae

Genera:
1. *Zizyphus*
2. *Rhamnus*
3. *Helinus*
4. *Ventilago*

Family: Vitaceae

Genera:
1. *Vitis*
2. *Leea*
3. *Cissus*
4. *Cayratia*

Family: Anacardiaceae

Tribes:
1. Dobineae - Dobinea
2. Rhoideae - Pistacia, Rhus
3. Spondiceae - Spondias
4. Mangiferae - Mangifera

Genera:
1. *Mangifera*
2. *Pistacia*
3. *Anacardium*
4. *Rhus*
5. *Spondias* (Hog plum)
6. *Buchania* (Cudapah almond)
7. *Schines*

Family: Sapindaceae

Genera:
1. *Litchi*
2. *Euphorbia*

3. *Nephalium*
4. *Pautlinia*
5. *Sapindus*
6. *Dodoneae*

Sub-families:
1. Eusapindaceae
2. Dyssapindaceae

Family: Rosaceae

Tribes:
1. Pomeae
2. Pruneae
3. Rubeae
4. Potentilleae

Genera:
1. *Rosa*
2. *Rubus*
3. *Malus*
4. *Pyrus*
5. *Prunus*
6. *Fragaria*
7. *Potentilla*
8. *Eriobotrya*
9. *Cydonia*
10. *Crataegus*
11. *Spiraea*
12. *Prisepia*

Family: Myrtaceae

Tribes:
1. Myrteae - *Psidium, Syzygium, Eugenia*
2. Leptospermeae - *Callistemon*
3. Chamaelancieae - *Verticordia*

Genera:
1. *Psidium*
2. *Eugenia*
3. *Syzygium*
4. *Eucalyptus*
5. *Callistemon*

Family: Sapotaceae

Genera:

1. *Achrus*
2. *Bassia*
3. *Palaquim* (Gutta-percha)
4. *Chrysophyllum* (G. Parcha)
5. *Sideroxylon* (Hardwood timber)
6. *Payena* (Latex yielding G. parcha)

Family: Oleaceae

Genera:

1. *Olea*
2. *Jasminum* (Jasmine)
3. *Osmanthus* (Fragrant)
4. *Fraxinus* (Common ash)
5. *Syringa* (Lilae)
6. *Nyctanthes* (Night flower)

Family: Apocynaceae

Sub-families:

1. Echitoideae
2. Plumeroideae

Genera:

1. *Carissa*
2. *Nerium*
3. *Vina* (Sada bahar)
4. *Plumiera*
5. *Allamanda*
6. *Kopsis* (Decorative plam)
7. *Wrightia* (Ornamental)
8. *Rauwolfia* (Sarpgandha)
9. *Ervatamia* (Chandni)

Family: Solanaceae

Genera:

1. *Physalis* (Rasbhari)
2. *Cyphomandra* (Tree tomato)
3. *Atropa*
4. *Capsicum*
5. *Cestrum*

6. *Datura*
7. *Nicotiana*
8. *Solanum*
9. *Withania* (Aswagandha)

Family: Euphorbiaceae

Genera:

1. *Phyllanthus*
2. *Euphorbia*
3. *Croton*
4. *Manihot*
5. *Jatropa* (Ratanjot)
6. *Hevea*
7. *Ricinus* (Arind castor)
8. *Buxus*
9. *Poinsettia*
10. *Sapium* (Chinese tallow/pahari shisham)

Family: Punicaceae

Species:

1. *Granatum*
2. *Protopunica*

Family: Moraceae

Genera:

1. *Artocarpus*
2. *Ficus*
3. *Humulus* (Hop)
4. *Castilla* (Panama rubber)
5. *Canabis*
6. *Broussonetia* (Paper mulberry)
7. *Brosimium* (Cow tree)

Family: Musaceae

Genera:

1. *Musa*
2. *Ravenala* (Traveller's tree)
3. *Sterlitzia*
4. *Heliconia* (Ornamental)

Family: Palmaceae

Tribes:

1. Phoeniceae - phrenix
2. Sabaleae
3. Borasseae
4. Lepidocaryeae
5. Arecae - *Areca*
6. Cocoeae - *Cocus, Elaeis*
7. Phytelephanteae

Genera:

1. *Cocos*
2. *Phoenix*
3. *Areca*
4. *Borassus* (Delab-palm tar yields toddy)
5. *Caryota* (Toddy palm/yields palm wine)
6. *Elaeis* (Oil palm)

Family: Bromeliaceae

Genera:

1. *Ananas*
2. *Billbergia*
3. *Tillandsia*
4. *Catopsis*
5. *Puya*

Family: Annonaceae

Tribes:

1. Xylopeae (Annona, Artabotrys)
2. Uvarieae
3. Miliuseae
4. Hexalobeae
5. Monodoreae

Genera:

1. *Annona squamosa*
2. *A. reticulata*
3. *A. muricata*
4. *Asimina*
5. *Unona* (Lavender champa)
6. *Miliusa* (Gausal)
7. *Desmos*
8. *Artabotrys* (Kantli champa)
9. *Polyalthia* (Mast tree)

❑❑❑

Annexure-IV

Botanical Terms Chart

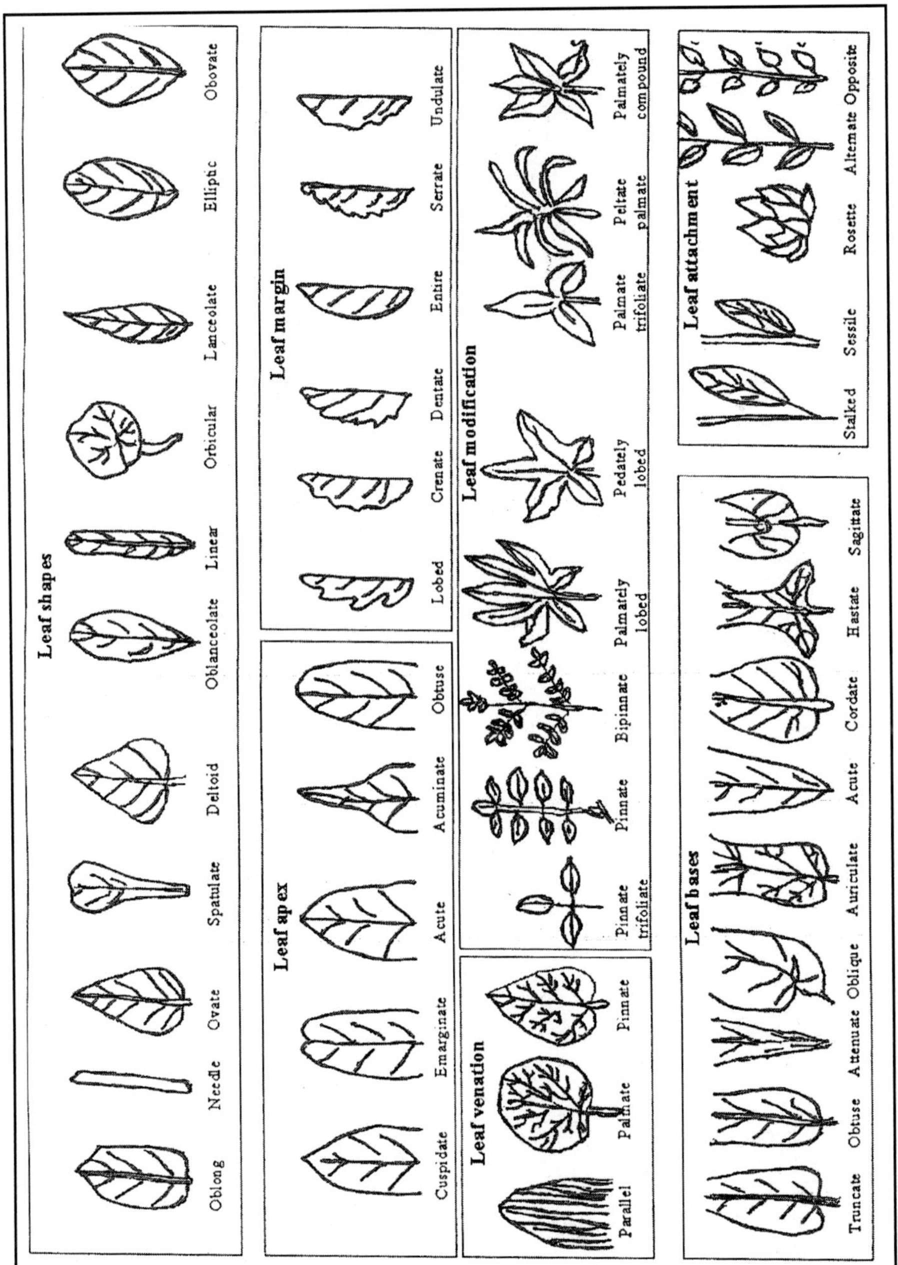

Floral Structure

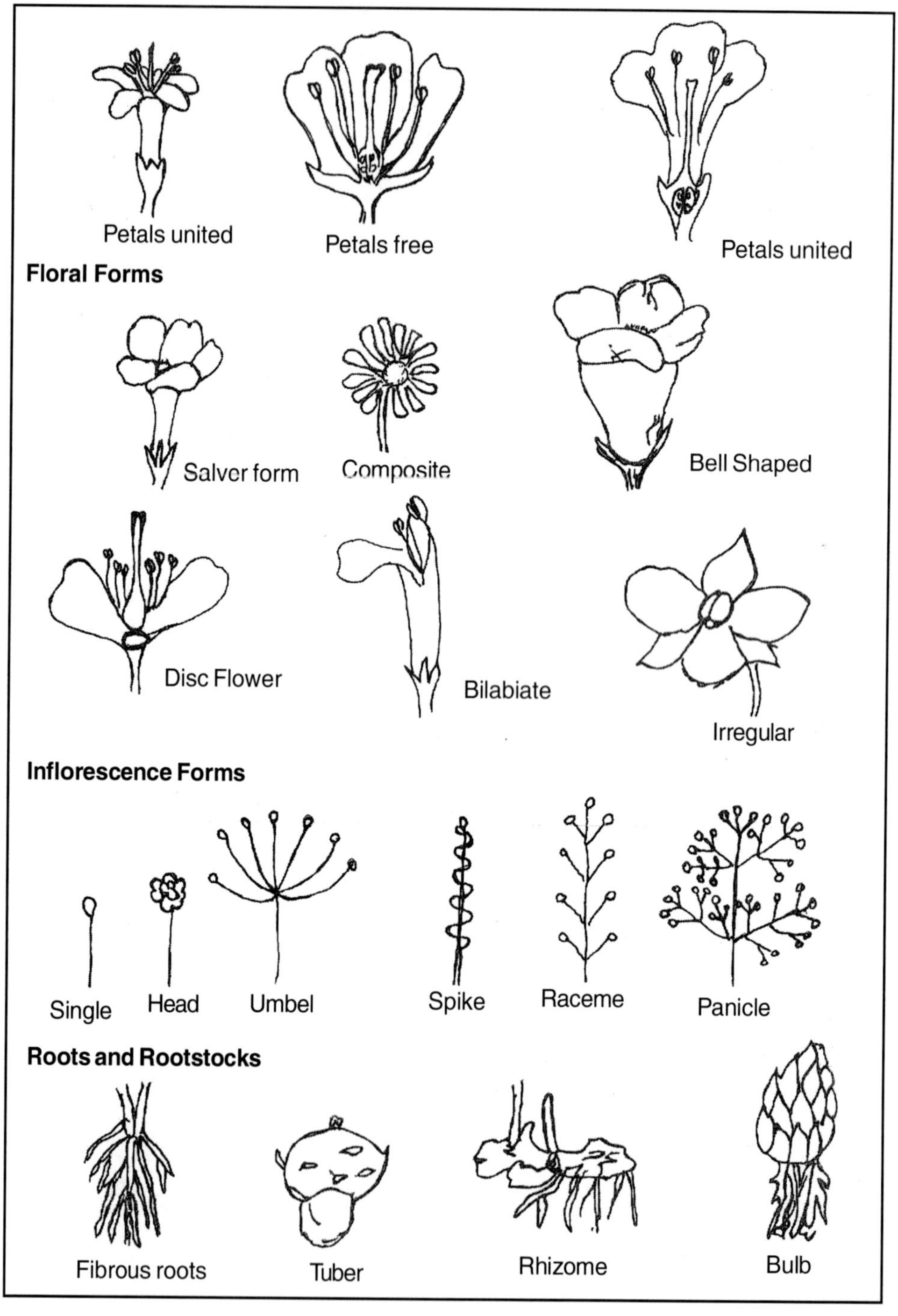

Leaf Shape

1.	Acicular/Needle	Needle shaped
2.	Oblong	Longer than broad
3.	Ovate	Egg-like rounded at both the ends, broadet in th ecenter or below
4.	Spatulate	Spatula spoon like, apical end rounded, basal and narrow
5.	Deltoid	Like; equilateral traingle, broader width above middle portion
6.	Oblanceolate	Tapering at base, broadest width above middle portion
7.	Linear	Long and narrow, sides more or less parallel as in most grasses
8.	Orbicular	Circular or nearly so
9.	Lanceolate	Lance shaped, narrow/tapering at both ends, maximum width in the center
10.	Elliptic	Narrow to rounded ends, maximum width near or at middle
11.	Obovate	Opposite of ovate, broadest above and not below the middle

Leaf apex

1.	Cuspidate	With an apical cusp
2.	Emerginate	Narrow notch at apex
3.	Acute	Sharp pointed, essentially straight
4.	Acuminate	Tapering to a protracted acute point
5.	Obtuse	Rounded, making wide angle, blunt

Leaf margin

1.	Lobed	Major segment of an organ, division halfway to the middle of the organ
2.	Crenate	Teeth rounded, shallow or obtuse
3.	Dentate	Teeth sharp, extended outwards perpendicular to the margin
4.	Entire	Margin continuous, without any indent serration or tooth
5.	Serrate	Margin with small, saw like teeth directed towards apex of the organ
6.	Undulate	Wavy surface

Leaf venation

1.	Parallel	Veins at equidistance from point of emergence to end point
2.	Palmate	Three or more lobes, nerves or leaflets emerging from basal point of attachment
3.	Pinnate	Some what-like feather with parts arranged along both the sides of an axis

Leaf types and its modification

1.	Pinnate trifoliate	Leaf arranged along the axis, and divided into three parts
2.	Pinnate	Made like a feather, i.e. parts like leaves veins branches arranged a long both sides of an axis
3.	Bipinnate	Double pinnate
4.	Palmately lobed	Divided in three or more nerves lobes or leflets arising from point of attachment
5.	Pedately lobed	In palmately lobed leaf the two outer sides lobes are again divided
6.	Plamate trifoliate	Divided in three distinct leaflets from single point of attachment
7.	Peltate palmate	Organ more or less circular, attached near its center or at least inside its margin
8.	Palmately	Divided into three or more leaflets from basal point of attachment

Leaf bases

1.	Truncate	Cut off straight across at the end
2.	Obtuse	Blunt, round, wide angled
3.	Attenuate	Long tapering, applied to bases or apices of parts
4.	Oblique	Slanting with unequal sides
5.	Auriculate	Having an auricle i.e. ear shaped lobe at the base of leaves or petals
6.	Acute	Sharp pointed, tapering straight to a point
7.	Cordate	Heart shaped, generally ovate in outline
8.	Hastate	Shaped like arrow head, with basal lobes turned outwards
9.	Sagittate	Triangular or arrow shaped, basal lobes directed towards the stalk

Leaf attachment

1.	Stalked	More or less elongated structure of any organ as of a petiole, pedicel, peduncle etc.
2.	Sessile	Without a stalk
3.	Rosette	Clustered at one point number of leaves or flowers emerging from single point
4.	Alternate	Arranged singly at different heights and sides of the axis or stem
5.	Opposite	Usually two at a node on opposite side of an axis

Floral structures

1.	Petals united	—
2.	Petals free	Ovary superior
3.	Petals united	Ovary inferior

Floral forms

1.	Salverform	Gamepetalours corolla, with a slender tube and an abruptly expanded limb
2.	Composite	Compount, organs or structure made up of distinct parts
3.	Bell shaped	Campanulate, shape like that of a bell
4.	Disc flower	Fleshy and raised development of the receptacle; circular, flattened organ
5.	Bilabiate	Two lipped, divided into upper and lower i.e. calyx or corolla
6.	Irregular	Cannot be divided into two equal parts, i.e. asymmetrical flower

Inflorescence forms

1.	Single	Comprising of one unit only - leaf, flower, stamen, pistil etc
2.	Head	A shor dense cluster of flowers
3.	Umbel	Generally flat headed an indeterminate inflorescence in which flower pedicels arise almost from a single point
4.	Spike	Usually unbranched, elongated indeterminate inflorescence, flowers are sessile

5.	Raceme	An unbranched, elongated indeterimate inflorescense, flowers are pedicelled
6.	Panicle	An indeterminate, branching inflorescence branches being racemes or corymbs

Roots and rootstocks

1.	Fibrous roots	Small tender succulent hair like usually below soil surface
2.	Tuber	A short thick stem or branch producing buds or eyes and serves as storage organ
3.	Rhizome	Rootstock, generally horizontal stem and give rise to leaves or stems at the apex
4.	Bulb	Modified leaf bud serves as storage organ comprising of a short thick stem, crowded leaf bases

❑❑❑

References

Bailey, L.H. (1963). The Standard Cyclopedia of Horticulture. Macmillan Company, New York, US.

Bailey, L.H. (1964). Manual of cultivated plants. Macmillan Company, New York, US.

Bailey, L.H. and E.L. Bailey (1976). A Concise Dictionary of Plants Cultivated in US and Canada.

Bessey, C.E. 1893. Evolution and Classification. Bot. Gaz. 18: 329-332.

Chopra, G.L. (1971). Angiosperms. Systematics and Life Cycle. 10th Edition. S. Nagin Sales Corporation.

Galletta, G.J. and D.G. Himerlick (1990). Small Fruit Crop Management. Prentice Hall Inc., New Jersey 07632.

Gangolly, S.R., R. Singh, S.L. Katyal and Daljit Singh (1957). The Mango. Saraswati Press Ltd., 32, Upper Circular Road, Calcutta-9.

Hooker, J.D. 1884. Flora of British India, Vol. 6.

Hooker, J.D. 1884. Flora of British India. (A to Z Volumes).

Hutchinson, J. (1934). The families of flowering plants. 1st Edition, Vol. 2. Monocot. London, Macmillan.

Hutchinson, J. 1926. Families of flowering plants. Vol. I. Dicot. London 1959. Families of flowering plants, Vol. II. London.

Popenoe, W. (1974). Manual of Tropical and Subtropical Fruits. Macmillan Publishing Co., Inc. New York 10022.

Reginald, C. (1964). Coconut, Longmans, Great Britain.

Rendle, A.B. (1938). The Classification of flowering plants. 1st Edition. Vol. II. Dicotyledons. Cambridge Univ. Press.

Tippo, O. 1942. A modern classification of plant kingdom. Chron. Bot. 7: 203-206.

❑❑❑